Interpreting Voice-Data-Video System Drawings and Specifications

Limited Energy

Low Voltage

Structured Cabling

AMERICAN TECHNICAL PUBLISHERS
Orland Park, Illinois

Christopher D. Sherlock, RCDD

American Technical Publishers Editorial Staff

Editor in Chief:
Peter A. Zurlis
Director of Product Development:
Cathy A. Scruggs
Assistant Production Manager:
Nicole D. Burian
Technical Editor:
James T. Gresens
Supervising Copy Editor:
Catherine A. Mini
Copy Editor:
Dane K. Hamann
Editorial Assistant:
Lauren D. Bedillion

Art Supervisor:
Sarah E. Kaducak
Cover Design:
Bethany J. Fisher
Illustration/Layout:
Nick G. Doornbos
Darrell E. Dwyer
Bethany J. Fisher
Nick W. Basham
Kevin L. Cuasay
Digital Media Manager:
Adam T. Schuldt
Digital Resources:
Lauren M. Lenoir
Tim A. Miller
James V. Cashman

1 2 3 4 5 6 7 8 9 – 21 – 9 8 7 6 5 4 3 2 1

Printed in the United States of America

ISBN 978-0-8269-1830-7
eISBN 978-0-8269-9593-3

This book is printed on recycled paper.

Introduction

Interpreting Voice-Data-Video System Drawings and Specifications covers the essential interpretation skills needed to safely install and troubleshoot common residential, commercial, and industrial voice-data-and video (VDV) systems. This book focuses on the fundamentals of interpreting drawings and specifications, including symbols, print elements, and construction documentation, of VDV systems. Practical applications and procedures are presented throughout the book as they relate to estimating and project management. VDV standards and codes, as well as the grounding and bonding of VDV systems, are covered. Current and future electricians and technicians working with the following systems will find this book invaluable:

- Voice-Data-Video
- Fire Alarm
- Security
- Audiovisual
- Outside Plant
- Premises Distribution

The author and publisher are grateful for the photographs and technical information provided by the following companies and organizations:

The American Institute of Architects
Autodesk, Inc.
Beast Cabling Systems, Inc.
Carlon®-Thomas&Betts
CM Fusion LLC
FEMA
Hilti, Inc.
International Code Council, Inc.
James Hardie Building Products, Inc.
Linden Group Architects
NJATC
NREL
Panduit Corp.
Shieldlab USA
3M
U.S. Army Corps of Engineers by Marc Barnes
U.S. Army Corps of Engineers

Contents

Contents

Features

Chapter introductions preview content to be covered.

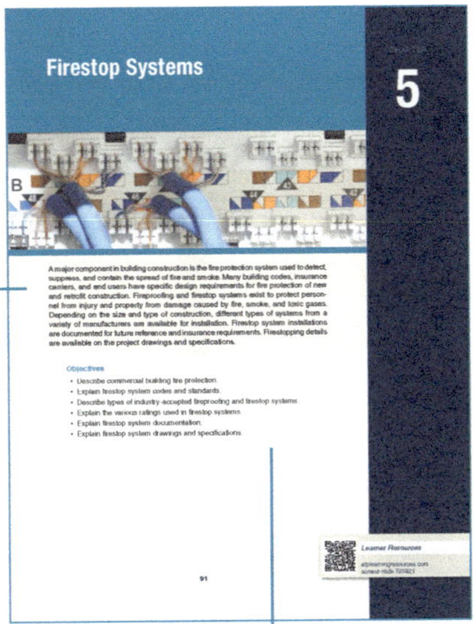

Objectives identify key concepts addressed in each chapter.

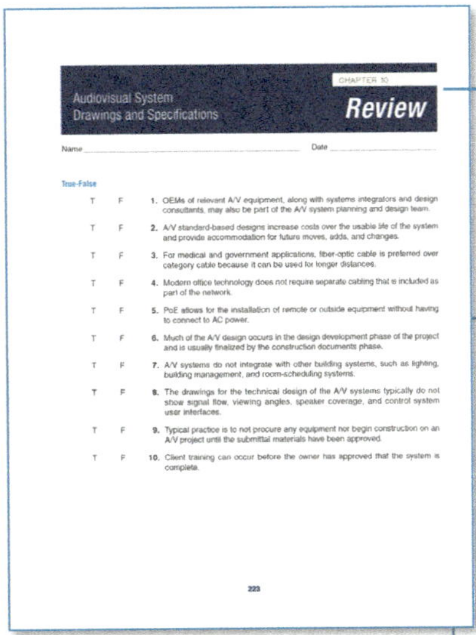

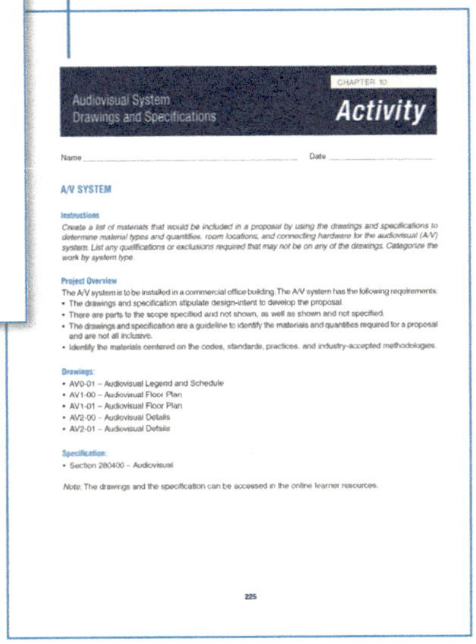

Chapter reviews and activities reinforce learner comprehension of chapter content.

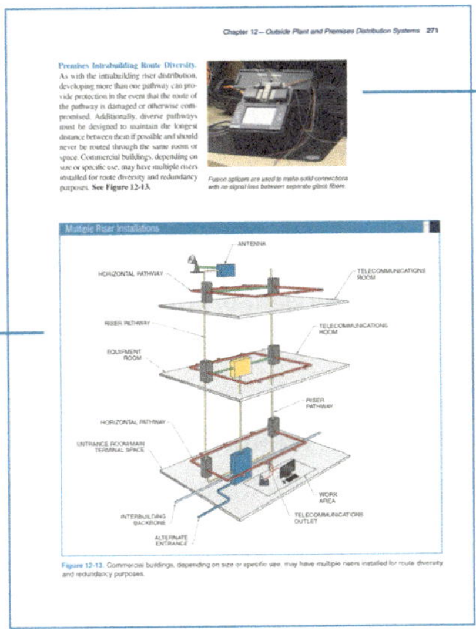

Photographs depict VDV equipment and systems.

Detailed illustrations depict major topics and concepts.

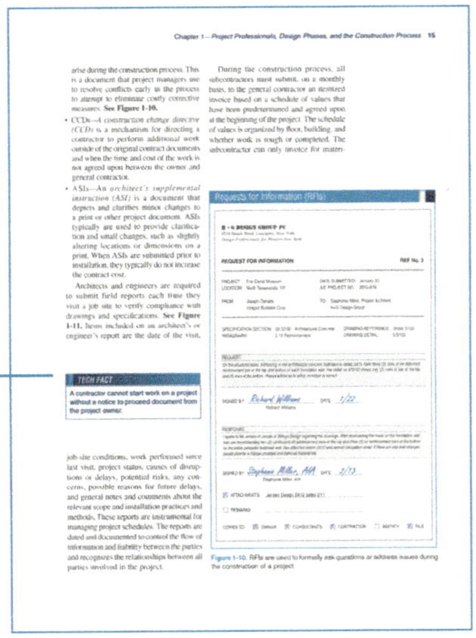

Tech Facts provide supplemental technical information.

Learner Resources

Interpreting Voice-Data-Video System Drawings and Specifications includes access to online learner resources that reinforce content and enhance learning. These online resources can be accessed using either of the following methods:

- Key atplearningresources.com into a web browser and enter QuickLinks™ Access code 783821.
- Use a Quick Response (QR) reader app to scan the QR Code with a mobile device.

The Learner Resources include the following:

- **Quick Quizzes®** that provide interactive questions for each chapter, with embedded links to highlighted content within the textbook and to the Illustrated Glossary
- **Flash Cards** that provide a self-study/review of common terms, definitions, symbols, VDV cable types, and connectors
- **Chapter Reviews** provide interactive PDFs of the chapter reviews
- **Activities** provide interactive PDFs of selected activities included in the textbook
- **Reference Prints** that can be accessed for the activities
- **Media Library** that consists of videos and animations that reinforce textbook content
- **Internet Resources** that provide access to additional resources to support continued learning

To obtain information on other related training material, visit the American Technical Publishers website at www.atplearning.com.

The Publisher

CHAPTER

1

Project Professionals, Design Phases, and the Construction Process

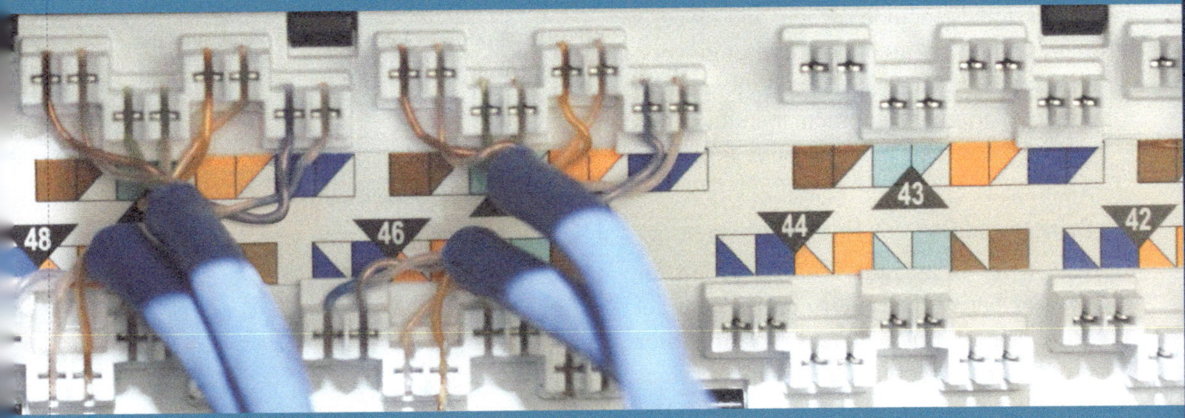

Every building project requires an accurate set of drawings and specifications that depict the owner's requirements for the facility to be constructed. Several different professionals and types of processes are involved in determining the owner's requirements, procuring the services of contractors or builders, and coordinating and completing the work. The time frame of the project schedule and costs from conception to completion will vary for each project, depending on the size, type, and complexity of the facility and the quality of the design and installation.

Learning Objectives

- Explain the roles of the different professionals involved with a project.
- Describe the design phases and how the project drawings and specifications are developed during each phase.
- Explain the different bidding methods that owners and general contractors (GCs) use to secure the service of subcontractors.
- Describe the considerations for the final product selection for a project.
- Describe the construction process and the different forms of documentation used on the job site.

Learner Resources

atplearningresources.com
access code 783821

PROJECT PROFESSIONALS

Project professionals not only develop the drawings and specifications used for a project but also work on the project to ensure that it is completed according to specifications in a timely manner. Project professionals include owners and owner's representatives, architects and engineers, general contractors (GCs) and subcontractors, and original equipment manufacturers (OEMs) and their representative firms.

Owners and Owner's Representatives

An *owner* is an individual, a group, or another entity that enters into a contract with designers, contractors, and other professionals to obtain goods and services provided by the construction industry. The owner, or the owner's representative, initiates a construction project and is present during all phases of planning, design, and construction. The owner has the authority to make decisions and is contractually tied to other parties. The owner may have several contractual obligations and responsibilities, such as making payments, attending meetings, and approving any changes to the work. **See Figure 1-1.**

A typical commercial construction project begins with an owner seeking to develop a new or existing property to meet specific needs. Quite often, owners lack expertise in the construction services field and require the assistance of construction professionals to not only identify properties but also negotiate building purchases and leases on the owner's behalf. Many of these construction professionals provide their services to owners, including project conception, site selection, design, construction, occupancy, and even the continued management of the facility in some cases.

An *owner's representative* is an individual who represents the business interests of the owner, especially if the owner is a group of individuals or a separate corporation. An owner's representative provides value to the owner by assisting with the engagement and evaluation of the architects and engineers that will provide design services. Upon design completion, the owner's representative will identify reputable general contractors and, in some cases, the various subcontractors. The owner's representative also develops and monitors the project budget, guards against cost overruns, and identifies scope changes that may save the owner money.

Owners

FEMA News Photo

Primary Responsibilities	
Obligation	**Description**
Project site information	During the design process, the owner is required to provide pertinent information about the property or any existing structures that might have an impact on planning or design.
Access to project site	The owner shall provide safe and legal access for designers and contractors to undertake their work.
Payment	The owner shall pay all fees for approvals, easements, assessments, use permits, and the work undertaken by consultants, designers, contractors, and suppliers.
Suspend or terminate work	Under the terms of the contract agreement, the owner has the right to stop or suspend the work under contract.

Figure 1-1. An owner may have several contractual obligations and responsibilities, such as making payments, attending meetings, and approving any changes to the work.

Architects and Engineers

An *architect* is a person licensed and qualified to perform architectural services such as analyzing a construction project, creating the project design and specifications, and managing the project. The architect determines the parameters of the necessary building space based on items such as property size and shape, desired interior and exterior spaces, and available project budget. Other factors that the architect reviews with the owner include the intended usage of the project, lifespan of the structure, length of a typical occupancy lease, and design and style preferences.

The architect reviews and approves all design elements for a project. The architect also specifies the materials, equipment, finishes (such as flooring and coatings), and fixtures. The architect is responsible for all drawings and specifications that will be used by the various professionals on the project. The architect will also work with the owner and/or the owner's representative(s) to ensure that zoning requirements and building codes are in compliance.

For example, when designing a building, an architect must comply with the Americans with Disabilities Act (ADA) of 1990. The *Americans with Disabilities Act (ADA) of 1990* is an antidiscrimination law that prohibits employers from denying employment to applicants based on their physical disabilities. Additionally, employers are required to provide reasonable accommodations and accessibility to the facilities. This law has made a significant impact on building design and code requirements for all new and rehabilitated building projects.

The architect, after determining the owner's requirements for a project, develops a conceptual design and retains and oversees various engineers and consultants that design the facility's systems. In doing so, the architect becomes the single point of contact for all design activities during construction. The architect's design process begins with the identification and allocation of space in the facility to be used for occupancy, elevators and stairways, tenant lavatories, and mechanical, electrical, and voice-data-video (VDV) systems services. Although engineers are also concerned with those functional considerations of a project, aesthetics are also taken into consideration.

The engineering professionals overseen by the architect may include professional engineers (PEs), mechanical engineers (MEs), electrical engineers (EEs), mechanical/electrical/plumbing engineers (MEPs), fire protection engineers (FPEs), and registered communications distribution designers (RCDDs). **See Figure 1-2.** A *registered communications distribution designer (RCDD)* is an individual with education, training, and expertise in the design, implementation, and integration of telecommunications/data transport systems and infrastructure. The VDV system is designed, reviewed, and approved by the RCDD.

Similar to architectural firms, and depending on the state where the company is located, at least one of the principal officers, or partners, of an engineering firm must be a licensed engineer. Engineering firms are required to maintain design and calculation records for all projects. Some engineering firms are multidisciplined and can offer a variety of professional services. Engineering firms are often retained and used for civil, geotechnical, structural, mechanical, electrical, and plumbing designs and analysis, as well as for consultation concerning the removal and transportation of hazardous materials. The need for specialized systems, such as life safety, solar heating, fiber-optic networks, and geothermal heating, have prompted engineering firms to become more highly specialized.

General Contractors and Subcontractors

A *general contractor (GC)* is an individual or commercial firm that is hired to provide goods and services for the undertaking of construction activities. Often, GCs offer construction management services and bring a high level of building expertise to the project team. A GC may either self-perform the work or manage subcontractors to complete the various trades work.

PROJECT ARCHITECTS AND ENGINEERS	
Architect	Licensed to perform architectural services; creates, designs, and administers the project
Professional engineer (PE)	State-licensed and qualified to prepare, sign and seal, and submit engineering plans and drawings to a public authority for approval, or seal engineering work for public and private clients; fulfills legal requirement of all construction projects
Mechanical engineer (ME)	Responsible for design and development of equipment such as HVAC systems, elevators, escalators, and motors
Electrical engineer (EE)	Responsible for the design and development of all electrical systems and equipment for a project
Mechanical/electrical/plumbing engineer (MEP)	Responsible for the planning and design of mechanical, electrical, and plumbing systems
Fire protection engineer (FPE)	Responsible for the design and development of fire suppression and safety systems for a project
Registered communications distribution designer (RCDD)	Responsible for the design, implementation, and integration of VDV systems infrastructure

Figure 1-2. An architect, after determining the owner's requirements for a project, develops a conceptual design and retains and oversees various engineers and consultants, including PEs, MEs, EEs, MEPs, FPEs, and RCDDs, that design the facility's systems.

A *subcontractor* is an individual or commercial firm that specializes in a particular trade or field of construction. The GC will provide the drawings and specifications to the subcontractor to determine the materials and labor required to complete the project, in order for the subcontractor to provide a cost estimate to complete the work. The GC will also develop a construction schedule illustrating project milestones and mutually agreed upon completion dates.

All subcontractors on a particular project are hired and managed by the GC. This arrangement provides a project hierarchy where all subcontractors and specialty contractors fall under the direction of the GC. This hierarchy also creates one point of contact (the GC) for communication with the architect/engineer and owner. **See Figure 1-3.** All project coordination between the subcontractors is overseen by the GC. This allows the subcontractors to work efficiently together and have access to work areas, which ensures a steady sequence of work so that they can complete their work in an efficient and timely manner.

Beyond the subcontractors, GCs also oversee project management meetings, jobsite documentation, changes in work scope, work activities, schedules, subcontractor billing, payments, project documentation, and project records. Another important responsibility that the GC fulfills is obtaining necessary permits and variances with municipal agencies. The GC is also responsible for all inspections, the final building occupancy approval sign-off, and the commissioning of all building systems. However, if there are issues that must be addressed after the building is occupied, the building owner typically hires the necessary vendors for service work.

Original Equipment Manufacturers and Representative Firms

The development of drawings and specifications for the VDV system installation on a project is often influenced by original equipment manufacturers (OEMs) and the design or distribution firms that represent them. Many OEMs employ design professionals who engage owners, architects, and engineers to assist in specifying and designing systems. These professionals assist the owner and design team in using their products for the project or substituting their products in place of competitors' products.

Project Hierarchy

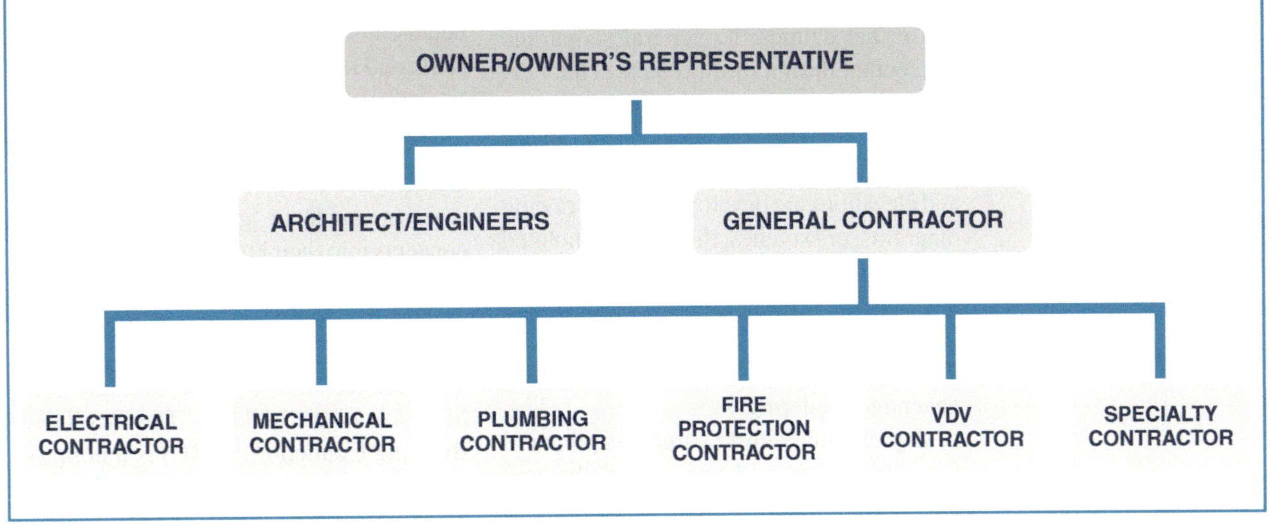

Figure 1-3. All subcontractors and specialty contractors fall under the direction of the general contractor (GC), creating a one-contact hierarchy for communication with the architect/engineers and owner.

Many OEMs have established their own standardized VDV systems that they specify for use. These OEM-standardized VDV systems are also known as "proprietary systems." These systems can be customized to an owner's specific requirements.

Sometimes OEMs will partner with distributors and product suppliers to secure competitive pricing for the equipment and materials that are required for a project. This is a concept used to provide discounted project materials to the owner in exchange for design assistance and exclusive access to the project. This partnership is also beneficial because it helps avoid delays in procuring specialty items that may have long lead times. For example, the supply of fiber-optic cables is dependent on the supply of optical glass. A partnership between a fiber-optic glass OEM and a distributor can secure the material needed to provide fiber-optic cables for a project in a timely manner, before the delivery of such material becomes delayed and the project deadline is missed.

DESIGN PHASES

Design is the process of taking an abstract idea and developing it into a working solution. A number of design phases are used by architects and engineers (the designers) in the development of a project. These design phases help both the owner and designers identify needs and compile the necessary information required to complete the construction documents. The needs of the project are based on communication with the owner. Items covered include everything from space requirements and building appearance to power and lighting, as well as any special occupancy requirements.

The methodology of design for each design firm varies. However, the process of schematic design to the completion of construction drawings and specifications is developed through a series of phases that refine the design concept. During the design phase, the administration and evaluation of a project includes the identification of project site conditions, evaluation of the project budget, and establishment of a preliminary project schedule. The designer is required to assess all information obtained from the owner, including property and topographical surveys, soils testing results, historic reviews, and any environmental concerns. The designers may also assist the owner in preparing and filing any permit or variance applications.

Schematic Design

Schematic design (SD) is the initial design phase that defines the general scope and conceptual design of a project. Schematic design begins with an architect developing sketches and conceptual drawings based on the owner's requirements. Floor plans and elevations are developed using drawing diagrams, or sketches, that show annotated room layouts and areas, general notes, and components. The graphic representations may also be conveyed to the owner in the form of scaled models or computer-generated renderings.

A preliminary cost estimate is prepared for comparisons with the owner's budget. The schematic design gives the owner the first opportunity to see the potential design and get an approximation of the costs. Costs are evaluated on a square foot basis. Project feasibility is assessed, design options are often pursued, and alternatives are presented to the owner. At the completion of the schematic design phase, the owner must approve the work before proceeding to the next phase, design development.

TECH FACT

Design fees can be a lump sum, hourly, or a percentage of the construction cost.

Design Development

Design development (DD) is the design phase that further develops and refines a schematic design. The layout of structural, mechanical, plumbing, electrical, and VDV systems are added and architectural details are selected. The selection and specification of products and materials also occur during this phase. Other design requirements, such as building- and energy-code compliance, structural calculations, and materials locations and quantities are further addressed. Design development includes refinements that establish the form and appearance of the project through plans, elevations, and details. Technical specifications sections are edited to include specific materials and equipment details, and the design is more accurately estimated for cost. **See Figure 1-4.**

Sustainable Building Design

Owners and architects will often incorporate sustainable, or "green," building design concepts into their buildings throughout the design phase. *Sustainable building* is the practice of constructing a building using processes that are environmentally responsible and resource-efficient throughout the life cycle of the building. The U.S. Green Building Council (USGBC) is a nonprofit organization that promotes sustainable building through the LEED rating systems. USGBC's mission is "to transform the methods in which buildings and communities are designed, built, and operated, enabling an environmentally and socially responsible, healthy, and prosperous environment that improves the quality of life."

The U.S. Green Building Council promotes sustainable building through the LEED rating system.

Design Development

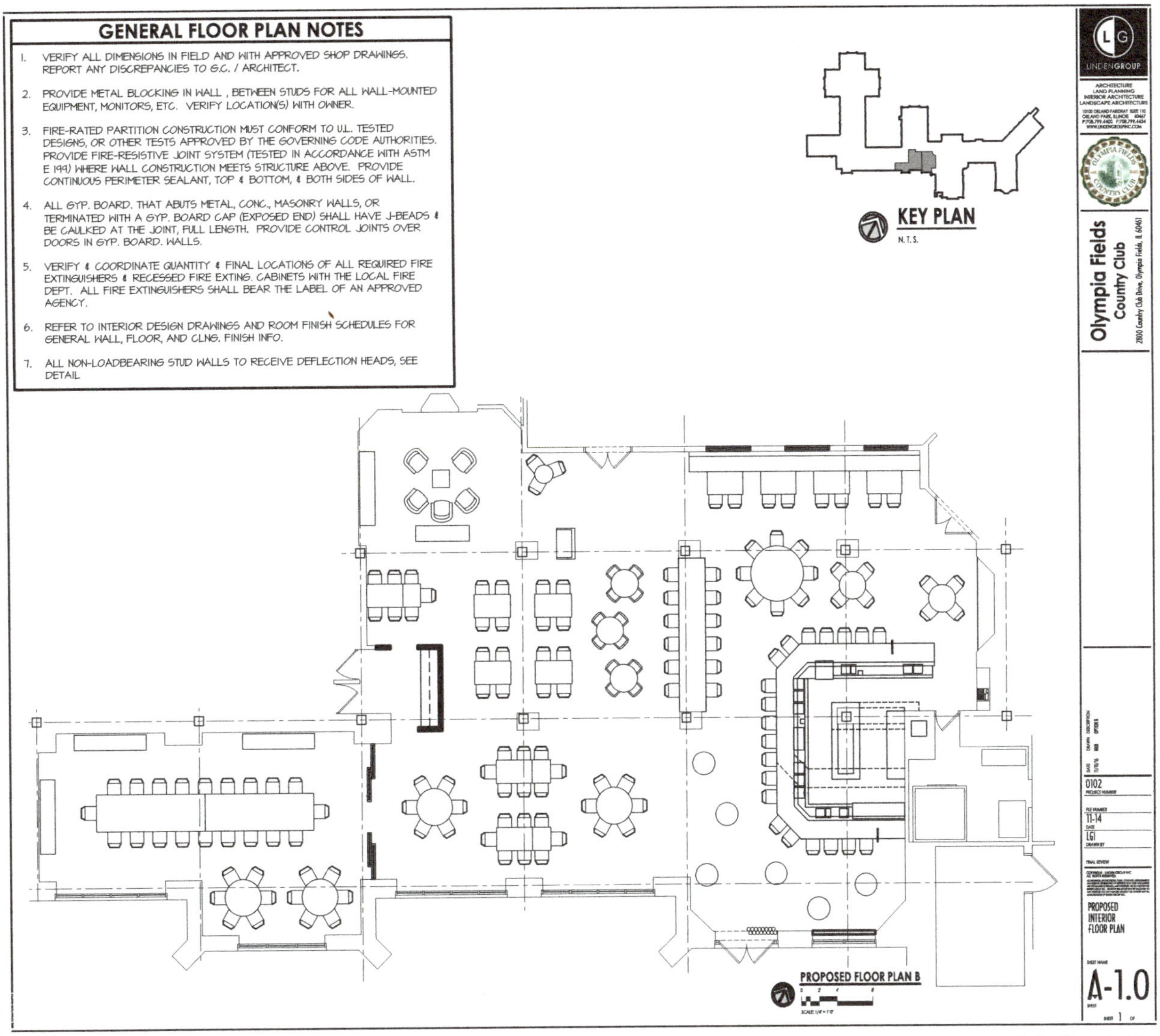

Linden Group Architects

Figure 1-4. In the design development phase, schematic designs are further refined and developed into construction drawings.

Leadership in Energy and Environmental Design™ (LEED®) is an international green building certification system developed by the USGBC. LEED acts as a guideline for building sustainability, utilization of recycled materials, water and energy efficiency, renewable energy, indoor environmental quality and comfort, lighting, and acoustics. LEED rating systems provide the framework for meeting these goals and assessing performance for many project types, such as new and existing buildings, commercial interiors, schools, health care facilities, retail spaces, and homes. **See Appendix.** The different variations of the LEED rating systems provide credits that buildings can earn to achieve one of four levels of certification: Certified, Silver, Gold, or Platinum. Owners who design and build projects with LEED certification are often awarded tax savings by municipalities and have lower maintenance costs over the life of the building.

Coordination of Requirements—VDV Systems

During the design phase, the VDV system designer, typically an RCDD, works with the architect's design team and the owner's information technology (IT), security, and facility engineers to determine the owner's requirements and coordinate the systems that operate together. The coordination of requirements phase begins with the architect and RCDD reviewing the overall floor plans to determine the allocation of space for VDV system wiring. **See Figure 1-5.** With commercial buildings, VDV installations have distance limitations. The designer must ensure that VDV systems cabling will not exceed the maximum allowable distance of 100 m (328′) from work area outlets to the network equipment.

For example, if the footprint of a building is 350′ × 250′ and only a single equipment room is placed in the middle of the floor, some of the system wiring may exceed the allowable distance of 100 m (328′). The quantity and amount of equipment to support the floor will also need to be factored into the design. There may be too much equipment and terminating hardware to effectively fit in a single rack or in one centrally located room. In this situation, a designer will have to locate two equipment rooms on the floor plan to strategically balance the load from one side of the floor to the other.

Information required to determine an owner's VDV system requirements include the following:

- type of telephone system—analog or Internet Protocol (IP)
- type of Ethernet network switching and routing equipment connecting to the physical layer, including quantity of LAN ports and network speeds
- types and quantity of network backbone cables connecting primary equipment rooms and equipment rooms that serve the floors
- types and quantity of network backbone cables connecting one building to another
- types and quantity of VDV cables required to be installed at work area outlets
- types and quantity of VDV cables required for wireless access points (WAPs)
- size and location of specialty equipment rooms to support network equipment, including server and storage equipment
- various systems designed to work on the IP platform and devices that utilize Power over Ethernet (PoE), such as lighting and building controls
- card access, security, and fire alarm interfaces
- audiovisual digital signage and wayfinding applications—A *wayfinding sign* is an interactive touch-screen display. Wayfinding signage is commonly found in lobbies of commercial buildings and provides a campus map or a building floor plan. It may note occupants by floor and may direct building visitors to various rooms for events. Digital wayfinding signs can easily be changed to suit particular or daily needs.
- identification and administration preferences for each system

The types and quantities of cables to be used with wireless access points (WAPs) are determined as part of the coordination of requirements phase.

Voice-Data-Video (VDV) System Floor Plans

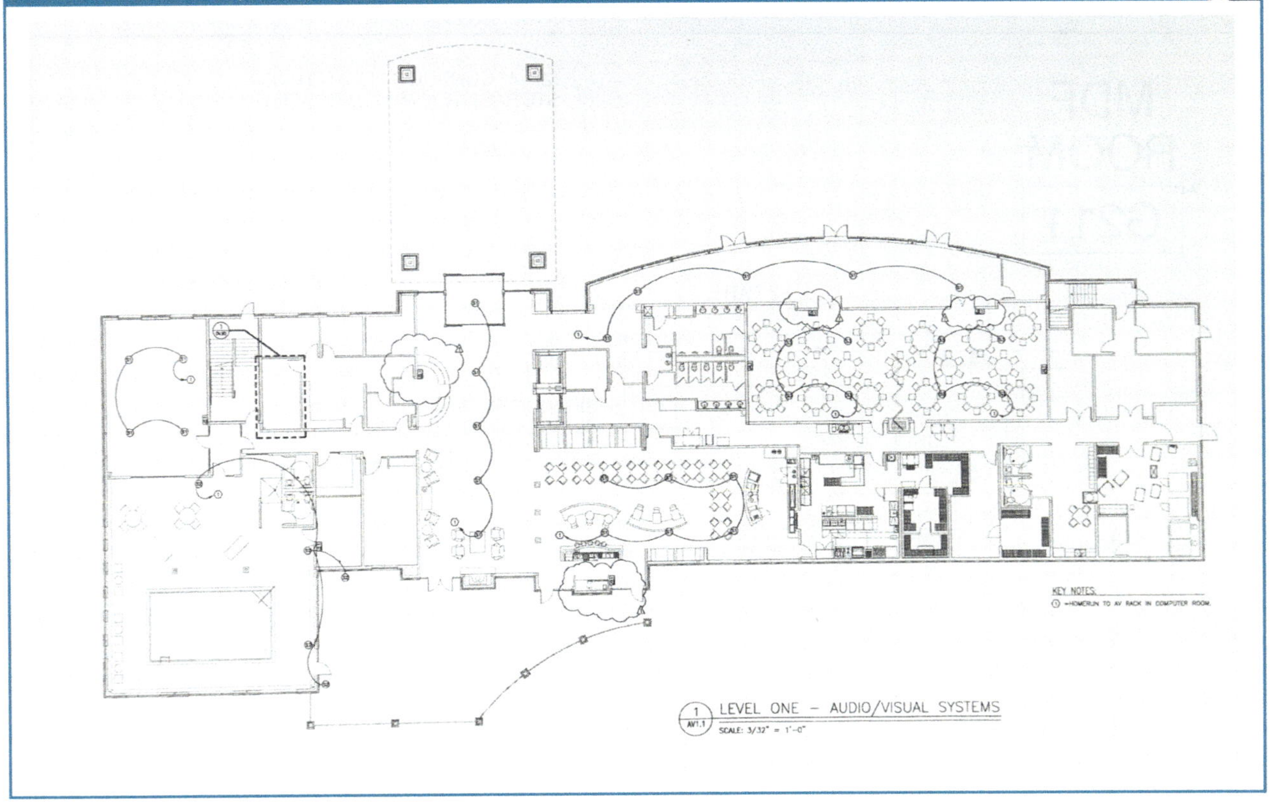

Figure 1-5. The coordination of requirements phase of the project begins with the architect and RCDD reviewing the overall floor plans to determine the allocation of space for the VDV system wiring.

By compiling the information on the owner's VDV system requirements, the VDV system designer can determine and coordinate with the architect the following criteria:

- location of all equipment rooms serving each floor
- size of the equipment rooms required to support the equipment
- size and location of backboards for wall-mounted equipment
- required lighting, grounding system, floor type, and color of the equipment rooms
- quantity, type, and location of equipment racks and cabinets in equipment rooms **(See Figure 1-6)**
- size and placement of overhead cable runways/trays to support the VDV cables within the equipment room
- size, type, and quantity of the pathways required to support VDV cables and wires above the ceiling and throughout the facility
- size, type, and quantity of pathways required to support backbone cabling between equipment rooms within the building and between other buildings
- National Electrical Manufacturers Association (NEMA) standardized configurations for the electrical receptacles to power the network equipment in the equipment rooms
- HVAC (cooling) requirements to maintain the network equipment within the rooms—This information is found on the nameplate of the equipment and provided to the mechanical engineer for cooling equipment sizing. *Note:* HVAC control system prints consist of abbreviations and symbols, wiring diagrams, detail drawings, and sequence-of-operation narratives.

Equipment Rooms

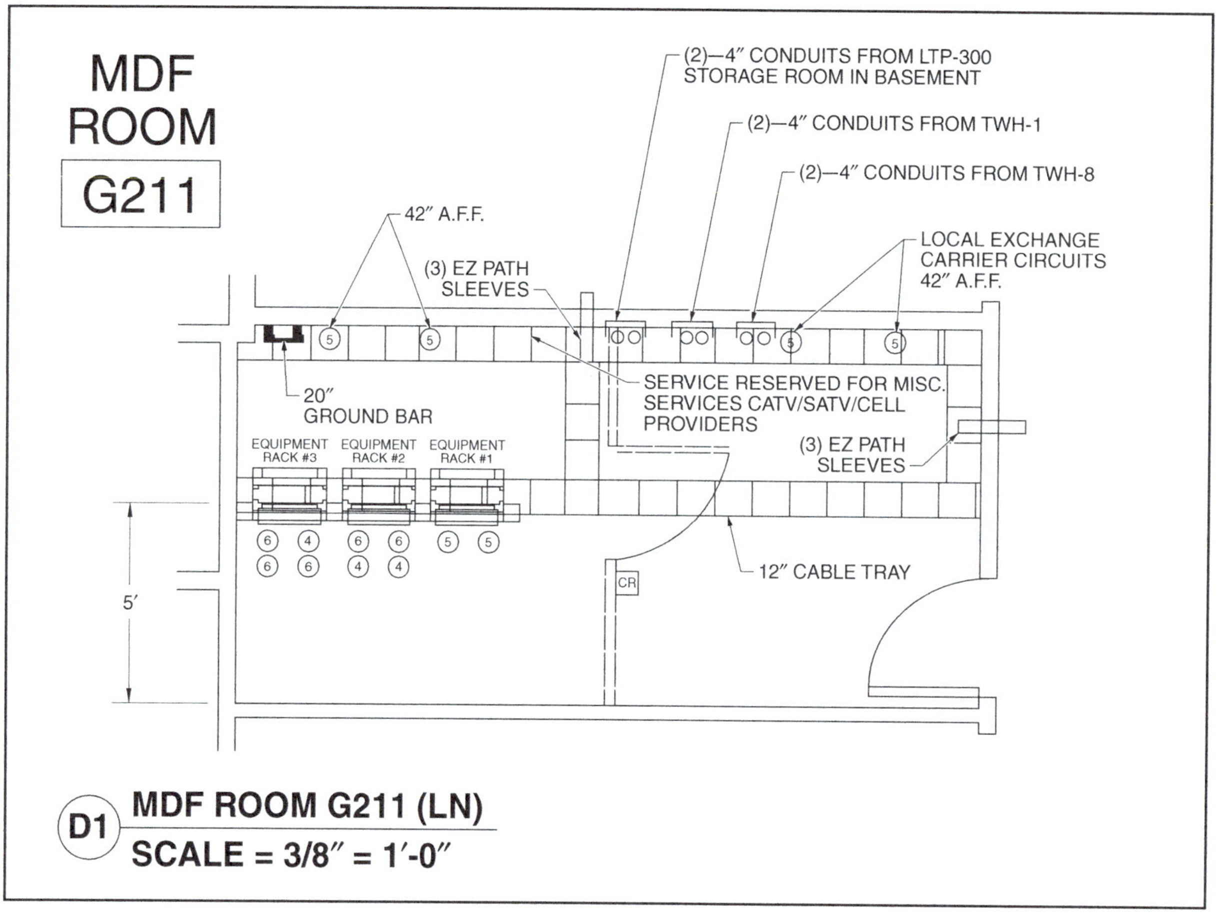

Figure 1-6. Equipment rooms are detailed on prints to show locations of equipment such as equipment racks, conduits, and cable trays.

A designer will also provide the architect with the type, mounting heights, and locations of work area outlets for network access. This includes the details of electrical wall boxes and conduits required and any specialty outlets such as floor boxes, poke-thru devices, and modular furniture outlets. **See Figure 1-7.**

Coordination between the architect and designer is necessary to ensure that outlets, such as for wireway-mounted devices, will fit correctly with equipment specified under other disciplines. For example, if a VDV system drawing indicates that the mounting height of a wall-mounted telecom outlet is to be installed at 54″ but the electrical contractor has a drawing with an outlet symbol that indicates the mounting height of a wall-mounted telecom outlet is to be installed at 18″, the conduit and box will not be installed at the correct height. Symbol uniformity must be maintained between drawings so that all devices are installed in the correct locations.

During the coordination of requirements, items such as plenum-rated and non-plenum-rated areas are identified. Locations for fire-rated walls, barriers, and other fire-stop systems are determined and specified to maintain the ratings of the walls that have been previously established. The designer will work with the architect to ensure that all of these requirements are shown on the architectural, electrical, VDV, and mechanical drawings for the project.

Wall Box Details

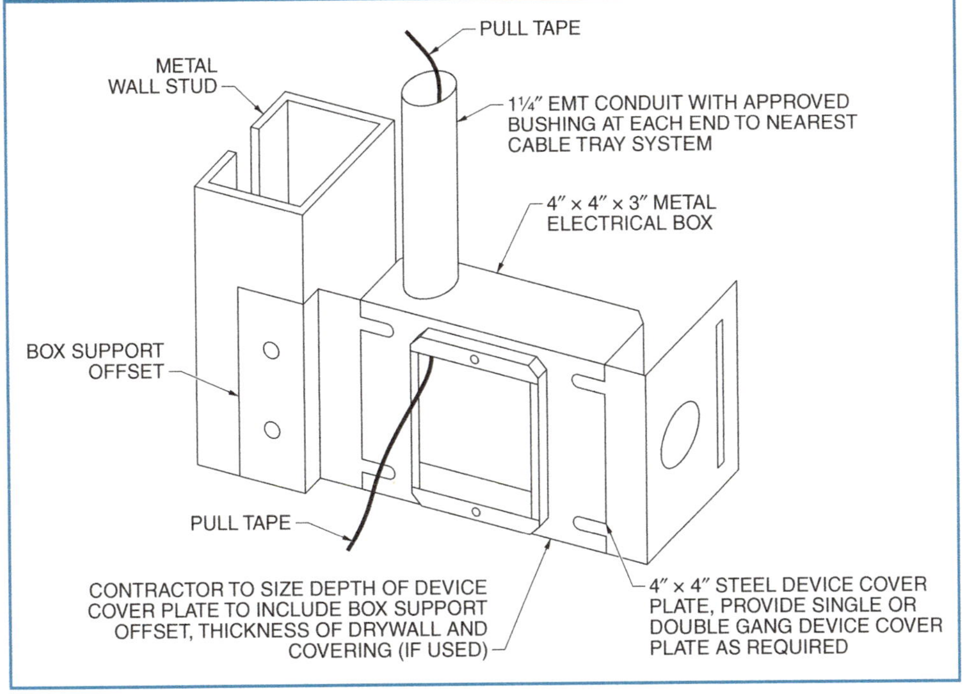

Figure 1-7. Drawings and specifications include the details of required electrical wall boxes.

Construction Documents

The construction document phase is the final design phase of a project. The *construction documents* are a set of instructions created by a designer and used by a contractor to estimate project costs and build a project. The construction documents are the fully completed design that complies with building codes and industry standards and are ready for bidding and, ultimately, the construction of the facility.

The drawings and specifications developed during the design phases are included in the construction documents. The number of drawings and volumes of specifications required depends on the size and complexity of the project. It is not uncommon for large projects to have a hundred or more drawings in a construction document set and multiple volumes of technical specifications.

During the bidding process, addenda are issued to amend any incomplete drawings and/or technical specifications or to make required modifications or additions. An *addendum* is the addition of omitted or updated information to a set of construction drawings, technical specifications, or a contract.

Building-code enforcement officials require that construction documents be signed and sealed by the design professional, usually the architect or PE associated with the project. This is to certify that all design work has been completed under the supervision of a licensed professional. The submission of signed and sealed construction documents to the local building department is required in order to obtain building permits. **See Figure 1-8.**

Signed and Sealed Construction Documents

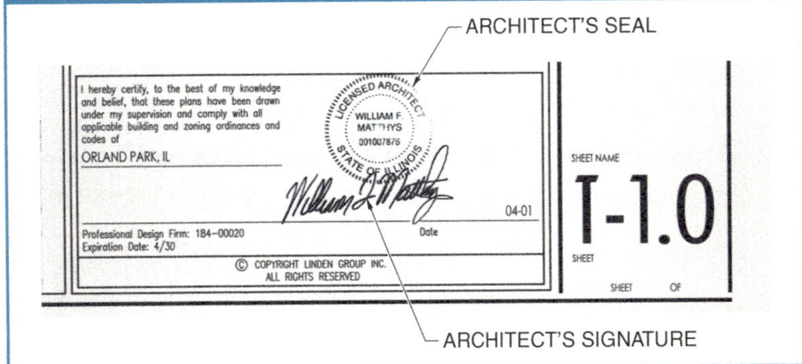

Figure 1-8. Construction documents are signed and sealed by the design professional before they are submitted to the local building department for review.

PROCUREMENT

The owner, owner's employees, representatives, OEMs, and the GC identify suitable subcontractors to bid on the work for the project. The subcontractors identified may have an existing relationship, based on past performances, with one of the parties or the ability to certify or provide a warranty for the work to be installed. There are several bidding methods used to procure the services of subcontractors and the necessary materials and equipment, as well as several considerations for selecting the final products, and their vendors, for use in a project.

Bidding Methods

The procurement of VDV systems and related equipment is typically pursued in the same manner as the other disciplines for a construction project through the competitive bid process. In addition to the competitive bid process, other procurement methods are the cost-plus, negotiated/design-assist, and labor-only methods. Bid documents are integral to these different methods of bidding on a project.

The bid documents for a project include a complete set of construction documents, including specifications, the project schedule, and other pertinent information for calculating an estimate to complete the work. The bid documents may request a breakdown of work by type, labor required, and materials. The bid documents may also include unit pricing for specific items to be added during the project. After careful review and calculation of materials, labor, and project schedule, the subcontractor provides a lump sum bid proposal to complete the work.

The competitive bid process is typically managed by the GC under the hierarchy of the one-contractor model. A pre-bid meeting at the job site is usually conducted so that all potential bidders may review the site logistics and the current state of the work. Pre-bid meetings are especially crucial when preparing an estimate for buildings that are being retrofitted. The prospective bidder will be able to visually inspect and review the challenges that the retrofit project will require, which are not often clearly depicted in the construction documents, such as the existing building materials and size of mechanical spaces.

The cost-plus method for bidding on a project is used when the owner is uncertain as to the cost of the project but knows it is likely to be higher than expected. In this case, the contractor has no reason to try and keep the costs down, as in the competitive bid process. However, there are many reasons to use the cost-plus approach. For example, some projects do not provide enough lead time to determine when it is required to start the work. This approach can often save time, provide flexibility, and avoid the risk of removing completed or partially completed work and starting over.

The negotiated/design-assist method is similar to the cost-plus method in that the price is negotiated but will most likely result in lower expenses due to the nature of the project or its fast-paced schedule. In some instances, the subcontractors will work directly with the owner and architect to provide design and engineering expertise and can move, add, and change work at a quicker pace than the traditional construction management processes. Cost oversight is scrutinized, and labor and material costs are usually pre-fixed.

The least common method for bidding on a project from a contractor's perspective is the labor-only method. This method is used when the owner furnishes the materials and a bid proposal requests the labor costs for installation services only. The labor-only method is often used for the electrical and VDV system portions of a project. Owners use the labor-only method to reduce costs and avoid markups on contractor-acquired materials.

Under the labor-only method, the contractor must account for handling the material, often multiple times. The project can be subject to delays in ordering and receiving the proper materials, which often delays work completion. Besides the loss of efficiency, another consideration when using the labor-only method is that the contractor can usually purchase materials at lower cost than most owners. The contractor and suppliers can also schedule deliveries so that the correct amount of materials is delivered per the schedule. This avoids either a shortage of material or an excess of material that would need to be stored and handled multiple times.

Although these bidding methods are all competitive, they still have differences. One method (cost-plus method) is by price only, while the others rely on professionalism, industry reputation, experience, efficiency, and integrity. This does not mean that one method is better than another. The owners ultimately decide the value for the work they are paying for. Some owners will obtain three competitive bids, remove the low and high bids, and take the middle bid. Or an owner may decide to pay a premium for a subcontractor based on their past performance of completing projects correctly and on time.

Ultimately, the procurement of services should not be based solely on the bid price. Procurement should account for the industry reputation, skill, efficiency, experience, and professionalism of the contractors as well. With those qualities in place, a building owner or GC can be assured that the project will be constructed at a fair price and completed on time. **See Figure 1-9.**

Procurement of Services

Figure 1-9. To ensure a successfully completed project, the procurement of services is not only based on the bid price but also the industry reputation, skill, efficiency, experience, and professionalism of the contractors that are hired.

Product Selection

During the final product selection for a project, the various types of products selected for use in a facility are usually determined by their performance and cost. Other considerations include application assurance warranties and ease of use. An *application assurance warranty* is an inclusion to a standard product warranty that is designed to provide the end user with additional assurances of performance. This means that the products are under warranty to support the current and approved applications for which they were initially designed. These applications must be designed to adhere to industry recognized standards, codes, and practices.

Many OEMs typically provide warranties for VDV systems for a period of 5 years to 25 years. Typically, all components, termination hardware, faceplates, outlets, and connecting cables must be manufactured by the company providing the warranty. The installation contractor must be a certified installer of the system and their employees must all attend and maintain certified manufacturer training. Most OEM warranties require installation and maintenance records and proof that all work has been completed by a certified contractor for that OEM in the event that a product malfunctions and requires repair or replacement under a warranty.

The cost and performance of the products specified can also drive product selection. With many purchases, cost is relative to quality. For example, the most economical option for a tenant office space with a two-year lease may be a cost-effective generic VDV system without a warranty. In this case, the VDV network requirements are basic and only normal office work is performed.

However, a facility with multiple buildings in a campus setting, such as a university or hospital, that performs heavy research and development and will serve large quantities of users with a high demand for network access may see the value in paying a premium for a higher quality product. A higher quality, standardized product along with an application assurance warranty provides security and value to owners. The return on investment (ROI) for the owner and users will be realized for many years. Standardized systems also provide structure to organizations. Standardized systems tend to be installed in a uniform manner and can be easily maintained and modified.

Another factor considered during product selection is whether the project is for a private, government, public, or municipal application. Most often, the product selection for private work is specified by specific OEMs. For government, public, and municipal projects, product selection is usually determined by performance specifications based on industry-recognized codes and standards. In this case, specific OEM part numbers are intentionally omitted as to maintain a fair and balanced selection process and to avoid conflicts with specifying any one OEM. This ensures that the consumer does not pay a premium, avoids price fixing, and receives a competitive and cost-effective price for the products.

CONSTRUCTION PROCESS

The construction process, from the beginning of the design concept until the end when users occupy the building, includes a variety of professionals with specific skill sets working together toward a common goal. The drawings and specifications developed during the design phase are the guidelines for the construction process. They are developed by the architect and engineers through analysis of the owner's and end users' specific requirements and require detailed planning. The final contract documents must be clear, concise, complete and correct.

The GC, through a procurement process, employs and manages subcontractors. The GC also puts processes and schedules into place to ensure that work is constructed in sequence by skilled professionals in an efficient manner. General contractors and subcontractors, through proper training and experience, interpret the drawings and specifications to complete the project.

During the construction process, the general contractor and the architect provide individually numbered logs for items such as equipment submittals, requests for information (RFIs), construction change directives (CCDs), architect's supplemental instructions (ASIs), and architect/engineer field reports. When the one-contractor model is used, the information flows from the subcontractors to the general contractor and then to the architect and engineers, accordingly. This approach prevents direct contact between the architect or engineer and the subcontractors, maintaining one-contractor communication.

All equipment, whether specified or shown, is subject to the submittal approval process. The submittal approval process provides the architect time to review and approve the products by detailing the OEM specifications and product data prior to purchase and installation. Additional documents and tools used by general contractors and other project managers to provide job-site documentation and recordkeeping include the following:

- RFIs—A *request for information (RFI)* is a document used for clarifying scope and the coordination of project elements that arise during the construction process. This is a document that project managers use to resolve conflicts early in the process to attempt to eliminate costly corrective measures. **See Figure 1-10.**
- CCDs—*A construction change directive (CCD)* is a mechanism for directing a contractor to perform additional work outside of the original contract documents and when the time and cost of the work is not agreed upon between the owner and general contractor.

- ASIs—An *architect's supplemental instruction (ASI)* is a document that depicts and clarifies minor changes to a print or other project document. ASIs typically are used to provide clarification and small changes, such as slightly altering locations or dimensions on a print. When ASIs are submitted prior to installation, they typically do not increase the contract cost.

Architects and engineers are required to submit field reports each time they visit a job site to verify compliance with drawings and specifications. **See Figure 1-11.** Items included on an architect's or engineer's report are the date of the visit, job-site conditions, work performed since last visit, project status, causes of disruptions or delays, potential risks, any concerns, possible reasons for future delays, and general notes and comments about the relevant scope and installation practices and methods. These reports are instrumental for managing project schedules. The reports are dated and documented to control the flow of information and liability between the parties and recognizes the relationships between all parties involved in the project.

TECH FACT

A contractor cannot start work on a project without a notice to proceed document from the project owner.

During the construction process, all subcontractors must submit, on a monthly basis, to the general contractor an itemized invoice based on a schedule of values that have been predetermined and agreed upon at the beginning of the project. The schedule of values is organized by floor, building, and whether work is rough or completed. The subcontractor can only invoice for materials and labor completed through the end of the invoice period, typically on a month-to-month basis. If specialized equipment or items with longer-than-normal lead times are required, a separate invoice may be sent in addition. The general contractor assembles all subcontractor invoices and submits one total lump sum to the owner each month for final approval and payment. Upon approval and payment by the owner, the general contractor then compensates the subcontractors per their monthly invoice.

Requests for Information (RFIs)

H + G DESIGN GROUP PC
4334 Beach Road, Lancaster, New York
Design Professionals for Western New York

REQUEST FOR INFORMATION **REF No. 3**

PROJECT: Erie Canal Museum DATE SUBMITTED: January 22
LOCATION: North Tonawanda, NY A/E PROJECT NO.: 2010-876

FROM: Joseph Daniels, Intrepid Builders Corp TO: Stephanie Miller, Project Architect, H+G Design Group

SPECIFICATION SECTION: 03 33 00 - Architectural Concrete DRAWING REFERENCE: Sheet S103
PARAGRAPH: 2.12 Reinforcement DRAWING DETAIL: 5/S103

REQUEST

On the structural plans, reinforcing in the architectural concrete bulkhead is called out to have three (3) rows of #4 deformed reinforcement bar at the top and bottom of each foundation wall. The detail on 5/S103 shows only (2) rows of bar at the top and (3) rows at the bottom. Please advise as to which condition is correct.

SIGNED BY: Richard Williams DATE: 1/22
Richard Williams

RESPONSE

I spoke to Mr. Jensen of Jensen & Billings Design regarding the drawings. After recalculating the loads on the foundation wall, they are recommending two (2) continuous #6 reinforcement bars at the top and three (3) #4 reinforcement bars at the bottom for the entire perimeter bulkhead wall. See attached sketch SK12 and revised calculation sheet. If there are any cost changes, please provide a change proposal and itemized material list.

SIGNED BY: Stephanie Miller, AIA DATE: 2/13
Stephanie Miller, AIA

☒ ATTACHMENTS Jensen Design SK12 dated 2/11

☐ REMARKS

COPIES TO: ☒ OWNER ☒ CONSULTANTS ☒ CONTRACTOR ☐ AGENCY ☒ FILE

Figure 1-10. RFIs are used to formally ask questions or address issues during the construction of a project.

Daily Field Reports

DAILY CONSTRUCTION FIELD REPORT

DAILY REPORT NO.: 14 DATE: 6-19
CONTRACTOR: LPT Construction PROJECT NO.: 0134
PROJECT NAME: Oak Hills Municipal Facility

WEATHER:			SITE CONDITIONS:		DAY:	
☒ Clear	☐ Windy	☐ Cool	☐ Clear	☒ Dusty	☐ Monday	☐ Thursday
☐ Overcast	☐ Fog	☐ Warm	☐ Muddy	☐	☒ Tuesday	☐ Friday
☐ Rain	☐ Cold	☐ Hot	Temperature Range: 74°-78°		☐ Wednesday	☐

REMARKS: (Describe significant events, work accomplished, materials & equipment on site, special instructions of contractors, and number count of personnel at the site):
-Foundation work complete
-Under slab utilites rough-in continues.
-Floor slab pour scheduled for 6-22

PROJECT LEADER COMMENTS:
-Steel columns on site. Beams scheduled for 6-22 delivery.
-Brick delivery scheduled for 6-29.
-Under slab utilities inspection called in.

NATURE OF EXCEPTION/DEFECTS (include specifications and/or drawing numbers, location, description):
NONE

CONTRACT DATE: 5-20 COMPLETION DATE: 8-1
ACTUAL WORK COMPLETED %: 10 PROBABLE COMPLETION DATE: 8-1
NUMBER OF WORKERS: 8
SUPERVISORS SIGNATURE: Joe Davy
☐ ATTACHMENTS

page of pages

Figure 1-11. Maintaining accurate daily field reports is crucial for providing a written history of daily activities on a job site.

A *punch list* is a list compiled by the owner and/or lead designer during the walk-through of a completed project that indicates items that must be repaired or completed before project finalization. **See Figure 1-12.** The end of the construction process includes a punch list generated by the architect and the architect's design team outlining work that remains to be completed, work not installed as specified, and work requiring adjustment or needing to be cleaned.

The generation of a punch list usually occurs prior to inspections by the authorities having jurisdiction (AHJs). These inspections typically involve structural building and electrical wiring inspectors as well as the fire marshal for the municipality where the project is located. The inspectors are to ensure that all life-safety requirements and building codes have been adhered to. Once all inspections are completed and any corrective action issues addressed, corrected, and re-inspected, the owner will receive an occupancy permit to allow personnel and related equipment into the building.

The final stage of the construction process for the contractors is project closeout and final payment. Closeout items include record drawings or as-built drawings that depict the final installation of each system. Operation and maintenance manuals are also required to provide a guide of proper use and maintenance of the installed equipment, as well as test reports and any warranties that are applicable to the completed systems. It is beneficial for the owner to have all operation and maintenance manuals bound together to prevent the loss of valuable equipment information. Upon receipt of all closeout documents, the GC submits a request for a final invoice to the owner. **See Figure 1-13.**

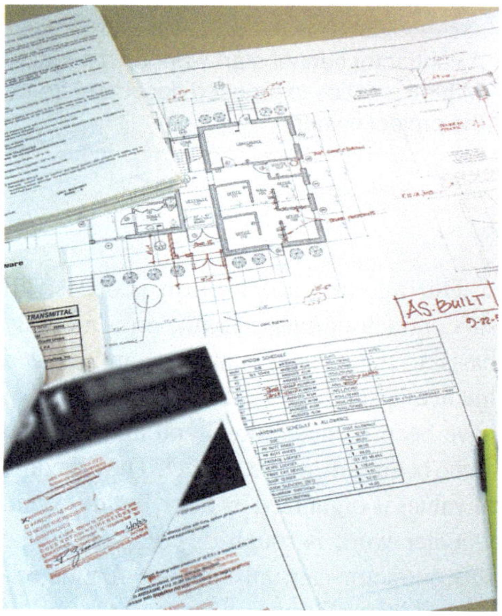

Closeout items include as-built drawings that depict the final installation of each system.

Punch Lists

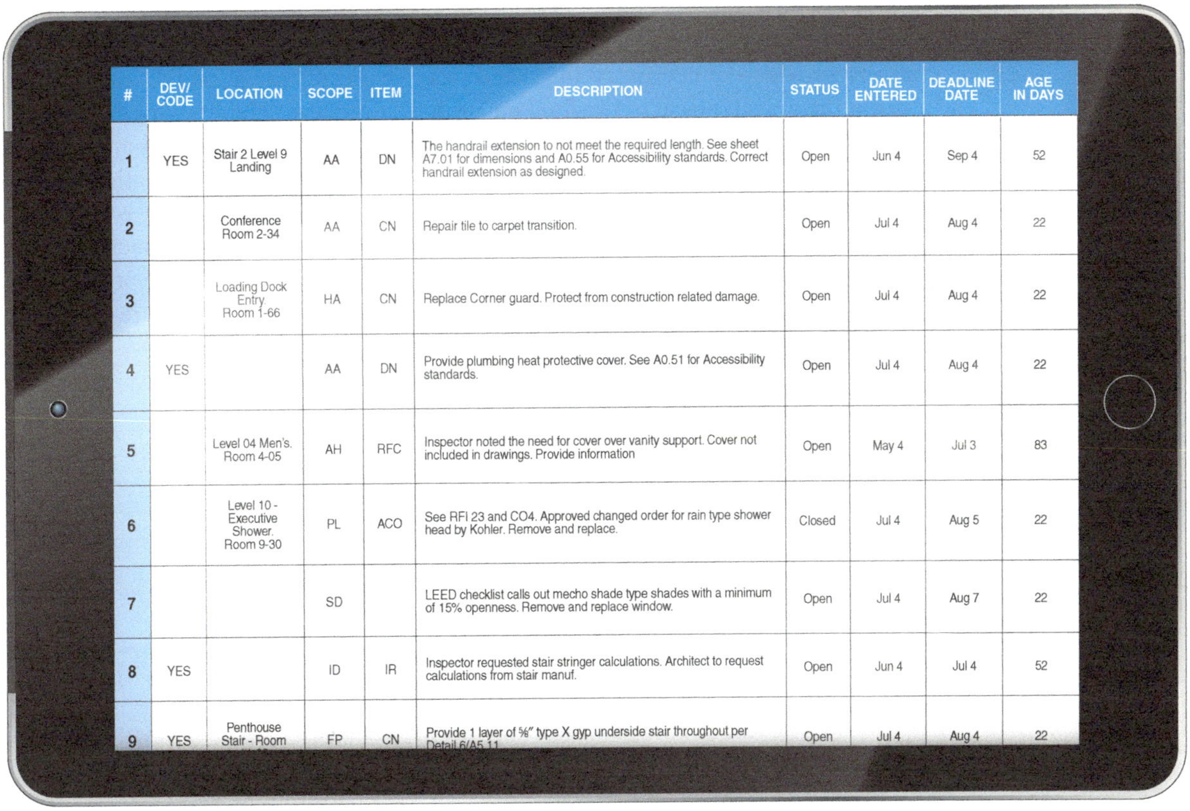

#	DEV/ CODE	LOCATION	SCOPE	ITEM	DESCRIPTION	STATUS	DATE ENTERED	DEADLINE DATE	AGE IN DAYS
1	YES	Stair 2 Level 9 Landing	AA	DN	The handrail extension to not meet the required length. See sheet A7.01 for dimensions and A0.55 for Accessibility standards. Correct handrail extension as designed.	Open	Jun 4	Sep 4	52
2		Conference Room 2-34	AA	CN	Repair tile to carpet transition.	Open	Jul 4	Aug 4	22
3		Loading Dock Entry. Room 1-66	HA	CN	Replace Corner guard. Protect from construction related damage.	Open	Jul 4	Aug 4	22
4	YES		AA	DN	Provide plumbing heat protective cover. See A0.51 for Accessibility standards.	Open	Jul 4	Aug 4	22
5		Level 04 Men's. Room 4-05	AH	RFC	Inspector noted the need for cover over vanity support. Cover not included in drawings. Provide information	Open	May 4	Jul 3	83
6		Level 10 - Executive Shower. Room 9-30	PL	ACO	See RFI 23 and CO4. Approved changed order for rain type shower head by Kohler. Remove and replace.	Closed	Jul 4	Aug 5	22
7			SD		LEED checklist calls out mecho shade type shades with a minimum of 15% openness. Remove and replace window.	Open	Jul 4	Aug 7	22
8	YES		ID	IR	Inspector requested stair stringer calculations. Architect to request calculations from stair manuf.	Open	Jun 4	Jul 4	52
9	YES	Penthouse Stair - Room	FP	CN	Provide 1 layer of ⅝" type X gyp underside stair throughout per Detail 6/A5.11	Open	Jul 4	Aug 4	22

Figure 1-12. A punch list is used to indicate items that need to be repaired or work that needs to be completed.

Operation and Maintenance Manuals

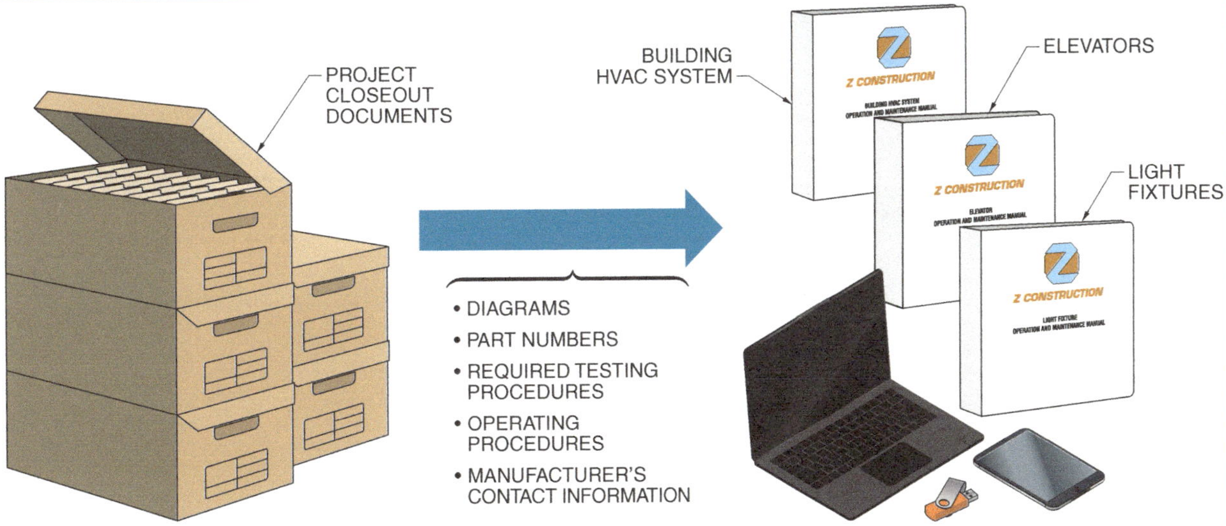

Figure 1-13. Operation and maintenance manuals are kept together for easy reference.

SUMMARY

Drawings and specifications developed by architects and engineers are used by the professionals on a project and are created during the design phase. They provide the necessary information for building based on the owner's needs, from the design phase to the final construction process. The drawings illustrate the scale and proportion of the building as well as how the materials are assembled and connected. The specifications provide detailed written information regarding materials and equipment for producing, testing, and executing the work. General contractors and subcontractors perform the construction of the project as specified in the drawings and specifications after successfully bidding on the project. OEMs and representative firms work with owners and architects to procure the required materials and equipment for the project.

CHAPTER 1

Project Professionals, Design Phases, and the Construction Process

Review

Name ______________________ Date ______________

True-False

T F **1.** The owner of a project determines the parameters of the necessary building space based on items such as property size and shape, desired interior and exterior spaces, and available project budget.

T F **2.** A professional engineer (PE) is an individual or commercial firm that specializes in a particular trade or field of construction.

T F **3.** OEM-standardized VDV systems are also known as "proprietary systems."

T F **4.** The most common method for bidding on a project from a contractor's perspective is the labor-only method.

T F **5.** Many OEMs typically provide warranties for VDV systems for a period of 5 years to 25 years.

Multiple Choice

______________ **6.** When designing a building, an architect must comply with ___.

A. the Americans with Disabilities Act (ADA) of 1990
B. Environmental Protection Agency (EPA) lead-removal rules
C. Occupational Safety and Health Administration (OSHA) regulations
D. Underwriter's Laboratories (UL) guidelines

______________ **7.** A(n) ___ oversees all project coordination between subcontractors so they may work efficiently together and have access to work areas, which ensures a steady sequence of work.

A. architect
B. general contractor
C. subcontractor
D. owner

______________ **8.** ___ is the initial design phase that defines the general scope and conceptual design of a project.

A. Coordination of requirements
B. Design development
C. Drawing development
D. Schematic design

_______________ **9.** VDV systems cabling must not exceed the maximum allowable distance of ___ from work area outlets to the network equipment.
   A. 25 m (82′)
   B. 50 m (164′)
   C. 75 m (246′)
   D. 100 m (328′)

_______________ **10.** A ___ would be best served by a cost-effective generic VDV system.
   A. multiple-building college campus
   B. hospital
   C. single-tenant office campus
   D. tenant office space with a two-year lease

Completion

_______________ **11.** A(n) ___ is an individual, a group, or another entity that enters into a contract with designers, contractors, and other professionals to obtain goods and services provided by the construction industry.

_______________ **12.** A(n) ___ is an individual or commercial firm that is hired to provide goods and services for the undertaking of construction activities.

_______________ **13.** A(n) ___ is an individual with education, training, and expertise in the design, implementation, and integration of VDV systems and infrastructure.

_______________ **14.** ___ is the process of taking an abstract idea and developing it into a working solution.

_______________ **15.** ___ is the practice of constructing a building using processes that are environmentally responsible and resource-efficient throughout the life cycle of the building.

_______________ **16.** ___ signage is commonly found in lobbies of commercial buildings and provides a campus map or a building floor plan.

_______________ **17.** A(n) ___ is the addition of omitted or updated information to a set of construction drawings, technical specifications, or a contract.

_______________ **18.** A(n) ___ is a document used for clarifying scope and the coordination of project elements that arise during the construction process.

_______________ **19.** A(n) ___ is a document that depicts and clarifies minor changes to a print or other project document.

_______________ **20.** A(n) ___ is a list compiled by the owner and/or lead designer during the walk-through of a completed project that indicates items that must be repaired or completed before project finalization.

Short Answer

21. Explain the role of owner's representatives.

22. List the types of engineering professionals overseen by the architect when designing the facility's systems.

23. What is LEED?

24. What are construction documents?

25. Explain when the cost-plus approach would be used for bidding on a project.

CHAPTER

2

Industry Codes and Standards

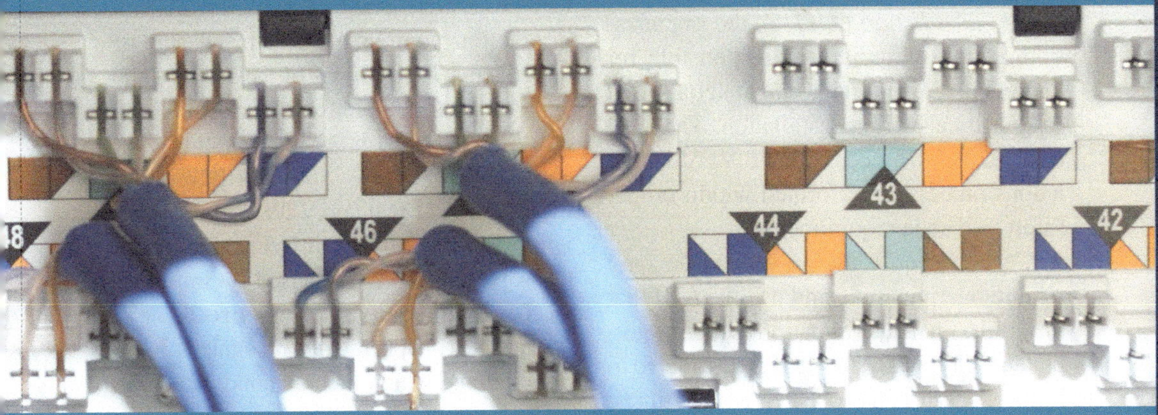

The architects and engineers that develop prints and specifications for building projects rely on industry codes and standards to meet life safety requirements as well as the requirements of the building owner. Industry codes and standards are developed by various industry associations to provide safety and performance guidelines for all trades concerning the design and construction of buildings and the systems installed within them.

Standards are sets of written definitions and guidelines for designers and manufacturers developed to provide the necessary requirements for products and their service and operation. Codes can be a legally enforceable standard when adopted by a government agency or municipality. Codes are also written into construction contracts. To meet the design and installation requirements, architects, engineers, and installers must be familiar with the codes and standards associated with their respective disciplines.

Codes and standards are developed and written by various industry associations. The authority having jurisdiction (AHJ) is responsible for the enforcement of codes. During the design and installation of VDV systems, codes and standards must be available and referenced as required. While a code may provide a safe and stable installation or design, the standard ensures that the finished construction adheres to specified performance requirements.

Objectives

- Explain the role of an authority having jurisdiction (AHJ) as it applies to codes and standards.
- Explain the purpose of the NEC® and how it is applied to VDV systems.
- Describe the development and differences between the various codes and standards.
- List the VDV industry associations.
- Describe how industry associations, codes, and standards influence VDV prints and specifications.
- List the codes and standards applicable to VDV systems projects and explain the meaning of the codes and standards.

Learner Resources

atplearningresources.com
access code 783821

THE AUTHORITY HAVING JURISDICTION

The process of designing, constructing, and occupying a new building requires the input and approval of the authority having jurisdiction (AHJ) on the project. An *authority having jurisdiction (AHJ)* is the organization, office, or individual responsible for approving the equipment and materials used for electrical wiring installation. The AHJ ensures that the electrical and VDV installations process is performed within proper safety and code requirements.

Certain requirements of a code may require interpretation, and it is the AHJ that makes the final determination as to the interpretation. Typically, the AHJ is considered to be the inspector on a job site. At other times, the on-the-job inspector is a designee of the AHJ. For example, the AHJ can be a federal, state, local, or other regional department; an individual such as a fire chief or fire marshal; or a labor department, health department, or tribal council. For example, with VDV installations, the AHJ is typically a building official, electrical inspector, or other person having statutory authority. Understanding the responsibilities of and cooperating with the AHJ is required for all electrical and VDV installations.

THE NATIONAL ELECTRICAL CODE® (NEC®)

A *code* is a regulation and requirement devised to maintain uniformity of work and proper standards of procedure. A code is designed to protect the health and welfare of the public. The *National Electrical Code® (NEC®)* is a standard on practices for the design and installation of electrical VDV systems.

The *National Fire Protection Association® (NFPA®)* is a national organization that provides guidance in assessing the hazards of the products of combustion. The NFPA sponsors the development of the NEC®, which indicates how electrical and VDV systems must be installed and how work must be performed. Fluency with the NEC® is also used as part of electrician licensing requirements. Electrician licensing requirements vary, depending on each state's governing body.

The NEC® is written to include specific definitions that apply to each of its documented sections. The NEC® specifies general installation requirements that installers must understand. The purpose of the NEC® is the "practical safeguarding of persons and property from the hazards arising from the use of electricity." The NEC® specifies the requirements for electrical and VDV wiring installation to help prevent injuries or damage to property from fires caused by improper electrical installations. **See Figure 2-1.**

The NEC® specifically recognizes VDV installations, which include conductors and related equipment, raceways, copper connectivity, and fiber-optic connectivity used for a variety of applications associated with VDV systems. With the constant convergence of Internet Protocol (IP) systems and other electrical systems such as IP lighting, building controls, building automation, security systems, and fire alarm systems, it is a requirement for architects, engineers, and installers to understand and apply the NEC® requirements to provide a legally compliant system.

NEC® Article 90 – Introduction

The NEC® is prepared by the National Electrical Code Committee, which consists of a correlating committee and 18 separate code-making panels. The code-making panels are composed of industry experts and have specific responsibilities relating to various topics. Article 90 of the NEC® is an introduction that explains how the different articles are applied to the entire NEC®.

NEC® Section 90.1 – Purpose. The purpose of the NEC® is the practical safeguarding of persons and property from hazards arising from the use of electricity and must be adhered to whether applying the NEC® or submitting new proposals to modify the NEC®. Per Subsection 90.1(A), the NEC® is not intended to be used as an instruction manual nor as a design specification. It contains the rules and necessary provisions for electrical and VDV systems but leaves the design and layout of those systems to others.

The NEC® does not provide reasons for the inclusion of its rules in its text. In order to understand the reason for a rule, it must be known in which version of the NEC® the rule was first placed. Also, the First Draft for Public Comment (previously the Report on Proposals) for that year's version of the NEC® must be reviewed. The code panel statements (or substantiations) for the related public inputs (proposals) should then be reviewed.

For example, Article 353 covers the use, installation, and construction specifications for high-density polyethylene (HDPE) conduit and its associated fittings. Article 353 was new to the 2005 NEC®. The Report on Proposals for the 2005 Code cycle, in the substantiation for the proposal to add the article, indicates that HDPE is a listed product that is restricted in its uses and that the new article is needed to provide the installation requirements and construction specifications for HDPE.

The NEC® only contains requirements that are necessary for safety. Subsection 90.1(B) states that following the provisions of the NEC® only results in an installation that is "essentially free from hazard." The NEC® rules and requirements are a minimum standard. Following the NEC® provisions does not ensure that electrical and VDV installations are suitable for future use or expansion.

NEC® Section 90.2 – Scope. In the 2017 NEC®, the scope of the NEC® in Subsection 90.2(A) states that the NEC® not only covers the installation of electrical and VDV wiring and systems but also the removal of wiring and systems. Often, abandoned, disconnected, or obsolete electrical and VDV wiring and equipment remain in place when new circuits and new systems are added in existing buildings. Leaving disconnected and obsolete equipment, wiring, and abandoned electrical and communications cables in place over time creates a much greater smoke and fuel load if a fire were to occur. The removal of abandoned electrical and VDV material reduces or limits the fuel load of a building and the resulting damage from a fire.

National Electrical Code® (NEC®)

Section	Title
Article 90	Introduction
Chapter 1	General
Chapter 2	Wiring and Protection
Chapter 3	Wiring Methods and Materials
Chapter 4	Equipment for General Use
Chapter 5	Special Occupancies
Chapter 6	Special Equipment
Chapter 7	Special Conditions
Chapter 8	Communication Systems
Chapter 9	Tables

Figure 2-1. The NFPA® sponsors the development of the NEC®, which indicates how electrical systems must be installed and how work must be performed.

Before the NEC® is applied, the scope must be reviewed to determine if the electrical installation must be compliant with the NEC®. Installations not covered are given in Subsection 90.2(B). Often, the distinction between what is covered and not covered is minimal, which can lead to different interpretations. For example, installations for electrical utilities are not covered if they apply to the metering, generation, transmission, or distribution of electric energy. However, electrical installations used by an electric utility that are not an integral part of the electrical energy transmission or distribution are covered.

NEC® Section 90.3 – Code Arrangement. The NEC® is divided into the introduction section and nine chapters. Chapters 1, 2, 3, and 4 are general applications for all systems and equipment. Chapters 6, 7, and 8 are applications for communications and information technology equipment, including VDV systems. Chapter 9 includes applicable reference tables.

NEC® Section 90.4 – Enforcement. All applications of the NEC® are ultimately subject to approval by the AHJ. The NEC® is designed to be used as a legal document. Any interpretations and approval of equipment rest with the AHJ. The AHJ can permit alternate methods or even waive specific requirements if they believe the installation is equally effective at meeting the intended objective. **See Figure 2-2.**

Authority Having Jurisdiction (AHJ)

Duties and responsibilities:

1. Make interpretations of NEC® rules.
2. Approve equipment and materials.
3. Grant special permission. (see Art. 100 definition)
4. Waive specific requirements. (must be in writing)
5. Permit alternate methods. (must be in writing)

Figure 2-2. The authority having jurisdiction (AHJ) is responsible for enforcement of the NEC® per Section 90.4.

NEC® Section 90.5 – Mandatory Rules, Permissive Rules, and Explanatory Material. The NEC® contains two types of rules: mandatory and permissive. Mandatory rules are designated by using the words "shall" or "shall not." For example, Section 110.23 is a mandatory rule stating that "any unused current transformer which may be associated with circuits that may be energized shall be short-circuited."

Permissive rules are rules that are not required but are permitted. Permissive rules are designated by using terms such as "shall be permitted" or "shall not be required." For example, Subsection 250.21(A) lists AC systems of 50 V to 1000 V that shall be permitted to be grounded but shall not be required to be grounded.

NEC® Section 90.7 – Examination of Equipment for Safety. In general, equipment and materials referred to in the NEC® are not required to be examined prior to use. It is intended that factory-installed internal wiring is sufficient and not within the scope of the installer's responsibility. The only exception to factory-installed equipment is where such equipment or materials have been damaged or altered to a degree that requires examination by the installer. However, installers of electrical equipment need to ensure that the equipment is listed, identified, or otherwise labeled as being suitable for the intended use.

In the 2017 NEC®, the term "suitability" is explained as "…shall be determined by the application of requirements that are compatible with this *Code*." Furthermore, Informational Note 3 explains that Informative Annex A contains a list of product safety standards that are compatible with the Code. Installers should also be aware that many state and local authorities require product listings for equipment used within their jurisdiction. In addition, OSHA includes requirements for listing and listing laboratories in its regulations. Article 100 provides definitions of "listed," "labeled," and "identified."

NEC® Section 90.8 – Wiring Planning. Subsection 90.1(B) and Section 90.8, taken together, indicate that installations completed in accordance with the NEC® merely meet a minimum established standard. The installations may not be adequate for future use, and while it is not required, allowance for future expansion is generally a preferred design method. Essentially, the NEC® is a minimum standard. The designer or installer can choose to provide for future use if desired. Installations in accordance with the NEC® are not necessarily efficient and may not meet the future needs and demands for electrical and VDV systems.

NEC® Section 90.9 – Units of Measurement. The NEC® utilizes a dual system of measurement. Metric units of measurement are shown utilizing the International System of Units (SI). The US customary units are shown using standard inch-pound units. Subsection 90.9(B) requires that the SI units appear first and the inch-pound units follow immediately and are shown in parentheses. Subsection 90.9(C) designates the rules for conversion between units and indicates where such conversion is required and where it is optional. Subsection 90.9(D) states that compliance with either unit of measurement is permitted.

NEC® Article 100 – Definitions

Article 100 – Definitions contains terms and definitions listed in alphabetical order that are required for the application of the code. Only terms that are referenced in two or more articles of the NEC® are listed in Article 100. Applying and interpreting the NEC® incorrectly can often be traced to a failure to properly understand the terms used.

In addition, terms may be defined in the particular article to which they apply. Many articles contain a separate section with the specific definitions for that article only. In most cases, specific definitions are placed in the "dot" 2 section of the article. For example, the definition for abandoned optical fiber cable is covered in Article 770.2. *Note:* Most definitions are included in Article 100, unless otherwise noted. **See Figure 2-3.**

Many electrical terms are also applicable to VDV applications and can be applied to the design specifications and drawings. Designers and installers of VDV equipment and systems must be familiar with the definitions for the following NEC® terms:

- approved
- authority having jurisdiction (AHJ)
- bonded (bonding)
- building
- cabinet
- cable, coaxial
- cable, optical fiber, composite
- cable, optical fiber, conductive
- cable, optical fiber, nonconductive
- cable routing assembly
- communications equipment
- grounded (grounding)
- grounded conductor
- grounding electrode conductor
- information technology equipment (ITE)
- innerduct
- intersystem bonding termination
- listed
- plenum
- premises wiring (system)
- raceway
- raceway, communications
- signaling circuit
- uninterruptable power supply
- weatherproof

NEC® Article 250 – Grounding and Bonding

Article 250 – Grounding and Bonding in NEC® chapter 2, Wiring and Protection, covers the general requirements for grounding and bonding of electrical and VDV installations. Table 250.3 – Additional Grounding and Bonding Requirements provides additional information or modifications to the grounding and bonding requirements shown in Article 250. The table includes a list of conductors and equipment and the relevant articles and/or sections of the NEC® in which the grounding and bonding requirements for those conductors and equipment can be found. For example, Section 770.100 is listed for optical fiber cables and provides specifications for grounding electrode conductors (GECs). **See Figure 2-4.** Section 250.94 provides detailed bonding requirements for communication systems, including intersystem bonding conductors and termination requirements.

NEC® Definitions	
100	Generally, only those terms that are used in two or more NEC® articles are defined in Article 100.
OTHER ARTICLES	Some articles contain a section that has definitions of terms used in that particular article only. For example, see 517.2, 550.2, 551.2, 680.2, and 690.2. Generally, article definitions will be located in the ".2" Sections.
INDIVIDUAL SECTIONS	Sometimes the definition is included in the specific NEC® section to which the definition is applicable. For example, see the definition of "point of entrance" in 770.2.

Figure 2-3. The definitions for terms used in the NEC® appear in Article 100 or in the particular article or section in which they are used.

Grounding Electrode Conductors (GECs)

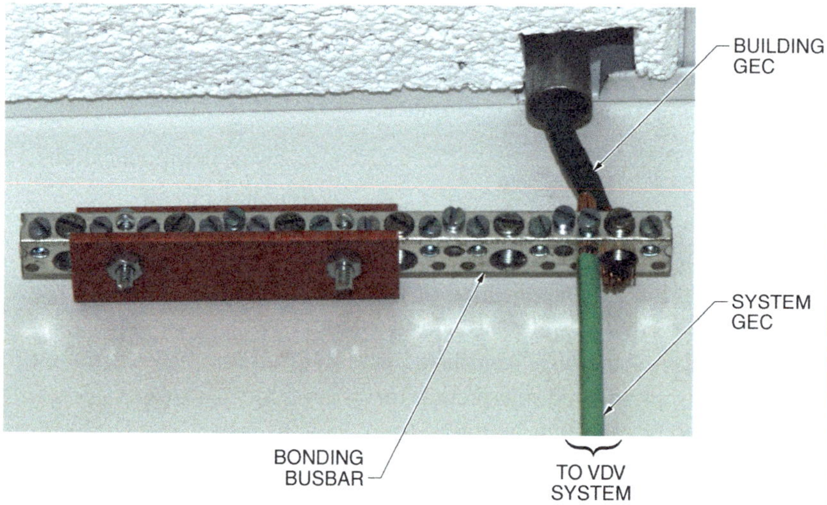

Figure 2-4. NEC® Article 250 covers the general requirements for the grounding and bonding of electrical and VDV installations.

NEC® Article 300 – General Requirements for Wiring Methods and Materials

Article 300 – General Requirements for Wiring Methods and Materials in NEC® chapter 3, Wiring Methods and Materials, covers the installation and wiring methods for conductors, cables, and raceways for applications up to 1000 V. The requirements are defined through the various descriptions and tables throughout the chapter. Comprehensive descriptions are provided for wiring methods and materials from the point of service to termination at the last receptacle in the electrical distribution system, which includes the conductors for general wiring and service entrance cables.

Other items outlined in chapter 3 include cabinets, cutout boxes, and meter socket enclosures; outlet, device, pull, and junction boxes; and various conduit and underground raceways. The requirements outlined in Article 300 provide the necessary pathways and spaces located within and between premises for providing the correct size and protection of VDV applications.

NEC® Article 645 – Information Technology Equipment

Article 645 – Information Technology Equipment in NEC® chapter 6, Special Equipment, covers equipment, power supply wiring, equipment interconnection wiring, and the grounding of information technology equipment (ITE) and systems. Typical design considerations for the type of VDV cables installed in spaces used for environmental air, or plenums, is detailed. The term "plenum," which is defined in Article 100 as a compartment or chamber in which one or more air ducts are connected and that forms part of the air distribution system, is a major design consideration for ITE rooms. **See Figure 2-5.**

ITE rooms, also referred to simply as equipment rooms, are commonly self-contained, air-conditioned spaces. The NEC® list of cables used in these spaces is required for reference purposes for proper installation. Cable selection and listing in these spaces is critical for meeting life safety requirements and is a major design consideration for developing drawings and specifications. The requirements for cable pathways above the celling and below the raised floors that are commonly installed in equipment rooms are also outlined in NEC® Article 645. Additional design items for ITE rooms are also provided, including grounding, special HVAC equipment, surge protection devices, and power distribution requirements.

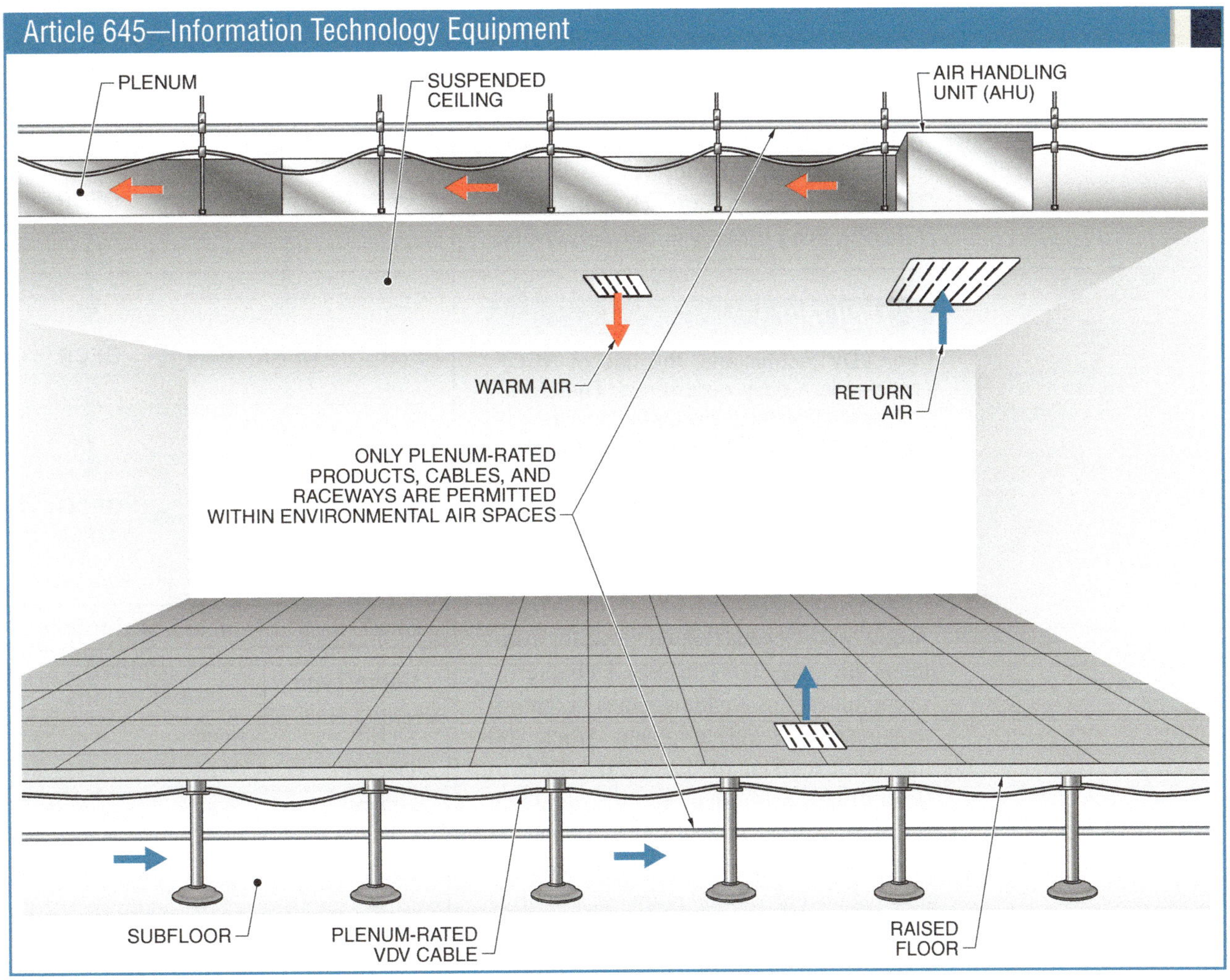

Figure 2-5. NEC® Article 645 covers equipment, power supply, wiring, equipment interconnection wiring, and the grounding of information technology equipment (ITE) and systems.

NEC® Article 770 – Optical Fiber Cables

Article 770 – Optical Fiber Cables in NEC® chapter 7, Special Conditions, covers the installation of optical fiber (fiber-optic) cables. Typical design considerations for optical fiber cable selection and pathways are detailed. The requirements for fiber-optic cables installed outdoors or underground and the method of entry when outside cables enter a commercial or residential premises are provided.

In many applications, fiber-optic cables entering a premises do not meet the requirements of the space used for environmental air. For such situations, Section 770.48 provides direction for the distance and raceway types allowed. The grounding methods for various listed fiber-optic cables are provided, including points of reference in fiber-optic cable selection and applications. Fiber-optic cable listings for conductive and nonconductive fiber-optic cables for use in plenum, riser, and general-purpose applications are detailed in Tables 770.154(a) and (b) and Table 770.179. **See Figure 2-6. See Appendix.**

TECH FACT

For VDV technicians, the use of NEC® articles is a daily occurrence. Therefore, all technicians must always be familiar with these articles and have access to them.

NEC® Article 800 – General Requirements for Communications Systems

Article 800 – General Requirements for Communications Systems in NEC® chapter 8, Communications Systems, covers communication circuits and equipment. Article 800 has been revised in the 2020 NEC® to consolidate redundant requirements in previous editions of the NEC® by placing them into one general article. Article 800 also details the various points of premises entry (underground or overhead), raceway types, terminations, grounding requirements, device protection, and installation methods. Grounding requirements and lightning protection for copper distribution are identified for various applications, including grounding electrode types, bonding busbars, intersystem bonding conductor terminations, and installation methods within buildings.

Article 800 also covers the bonding of grounding electrodes. In the event that there is a separate grounding electrode installed for the VDV system, it must be bonded to the grounding electrode system for the building electrical service. The minimum size of the bonding conductor must be 6 AWG copper. The electrodes must be bonded so that there is no difference of potential between them. **See Figure 2-7.** Tables 800.154 (a), (b), (c), and (d) provide routing directions, listings, and permitted substitutions for various communication cables, which are required for the development of drawings and specifications.

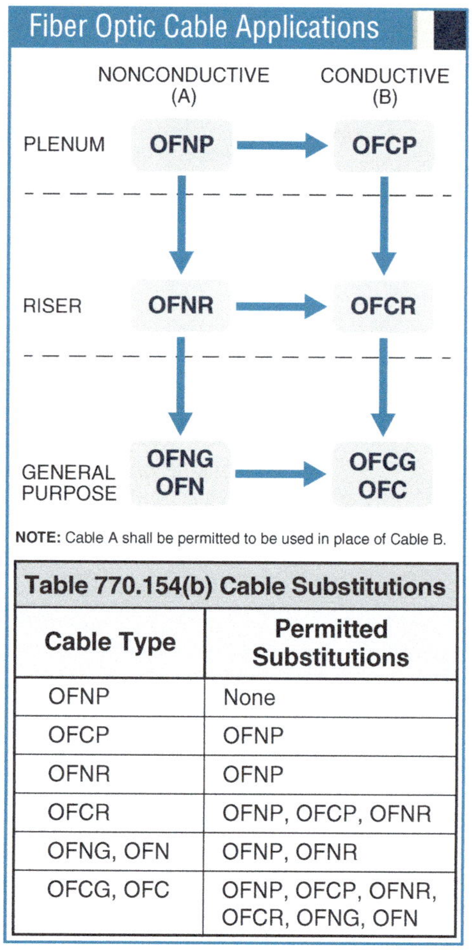

Table 770.154(b) Cable Substitutions	
Cable Type	**Permitted Substitutions**
OFNP	None
OFCP	OFNP
OFNR	OFNP
OFCR	OFNP, OFCP, OFNR
OFNG, OFN	OFNP, OFNR
OFCG, OFC	OFNP, OFCP, OFNR, OFCR, OFNG, OFN

Figure 2-6. NEC® Article 770 covers the installation of optical fiber (fiber optic) cables.

VDV Grounding Electrodes

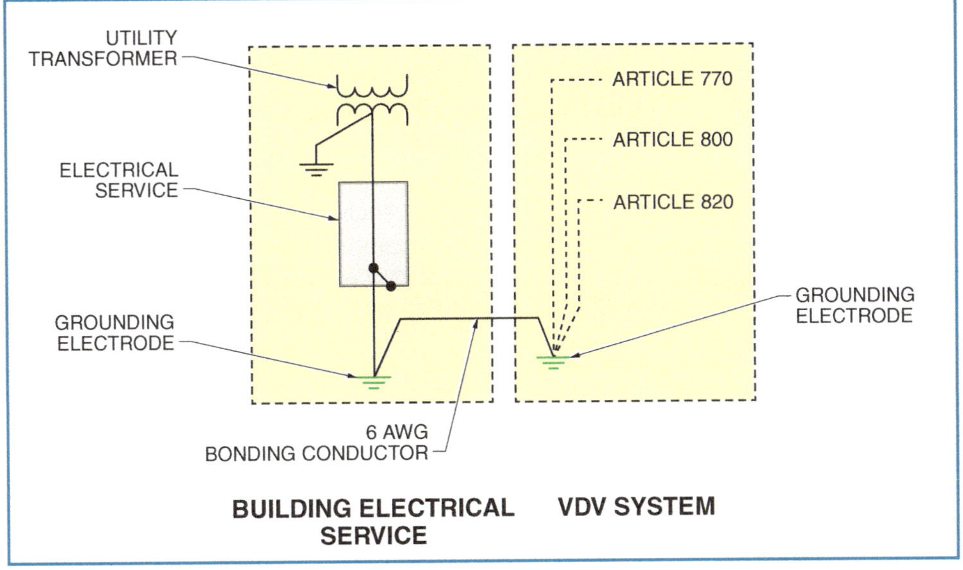

Figure 2-7. A separate grounding electrode for a VDV system shall be bonded to the grounding electrode system for the building electrical service.

NEC® Article 830 – Network-Powered Broadband Communication Systems. Article 830 – Network-Powered Broadband Communication Systems in NEC® chapter 8, Communications Systems, covers network-powered broadband communications systems that provide any combination of voice, audio, video, data, and interactive services through a network interface unit. As with the previous articles, cable types, methods of entry, grounding, raceways, and listings are provided. The interpretation of these requirements is critical to the development of drawings and specifications.

NEC® Article 840 – Premises-Powered Broadband Communication Systems. Article 840 – Premises-Powered Broadband Communication Systems in NEC® chapter 8, Communications Systems, recognizes optical, twisted-pair, and coaxial cables that provide a broadband signal (optical, electric, or wireless) to the premises via a network terminal. The network terminal converts the signal into voice, audio, data, wireless, and optical interactive services. The network terminal is a network device on the premises that is connected to a communications service provider and powered at the premises.

Section 840.160 addresses Power over Ethernet (PoE) on the IP platform and the powering of IP devices over communication cables. IP devices include voice, data, and video applications. As with previous articles, cable types, methods of entry, grounding, raceways, and listings are also provided in Article 840. The interpretation of the requirements is critical to the development of drawings and specifications.

VDV INDUSTRY ASSOCIATIONS

Every industry has trade associations and related organizations that collaborate with engineers, designers, and OEMs to provide standardization and educational resources for each specific discipline. Several associations provide guidelines and standards for VDV equipment and systems. **See Appendix.** The guidelines and standards provide design and installation practices. If any standards contradict each other, design and installation professionals must then determine the most practical, cost-effective solution that meets life-safety and performance requirements. Industry associations for VDV equipment and systems include the following:

- American National Standards Institute (ANSI)
- Building Industry Consulting Service International® (BICSI®)
- Construction Specifications Institute® (CSI®)
- International Electrotechnical Commission (IEC)
- Institute of Electrical and Electronics Engineers (IEEE)
- International Organization for Standardization® (ISO®)
- Telecommunications Industry Association® (TIA®)
- Underwriters Laboratories® (UL®)

American National Standards Institute (ANSI)

The *American National Standards Institute (ANSI)* is a private nonprofit organization that oversees the development of voluntary consensus standards for products, services, processes, systems, and personnel in the United States. ANSI also coordinates US standards with international standards so that US-made products can be used worldwide. ANSI standards are developed by representatives of other standards organizations, government agencies, consumer groups, companies, and other associations. ANSI standards ensure that the characteristics and performance of products are consistent, that definitions and terms are consistent, and that products are tested in the same manner.

Building Industry Consulting Service International (BICSI)

Building Industry Consulting Service International (BICSI) is a professional telecommunications standards and certification organization. Since the official breakup of AT&T in 1984, BICSI has assumed the role of AT&T's complimentary service, Building Industry Consulting Service (BICS). BICS was originally formed to assist with the development and design for information transport systems (ITS), such as the ANSI/TIA 568B structured cabling system standard.

BICSI professional credentials include the Registered Communications Distribution Designer (RCDD) designation, established in 1984. An RCDD is an individual with education, training, and expertise in the design, implementation, and integration of telecommunications/data transport systems and infrastructure. Other industry credential programs offered by BICSI include Registered Telecommunications Project Manager (RTPM), Data Center Design Consultant (DCDC), and Outside Plant Designer (OSP).

Construction Specifications Institute (CSI)

The *Construction Specifications Institute (CSI)* is an organization that develops standardized construction specifications. The CSI is a national association of construction industry professionals who are experts in building construction and the materials used therein. The work of the CSI is focused in the areas of standards and publications, construction industry professional certifications, and continuing education.

The CSI publishes the MasterFormat®, which is a standardized classification for construction information, such as detailed construction project cost estimates or construction specifications. The MasterFormat originally consisted of 16 divisions representing broad categories of construction, such as masonry, concrete, electrical systems, interior finishes, and mechanical systems. In November 2004, a revised edition was published that expanded the categories to contain front-end documents, 5 subgroups, and 50 divisions (including several unused divisions reserved for future use). **See Figure 2-8.**

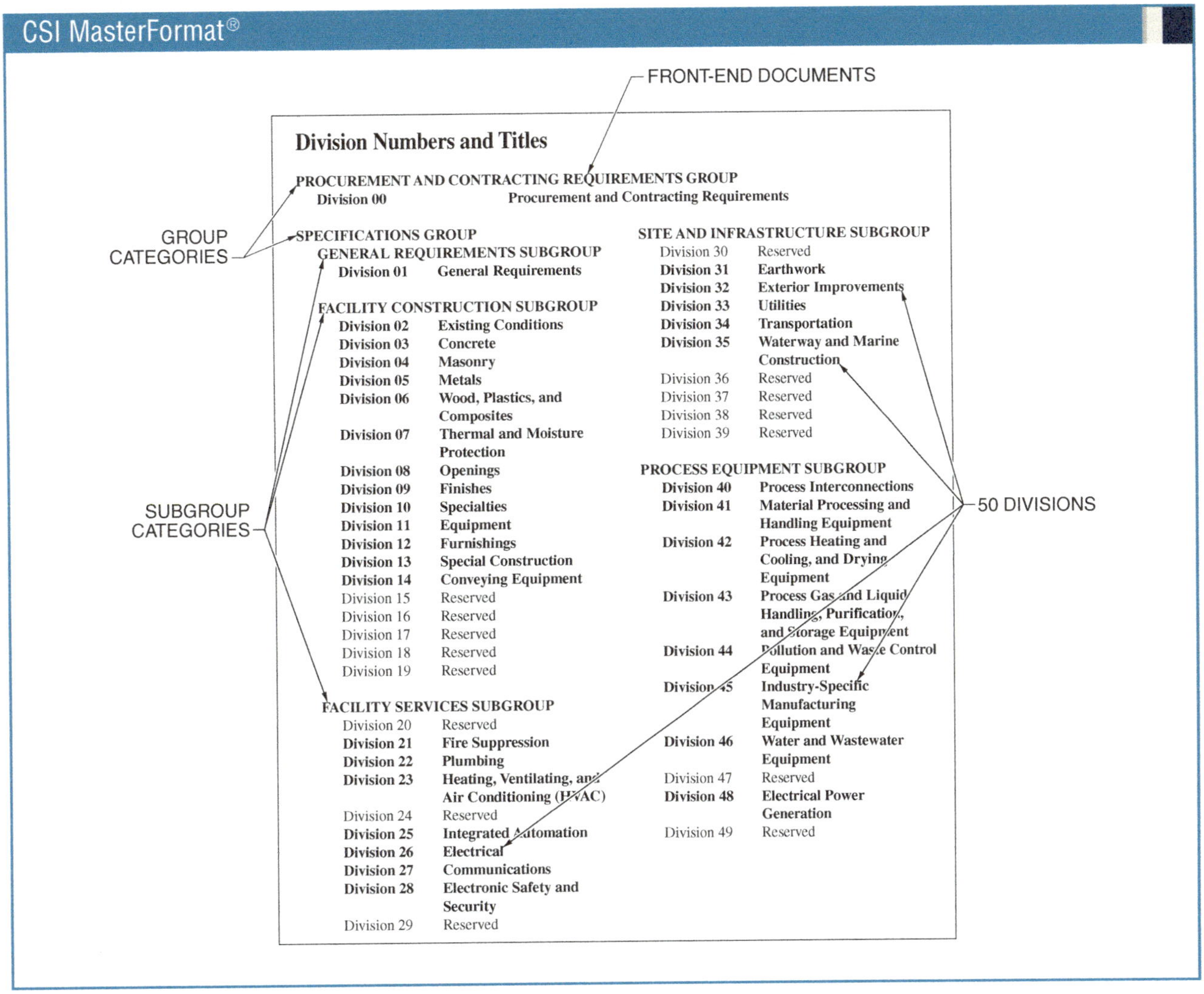

Figure 2-8. The CSI MasterFormat® contains front-end documents, 5 subgroups, and 50 divisions for use in organizing construction information.

International Electrotechnical Commission (IEC)

The *International Electrotechnical Commission (IEC)* is an organization that develops international safety standards for electrical equipment. The IEC standards reduce safety hazards that can occur from unpredictable circumstances when using electrical test equipment such as test instruments and meters. The IEC is a global organization that publishes consensus-based international standards and manages conformity assessment systems for electric and electronic products, systems, and services referred to as electrotechnology.

Institute of Electrical and Electronics Engineers (IEEE)

The *Institute of Electrical and Electronics Engineers (IEEE)* is a technical professional society that produces standards for telecommunications, local area networks, and computing. The IEEE is a professional association formed in 1963 from the combination of the American Institute of Electrical Engineers and the Institute of Radio Engineers and has more than 423,000 members in over 160 countries. The objective of the IEEE is the educational and technical advancement of electrical and electronic engineering, telecommunications, computer engineering, and allied disciplines. The IEEE provides numerous standards and publications and hosts educational conferences for industry professionals. The IEEE has developed and collaborated with numerous industry associations and societies to develop the most current industry standards.

International Organization for Standardization (ISO)

The *International Organization for Standardization (ISO)* is a developer of voluntary international standards and facilitates world trade by providing common standards between nations. The ISO is an independent, nongovernmental organization, of which the members are the standards organizations of the 164 member countries. The ISO had produced over 20,000 standards, ranging from manufactured products and technology to food safety, agriculture, and healthcare. The standards aid in the creation of products and services that are safe, reliable, and high quality. ISO standards also safeguard consumers and end-users of products and services, while ensuring that certified products conform to the minimum standards set internationally.

Telecommunications Industry Association (TIA)

The *Telecommunications Industry Association (TIA)* is an international trade group that represents several hundred telecommunications companies. The TIA is accredited by ANSI to develop voluntary, consensus-based industry standards for various information and information communication technologies (ICT) products. The TIA currently represents about 400 companies.

The TIA's Standards and Technology Department operates 12 engineering committees, which develop guidelines for private radio equipment, cellular towers, data terminals, satellites, telephone terminal equipment, accessibility, VoIP devices, structured cabling, data centers, mobile device communications, multimedia multicast, vehicular telematics, healthcare ICT, machine-to-machine communications, and smart utility networks. The TIA encourages engineers who represent the OEMs and/or users of network equipment technology products and services to become engaged in TIA's engineering committees, by voting and submitting technical contributions for inclusion in future standards.

Underwriters Laboratories (UL)

Underwriters Laboratories (UL) is an independent organization that tests equipment and products to see if they conform to national codes and standards. UL is a global safety consulting and certification company that maintains offices in 46 countries. Established in 1894 as the Underwriters Electrical Bureau (a bureau of the National Board of Fire Underwriters), it was known throughout the 20th century as Underwriters Laboratories and participated in the safety analysis of many of that century's new technologies.

UL is one of several companies approved to perform safety testing by the US Occupational Safety and Health Administration (OSHA). OSHA maintains a list of approved testing laboratories, known as Nationally Recognized Testing Laboratories, which includes UL. UL has expanded into an organization with 64 laboratories and testing and certification facilities serving customers in 104 countries. It has evolved from its roots in electrical and fire safety to address broader safety issues, such as hazardous substances, water quality, food safety, performance testing, safety and compliance education, and environmental sustainability.

VDV SYSTEMS STANDARDS

A *standard* is an accepted reference or practice. Standard for VDV installations are means by which installations are measured, ensuring that all other work meets the performance criteria. Codes and standards differ and sometimes may contradict each other. Codes exist to protect life and property. Codes are a legal mandate, are legally enforceable, and carry civil penalties if not followed. Standards exist to ensure consistency of performance and interoperability of components. Standards are not enforceable and carry no penalties for noncompliance. Standards are voluntary, but advisable.

The VDV industry standards are relatively new compared to the standards used for other trades. The standards-making bodies saw a need for standardization in the telecommunications industry in the early 1980s. At the time, there was only one telephone company (AT&T/Bell System), and they were responsible for the installation and maintenance of all telecommunications systems and equipment. Although AT&T/Bell System was once a government-sanctioned monopoly, the federal government determined, at the urging of numerous prospective competitors, that the AT&T/Bell System monopoly no longer had any valid purpose. In 1984, the US Department of Justice (DOJ) mandated the divestiture of AT&T/Bell System, which was broken up into smaller companies known as Regional Bell Operating Companies (RBOCs).

Interconnection between independent operating companies and the newly formed RBOCs was unclear and difficult. The cabling in an individual building was not always compatible with the type of cabling or signaling being delivered by the RBOC. In 1985, many service providers representing the telecommunications and computer industries expressed concern over the lack of a standardization for building telecommunication systems. There were local telephone service providers, formerly part of the Bell System and referred to as "baby Bells," who had always operated under their own self-written procedures.

OEMs of cable and components provided further confusion by designing proprietary cabling systems that were not interchangeable with other OEM systems, requiring replacement whenever a new version of their system was released. The release of new systems required re-cabling facilities to accommodate new features and equipment, which resulted in a large amount of abandoned cables and necessitated the re-training of facility managers. Since cable upgrades were constant, a form of standardization needed to be developed.

ANSI/TIA was soon formed and assumed the task of taking each phase or segment of the telecommunications industry and refining that segment of technology into a set of best practices. These best practices defined how components should be manufactured and installed. The beginning of the VDV technology currently in use can be traced back to the ANSI/TIA merger and alliance.

As various OEMs develop new performance standards for their equipment, the performance standards are submitted to ANSI/TIA and considered for issuance as new or updated standards. Typically, standards are updated at five- to seven-year intervals. New standards are developed for the changes in technology.

ANSI/TIA-568, Commercial Building Telecommunications Standard

The ANSI/TIA-568 standard was the first standard developed by the EIA/TIA in 1991. Prior to ANSI, the Electronics Industry Association (EIA), later known as the Electronics Industries Alliance, developed the standard with the TIA. As the VDV industry grew, technology developed, and more information and traffic were placed on data networks. The complete EIA/TIA-568 standard needed to be separated into more specific, manageable standard sections to effectively address the needs of the VDV industry. Also, as technology evolved, the older standards were no longer relevant. Cable transmission capacity was exceeded by the speeds of newer data transmission and equipment became more robust, requiring a more resilient infrastructure. Additionally, the standards grew and spread out into new areas of technology. The current ANSI/TIA-568 set of standards covers a wide range of performance criteria applied to telecommunications systems, including the following:

- generic telecommunications cabling for customer premises
- commercial building telecommunications cabling
- balanced twisted-pair telecommunications cabling and components
- optic fiber cabling and components
- performance testing

For the components of balanced twisted-pair cable systems, ANSI/TIA-568 defines the different work area outlet pin-to-pair versions of T568A and T568B as wiring standards. Functionally, there is no difference between the T568A and T568B standards; the only difference between the two is that the positions of green and orange wires are switched in the connector termination pattern. ANSI/TIA-568 also establishes cable distance limitations, with the breakdown of 295′ (90 m) permanent link length for installed horizontal cabling, an aggregate limit of 32.8′ (10 m) of apparatus cabling or patch cords, and 328′ (100 m) permanent channel link for horizontal work area outlets. **See Figure 2-9.**

Permanent Links

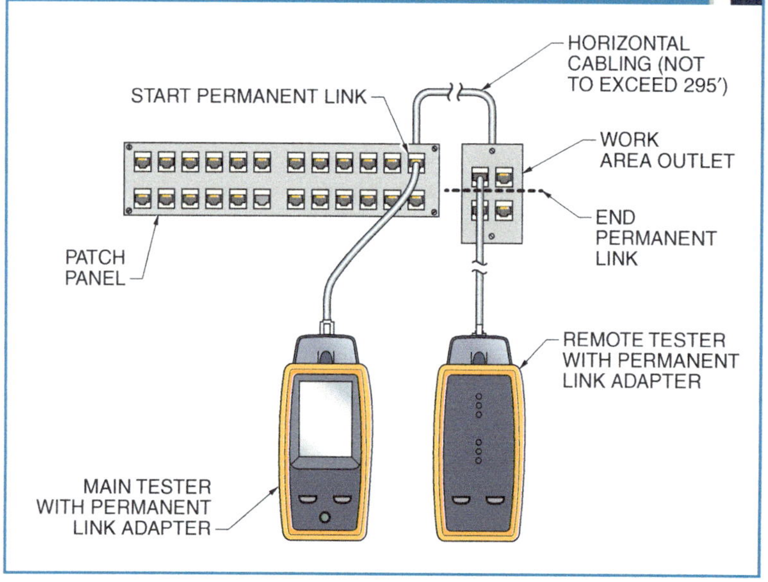

Figure 2-9. ANSI/TIA-568 defines the T568A and T568B wiring standards and establishes cable distance limitations, with 295′ (90 m) permanent link length for installed horizontal cabling.

Currently, ANSI/TIA-568 is on Revision D, which began development in 2017. The standards developed in 1991 (original standard), 1995 (Revision B), and 2009 (Revision C) are considered obsolete. In addition to ANSI/TIA-568, the other primary ANSI/TIA standards associated with VDV applications, systems, and equipment include standard numbers 569, 570, 606, 607, 758, 862, and 942.

ANSI/TIA-569, Telecommunications Pathways and Spaces Standard

The ANSI/TIA-569 standard was developed shortly after the development of the ANSI/TIA-568 standard. The scope of ANSI/TIA-569 provides the necessary telecommunications design requirements as part of the overall architectural design of a building during the construction process. ANSI/TIA-569 was developed to provide guidance and a standard specification for the typical equipment rooms and other pathways and spaces that house telecommunications equipment. The spaces covered under ANSI/TIA-569 include the following:

- entrance facilities
- service entrance pathways
- equipment rooms
- intrabuilding backbone pathways
- telecommunications rooms
- horizontal pathways
- consolidation points and multiuser telecommunications outlet assemblies (MuTOAs)
- areas with electromagnetic interference (EMI)
- firestops

Pertaining to telecommunications spaces, the ANSI/TIA-569 standard requirements define the following:

- environmental conditions, including heat, cooling, and humidity
- room or closet sizing minimums
- door sizes
- sleeves or entrance conduits
- lighting levels

- power levels
- locations within a building
- adjacent space limitations

Pertaining to pathways, the ANSI/TIA-569 standard requirements define the following:

- sizing and capacity limits of conduit and cable tray or ladder tray
- clearance from power
- cohabitation with other technologies
- interbuilding vs. intrabuilding pathways and their use and construction
- underfloor raceways or ducts
- access floors
- perimeter pathways
- surface-mount raceways
- plenum and ceiling spaces
- trench ducts

ANSI/TIA-569 recognizes that tenant and building requirements can change over the life cycle of the building. If proper pathways and spaces are incorporated into the architectural design for the allowance of future modifications, the building and its equipment rooms will be adaptable to the changing requirements and will not require major and costly space changes. **See Figure 2-10.**

ANSI/TIA-569 provides definitions and reference tables detailing conduit sizes and temperature and humidity requirements for telecommunications spaces and access/service provider spaces. The architectural, mechanical, and electrical requirements detailed in the standard must be referenced during the design process when coordinating requirements between various trade disciplines.

ANSI/TIA-570, Residential Telecommunications Infrastructure Standard

The ANSI/TIA-570 standard was developed and released in 1999 to address standardization for residential premises and to detail the requirements for residential telecommunications cabling. The cabling infrastructure specifications within ANSI/TIA-570 include support for security, audio, television, sensor, alarm, and intercom systems. The ANSI/TIA-570 standard is implemented for new construction, additions, and remodeled single-tenant and multitenant buildings.

Future VDV Sizing Requirements		
Number of Equipment Outlets Served	**Minimum Floor Space Requirement***	**Typical Dimensions†**
1 to 200	150	(10 × 15)
201 to 800	400	(20 × 20)
801 to 1600	800	(20 × 40)
1601 to 2400	1200	(30 × 40)

* in sq ft
† in ft

Figure 2-10. The proper sizing of equipment rooms is necessary to house current and future technology requirements.

The scope of ANSI/TIA-570 covers the numerous applications evolving for use in the residence, such as satellite and wireless television, high-speed internet (data and telephone), utility monitoring, and smart home appliances and lighting. Other applications covered by ANSI/TIA-570 include the following:

- security and fire alarm systems
- telecommunications
- energy management (lighting, climate control, and home automation)
- whole-home entertainment systems
- VDV infrastructure
- cable and connecting hardware
- installation requirements
- field test requirements

The ANSI/TIA-570 standard outlines the two following grades of VDV service that could be required for a residential premises:

- Grade 1 provides a generic cabling system that meets the minimum requirements for telecommunications services. Grade 1 provides for telephone, satellite, community antenna television (CATV), and data services. Grade 1 also specifies the minimum requirements for twisted-pair and coaxial cabling.

- Grade 2 provides a generic cabling system that meets the requirements for basic, advanced, and multimedia telecommunications services. Grade 2 covers current telecommunications services and any services that are in development. Grade 2 also specifies minimum requirements for twisted-pair cable, coaxial cable, and optical fiber cabling.

As with all standards, ANSI/TIA-570 provides acronyms, definitions, tables, and figures to assist in design parameters. Acronyms include ADO, or auxiliary disconnect outlet, and DD, or distribution device. An *auxiliary disconnect outlet (ADO)* is a device located within the tenant or living unit used to terminate the backbone cable. A *distribution device (DD)* is a hardware device used for the distribution of VDV systems throughout a dwelling or commercial structure. ANSI/TIA-568 and ANSI/TIA-569 are referenced in ANSI/TIA-570 for cable type, distance requirements, testing requirements, pathways, and spaces required to support residential applications.

ANSI/TIA-606-C, Administration Standard for Telecommunications Infrastructure

The ANSI/TIA-606 standard was developed to provide uniform guidelines for the administration of applications that may change over the life of the telecommunications infrastructure. ANSI/TIA-606 defines classes of administration so that the installed infrastructure is properly labeled and administered. The standard specifies the administration of a generic system that supports multiproduct and multi-OEM environments. The four classes of administration specified in ANSI/TIA-606 include the following:

- Class 1, which covers a single equipment room
- Class 2, which covers one or more equipment rooms in a single building
- Class 3, which covers multiple buildings on a single site
- Class 4, which covers multiple sites or campuses

TECH FACT

An intersystem bonding termination is a device that provides a means of connecting bonding conductors from communications systems to a grounding electrode system.

ANSI/TIA-606-C outlines identification guidelines for telecommunications infrastructure within and between buildings. The standard also provides identification requirements for outlets, grounding conductors, busbars, and firestop locations. In addition, color codes for various applications such as cables, jumpers, termination fields, and pathways are provided. **See Figure 2-11.**

Proper labeling and identification increases the value of an infrastructure by reducing system maintenance expense, extending the life of the system, and providing effective service to the users. ANSI/TIA-606-C is a reference that can be used during the design process to provide clear and concise customized identification for drawings and specifications. Many end users have standardized their identification means to meet their needs and often utilize software programs to track moves, adds, and changes (MACs) to their infrastructure.

ANSI/TIA-607-D, Generic Telecommunications Bonding and Grounding (Earthing) for Customer Premises

The ANSI/TIA-607-D standard specifies a uniform telecommunications grounding and bonding infrastructure for commercial buildings. The bonding and grounding of telecommunications infrastructure is an area where codes and standards can possibly overlap. While the bonding and grounding system that is put in place to protect the telecommunications spaces and equipment is separate from the building grounding system, the two systems must meet the same standards of NEC® compliance and be interconnected so that the building has one cohesive bonding and grounding system.

ANSI/TIA-606-C Administration Standard for Telecommunications Infrastructure

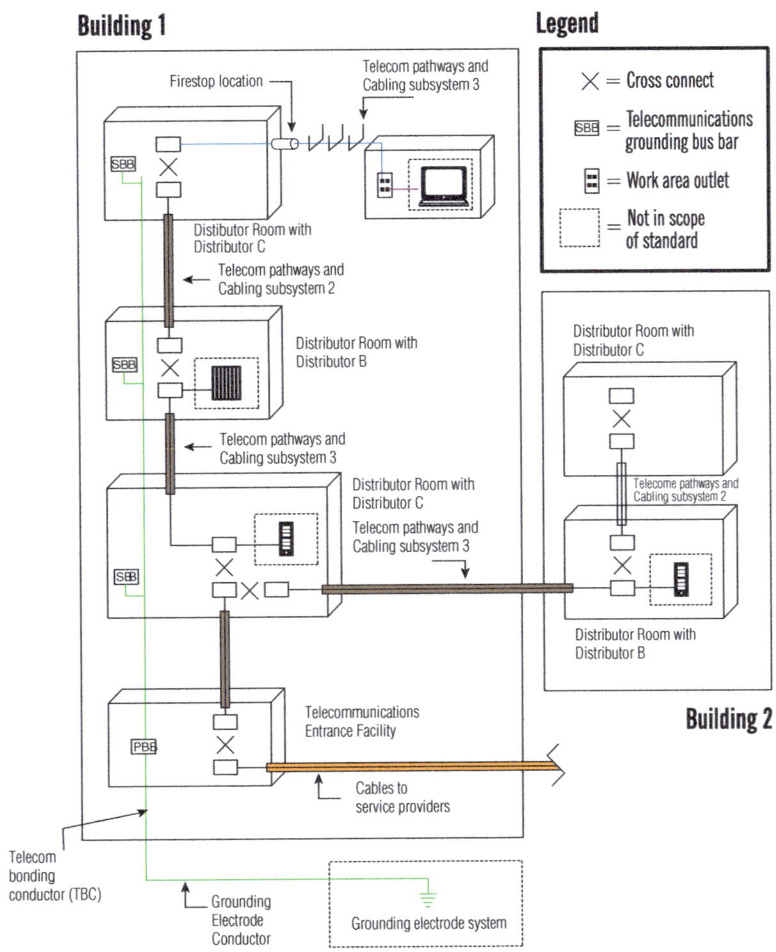

ADMINISTRATIVE SYSTEM INFRASTRUCTURE ELEMENTS

VDV Application Color Codes		
Termination Type	**Color**	**Typical Application**
Demarcation Point	Orange	Local exchange carrier (LEC) connection
Network Connection	Green	Bonding connection, GEC
Common Equipment	Purple	Phone/LAN
Cabling Subsystem 1	Blue	Cabling between outlet and equipment room
Cabling Subsystem 2	Gray	Second level riser
Cabling Subsystem 3	White	First level riser
Campus Cabling	Brown	Interbuilding riser
Miscellaneous	Yellow	Alarms, security, or building automation

Figure 2-11. ANSI/TIA-606-C outlines identification guidelines for telecommunications spaces, racks and cabinets, terminating hardware, and cables within and between premises (building, campus, or site), as well as VDV color codes.

Bonding and grounding of telecommunications equipment and systems is put in place for three primary reasons: to protect life and property, to protect equipment, and for electrical equipment operational purposes. Proper grounding is an NEC® requirement to protect life and property that ensures adequate conductors are in place to divert hazardous, unintended electrical current to ground, where it can be safely dissipated. Proper grounding is also a requirement of the electrical grounding system in a structure. To achieve uniformity and to provide a total, safe grounding system, the bonding conductors in the telecommunications system must be of the same gauge and type of conductor as the electrical system grounding and also must be bonded to the electrical ground, which creates a cohesive grounding system.

To achieve a protective grounding system for the installed telecommunications equipment, permanent bonding needs to be in place between apparatus cabinets, cable trays, ladder trays, the grounding busbar in a telecommunications room, and any other equipment in the system. Through the installation of a proper grounding system, any stray electrical current is dissipated to the building grounding system rather than damaging an equipment cabinet full of costly network switches or servers.

The components of a telecommunications ground system are also standardized by ANSI/TIA-607. Each component of a grounding system has an acronym associated with its specific intended purpose. The process of bonding equipment and components within telecommunications spaces is also critical. Equipment cabinets, ladder trays, cable trays, and raised access floor support columns are included in the bonding system. **See Figure 2-12.**

ANSI/TIA-758-B, Customer-Owned Outside Plant Telecommunications Infrastructure

The ANSI/TIA-758-B standard specifies the minimum requirements for customer-owned outside plant (OSP) system facilities in a campus environment. ANSI/TIA-758-B applies to any campus, regardless of the size of the structures, layout, or occupancy level. The standard specifies the cabling, pathways, and spaces to support the infrastructure to be installed outside of a structure. **See Figure 2-13.** ANSI/TIA-758-B also provides reference tables and figures depicting examples of entry methods, color-coding of excavation areas, and separation of services.

Customer-owned OSP cabling extends between separated structures, including the terminating connecting hardware at or within the structures. The OSP cabling specified by ANSI/TIA-758-B is intended to support applications such as VDV systems, alarms and environmental control systems, and security systems. The OSP pathway includes the pathway through the point of entry into the building space. Customer-owned OSP pathways may include aerial pathways, direct-buried pathways, underground pathways, and utility tunnels. In addition to pathways and spaces, other design recommendations and requirements include the following:

- topology type
- entrance point diversity
- entrance route diversity
- recognized cabling and related hardware
- bonding and grounding
- physical location and protection of below-ground-cable plant
- cabling enclosures
- common OSP symbols

ANSI/TIA-862, Structured Cabling Infrastructure Standard for Intelligent Building Systems

Typically applied to commercial structures, the ANSI/TIA-862 standard was developed to address the emerging technologies being placed on the IP platform. Similar to the development of ANSI/TIA-568, the ANSI/TIA-862 standard was needed due to proprietary systems and products. ANSI/TIA-862 addresses proprietary building automation systems (BASs) by specifying a generic VDV system with the capability of supporting multivendor and OEM infrastructures.

ANSI/TIA-607-D Generic Telecommunications Bonding and Grounding (Earthing) for Customer Premises

VDV Grounding and Bonding Acronyms	
Acronym	**Meaning**
TMGB	Telecommunications Main Grounding Busbar
TBB	Telecommunications Bonding Backbone
TGB	Telecommunications Grounding Busbar
TBC	Telecommunications Bonding Conductor
TBBIBC	Telecommunications Bonding Backbone Interconnecting Bonding Conductor

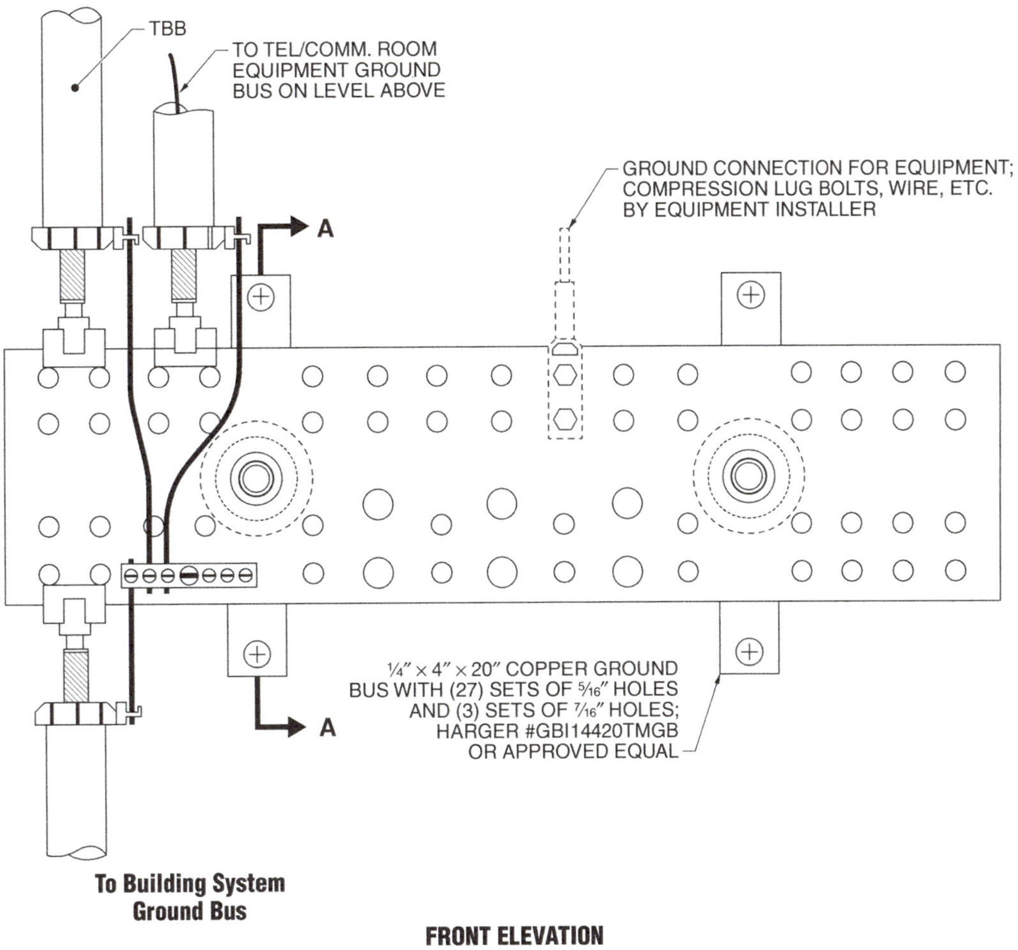

FRONT ELEVATION

NOTES:

1. Additional ground connections shall be by telecommunications equipment installer
2. 20″ for MDFs, 10″ for IDFs

BONDING BUSBAR DETAIL

Figure 2-12. The components of a telecommunications ground system are standardized by ANSI/TIA-607-D, and each component of the grounding system has an acronym associated with its specific intended purpose.

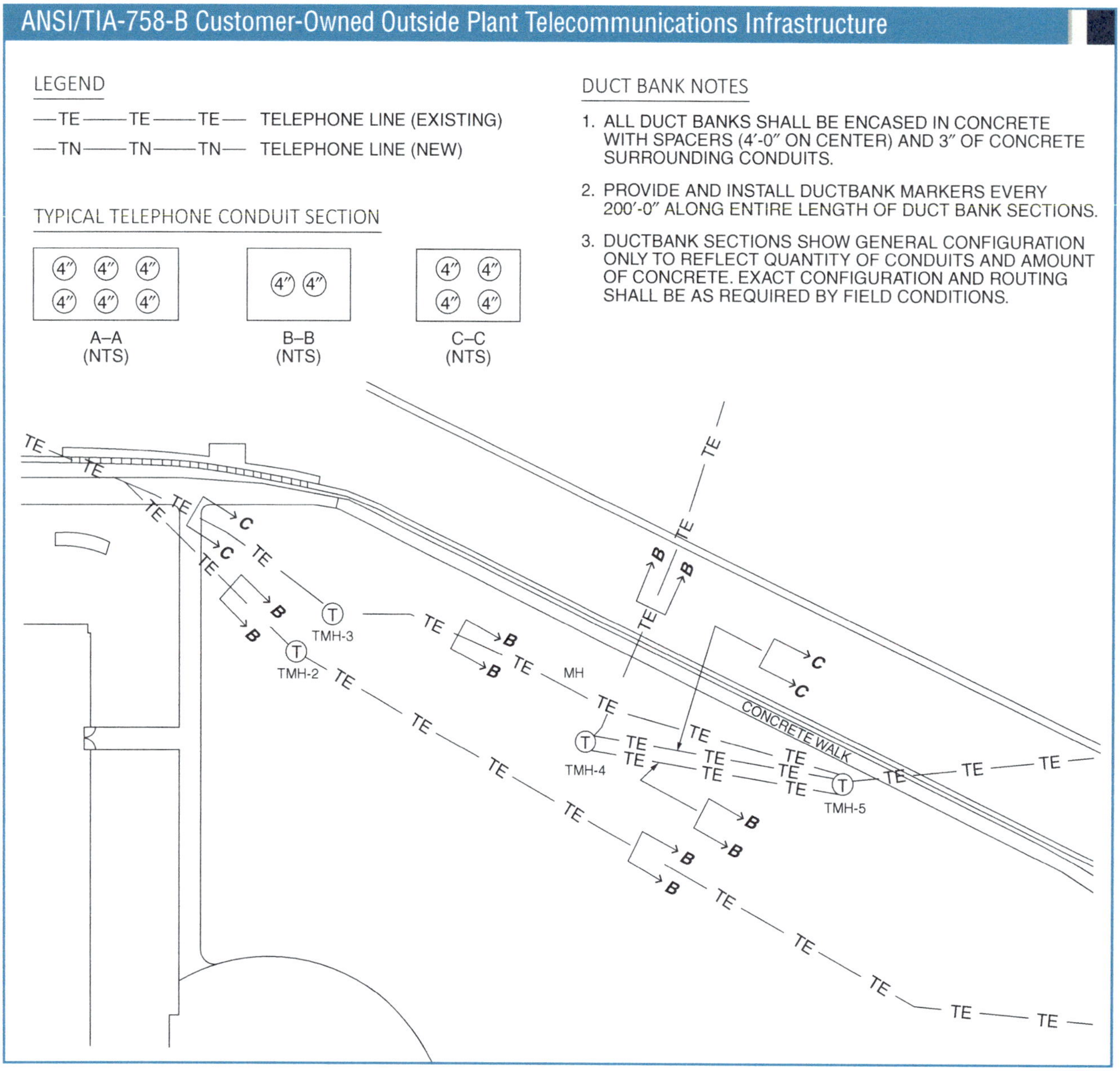

Figure 2-13. The ANSI/TIA-758-B standard specifies the cabling, pathways, and spaces to support the infrastructure to be installed outside of a structure.

ANSI/TIA-862 specifies the minimum requirements for a BAS cabling system, including topology, architecture, design and installation practices, test procedures, and components. BASs typically include fire protection, access control, elevator, security, HVAC control, electrical, plumbing, energy management and environmental control, lighting control, parking access and revenue control, and VDV systems. The standard specifies cabling that is intended to support applications related to building management and a wide variety of premises.

Additional systems covered by ANSI/TIA-862 include audio (intercommunication systems), data communications, and wireless access points (WAPs). A *wireless access point (WAP)* is a device connected to a wired network that connects wireless communication devices to the same network.

Wireless communication devices are typically smart phones, tablets, and laptop computers. ANSI/TIA-862 also specifies the requirements for the following hardware and spaces:

- horizontal cabling
- backbone cabling
- cabling subsystems
- telecommunications rooms, equipment rooms, mechanical rooms, and entrance facilities
- pathways and spaces
- system administration

The requirements of the ANSI/TIA-568, -569, -606, and -607 standards apply to ANSI/TIA-862 and should be referenced when determining structured cable subsystems, topology, connector types, distance limitations, pathways and spaces, grounding, bonding, and identification of systems. The standard also allows for zone cabling applications based on end user requirements to address accessibility and security concerns.

ANSI/TIA-942, Telecommunications Infrastructure Standard for Data Centers

The ANSI/TIA-942 standard was first published in 2005. The purpose of ANSI/TIA-942 is to establish a guideline for the various design and construction elements of both large- and small-scale data centers and to provide requirements and guidelines for the design and installation of computer rooms, server rooms, and similar spaces. ANSI/TIA-942 is primarily a telecommunications infrastructure standard; however, about half of the technical content covers facility specification requirements.

ANSI/TIA-942 is intended for use by designers for a comprehensive understanding of facility design, cabling design, and network design for data center applications. ANSI/TIA-942 enables data center design to be considered in the early stages of the building development process by contributing to the architectural considerations, providing evaluations of existing data centers, and promoting cooperation between the building disciplines during the design and construction phases.

Adequate planning during building construction or renovation is significantly less expensive and less disruptive than after the facility is operational. Data centers can benefit from infrastructure that is planned in advance to provide growth and changes in the VDV systems that they are designed to support. **See Figure 2-14.** The ANSI/TIA-942 standard covers multiple components and utilities installed in data centers, such as the following:

- electrical
- HVAC
- fire detection and suppression
- physical security
- physical space and access
- raised flooring

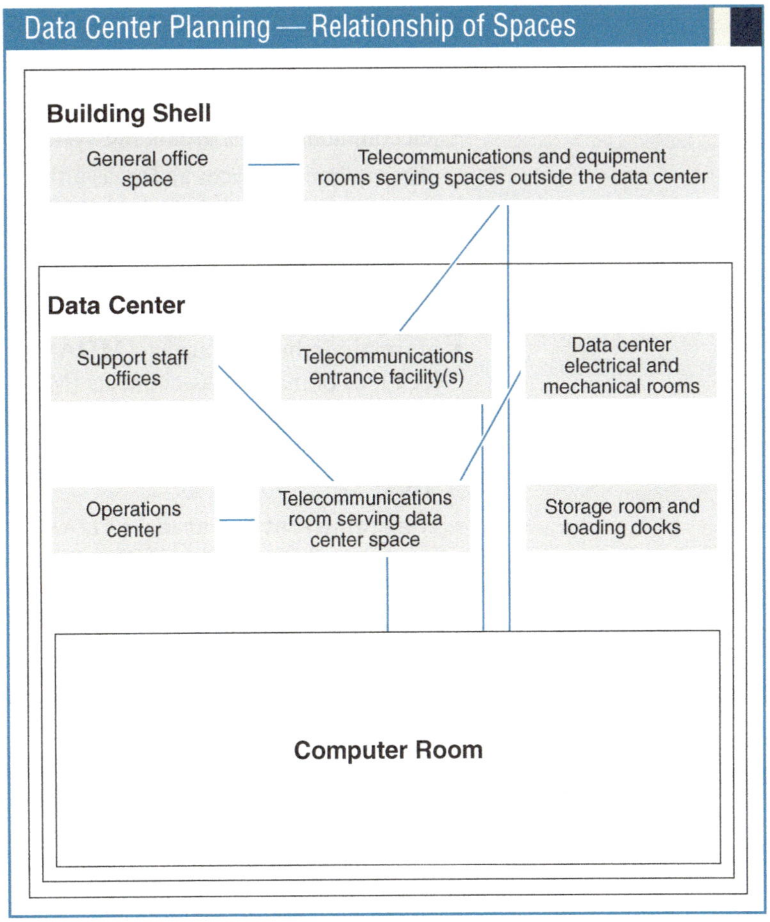

Figure 2-14. Data centers can benefit from infrastructure that is planned in advance to provide growth and changes in the VDV systems that they are designed to support.

General design elements identified in ANSI/TIA-942 for consideration that apply to most data centers include the following:

- standards-based open systems
- high-performance and high-bandwidth technologies (10+ GbE)
- support for storage devices
- support for convergence of new IP applications
- energy efficiency
- redundancy (continuance of function under catastrophic conditions)

ANSI/TIA-942 standard is the first standard that specifically addresses data center telecommunications infrastructure and includes the following:

- overhead and underfloor cable trays
- plenum and nonplenum cabling
- shielded and nonshielded cabling
- single-mode and multimode fiber cabling and termination methods
- telecommunications grounding systems

The equipment spaces and areas utilized in data center spaces include the following:

- entrance rooms (ERs)—locations of interface with campus and carrier entrance facilities
- main distribution areas (MDAs)—locations of main cross-connects (MCs)
- horizontal distribution areas (HDAs)—locations of horizontal cross-connects (HCs)
- zone distribution areas (ZDAs)—locations of zone outlets (ZOs) or consolidation points (CPs)
- equipment distribution areas (EDAs)—locations of horizontal cable outlet/patch panel and server/equipment cabinets and racks

ANSI/TIA-942 provides simplified, standardized guidelines for scalability and modularity so that future changes in technology can be achieved without replacing infrastructure. The standard also provides detailed descriptions of the data center tier levels. **See Figure 2-15.** Recommended specifications by tier are a uniform method to rate aspects of data center design and estimate costs.

The main difference in tier levels is the amount of uptime and downtime associated with each level. For example, a Tier 1 data center has an uptime of 99.671% and 28.8 hours of downtime, while a Tier 4 data center has 99.995% uptime and 26.3 minutes (0.44 hour) of downtime. Tier level certification is provided by the Uptime Institute®, which includes four certification levels:

- Tier 1—Tier 1 data center requirements are generally utilized by small commercial installations and feature 99.671% uptime, no redundancy, and 28.8 hours of downtime per year.
- Tier 2—Tier 2 data center requirements are generally utilized by small commercial installations and feature 99.749% uptime, partial redundancy in power and cooling, and 22 hours of downtime per year.
- Tier 3—Tier 3 data center requirements are generally utilized by large commercial installations and feature 99.982% uptime, 1.6 hours of downtime per year, and N+1 fault tolerance, which provides at least 72 hours of power outage protection.
- Tier 4—Tier 4 data center requirements are generally utilized by large enterprise installations and feature 99.995% uptime per year, 2N+1 fully redundant infrastructure (the main difference between Tier 3 and Tier 4 data centers), 96-hour power outage protection, and 26.3 minutes of annual downtime.

NREL

ANSI/TIA-942 identifies general design considerations for data centers.

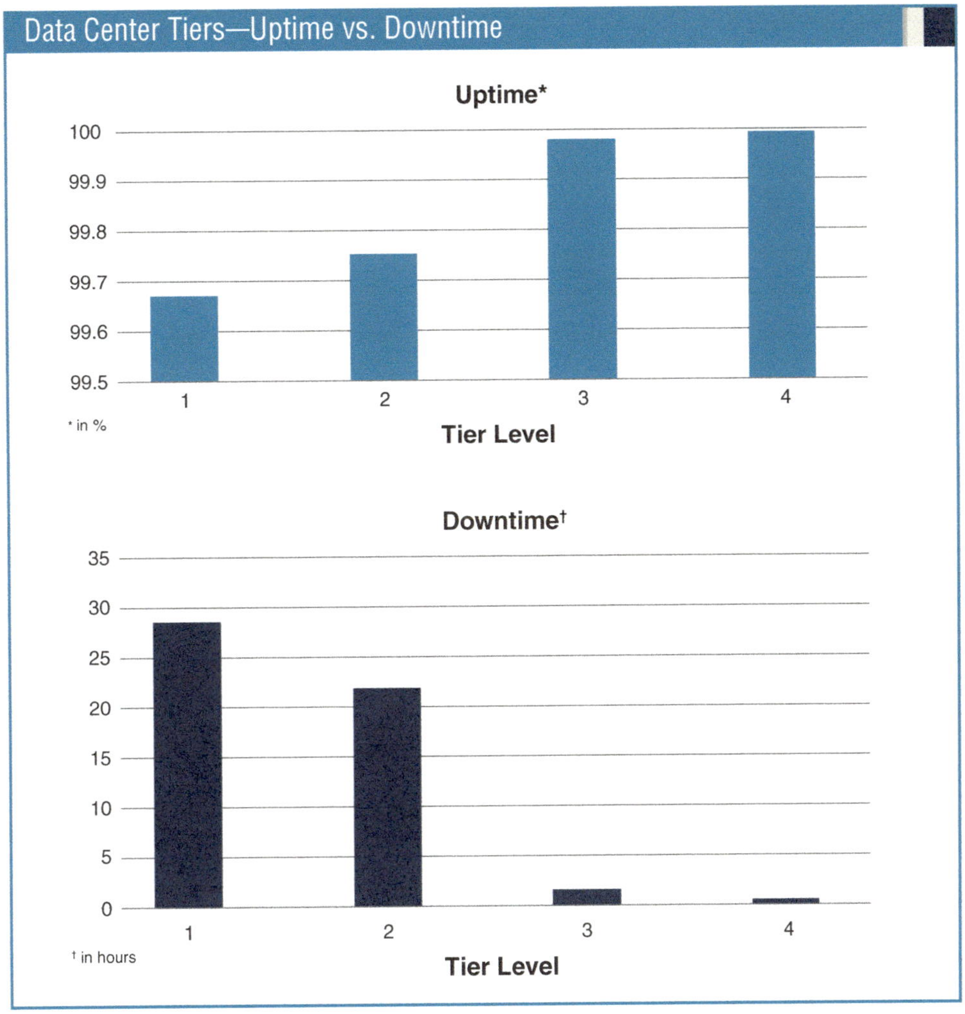

Figure 2-15. The main differences between data center tier levels is the amount of uptime and downtime associated with each level.

SUMMARY

VDV systems are installed in both new construction and existing construction as part of any retrofit projects. In order to troubleshoot, test, and use as desired, VDV systems must be installed according to the industry standards and codes that define proper installation. Codes, such as the NEC®, are legally enforceable by the AHJ on a project. Standards, such as those from ANSI/TIA, BICSI, CSI, ISO, and UL, are written as recommended practices. VDV industry associations contribute to the development of codes and standards.

CHAPTER 2

Industry Codes and Standards

Review

Name ______________________________ Date ______________

True-False

T F **1.** A standard is a regulation and requirement devised to maintain uniformity of work and proper standards of procedure.

T F **2.** The NEC® only contains requirements that are necessary for safety.

T F **3.** A code is an accepted reference or practice.

T F **4.** Only VDV drawings and specifications for copper systems reference standard ANSI/TIA-568.

T F **5.** ANSI/TIA-570 outlines three grades of VDV service that could be required for a residential premises.

T F **6.** Four classes of administration are specified in ANSI/TIA-606-C.

T F **7.** ANSI/TIA-862 was developed to address the emerging technologies being placed on the IP platform.

T F **8.** A wireless access outlet (WAO) is a device connected to a wired network that connects wireless communication devices to the same network.

T F **9.** Tier 1 data center requirements are generally utilized by small commercial installations and feature 99.671% uptime, no redundancy, and 28.8 hours of downtime per year.

T F **10.** The requirements of the ANSI/TIA-568, -569, -606, and -607 standards apply to ANSI/TIA-942.

Multiple Choice

______ **11.** Certain requirements of a code may require interpretation, and it is the ___ that makes the final determination as to the interpretation.
- A. architect
- B. authority having jurisdiction (AHJ)
- C. code committee
- D. project owner

______ **12.** The ___ is the organization, office, or individual responsible for approving the equipment and materials used for electrical wiring installation.
- A. architect
- B. authority having jurisdiction (AHJ)
- C. local IBEW Training Director
- D. project owner

______ **13.** NEC® Article 770 covers the installation of ___.
- A. copper cables
- B. coaxial cables
- C. optical fiber cables
- D. wireless access points

______ **14.** ___ is a professional telecommunications standards and certification organization.
- A. American National Standards Institute (ANSI)
- B. Building Industry Consulting Service International (BICSI)
- C. The Construction Specifications Institute (CSI)
- D. Telecommunications Industry Association (TIA)

______ **15.** A(n) ___ is a device located within the tenant or living unit used to terminate the backbone cable.
- A. auxiliary disconnect outlet (ADO)
- B. distribution device (DD)
- C. entrance room (ER)
- D. wireless access point (WAP)

______ **16.** A(n) ___ is a hardware device used for the distribution of VDV systems throughout a dwelling or commercial structure.
- A. auxiliary disconnect outlet (ADO)
- B. distribution device (DD)
- C. entrance room (ER)
- D. wireless access point (WAP)

______ **17.** Class ___ covers the administration of multiple sites or campuses in ANSI/TIA-606-C.
- A. 1
- B. 2
- C. 3
- D. 4

_______________ **18.** ANSI/TIA-___ is the first standard that specifically addresses data center telecommunications infrastructure.

A. 570
B. 606-C
C. 862
D. 942

_______________ **19.** The AHJ is typically a(n) ___, electrical inspector, or other person having statutory authority.

A. architect
B. building official
C. local IBEW Training Director
D. project owner

_______________ **20.** The NEC® is prepared by the National Electrical Code Committee, which consists of a correlating committee and ___ separate code-making panels.

A. 9
B. 12
C. 13
D. 18

Completion

_______________ **21.** The ___ sponsors the development of the NEC®.

_______________ **22.** The ___ ensures that the electrical installation process is performed within proper safety and code requirements.

_______________ **23.** Before the NEC® is applied, the ___ must be reviewed to determine if the electrical installation must be compliant with the NEC®.

_______________ **24.** ___ rules are rules that are not required but are permitted.

_______________ **25.** Only terms that are referenced in ___ or more articles of the NEC® are listed in Article 100.

_______________ **26.** The ___ is an organization that develops international safety standards for electrical equipment.

_______________ **27.** The ___ is a technical professional society that produces standards for telecommunications, local area networks, and computing.

_______________ **28.** The TIA is accredited by ___ to develop voluntary, consensus-based industry standards for various information and information communication technologies (ICT) products.

_______________ **29.** Typically, the ___ is considered to be the inspector on a job site.

_______________ **30.** NEC® Article 300 covers the installation and wiring methods for conductors, cables, and raceways for applications up to ___ V.

Short Answer

31. Explain the two types of rules used in the NEC®.

32. List the types of systems typically controlled by building automation systems (BASs).

33. Briefly explain the grades of VDV service outlined in the ANSI/TIA-570 required for a residential premises.

34. List the amounts of uptime and downtime for the four different data center tier levels as covered in ANSI/TIA 942.

35. List the terms from the NEC® that designers and installers of VDV equipment must be familiar with.

CHAPTER

3

Systems Drawings

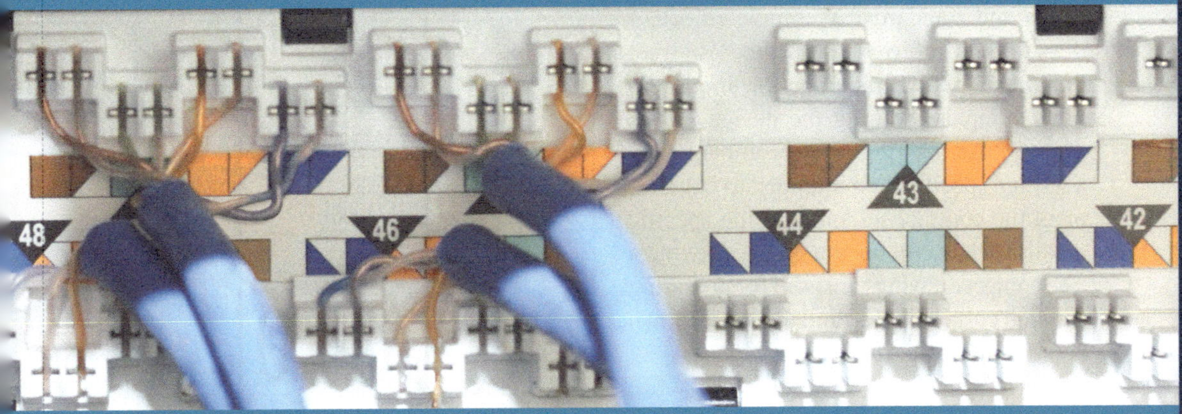

Construction design and installation professionals rely on drawings to illustrate the scope of the work to be performed. Drawings are created during the design development phase of a project. Designers and installers must understand individual drawing components and how to use drawing scales. The more detail provided on drawings, the more precise the cost estimate and the higher the savings to the owner; and with the proper amount of detail, the installer can work more efficiently with other trades in a timely manner. Interactive electronic devices and software programs are commonly used rather than hard-copy drawings.

Objectives

- Explain design development.
- List the different drawing types.
- Explain the purposes of the different components of a drawing.
- Describe how drawings are scaled and the differences between an architect's scale and an engineer's scale.
- Describe building information modeling (BIM) and the various software programs used for design and project management and implementation.

Learner Resources

atplearningresources.com
access code 783821

DESIGN DEVELOPMENT/CONTRACT DRAWINGS

Most drawings are developed after several revisions and updates based on owner input, coordination of work between the trades, the designer's expertise, and applicable codes and standards. A completed set of drawings is referred to as the "contract drawings," which includes the modifications completed during the design phase. The contract drawings are accompanied by a written specification intended to complement the contract set and provide specific details to the items shown on the drawings. The contact drawings are used for preparing detailed cost estimates, and the work shown is the basis of the contract scope.

DRAWING TYPES

Drawings are organized to provide a universally accepted order of structure by system. Most architects follow the American Institute of Architects (AIA) standard on drawing organization. The AIA standard incorporates letters, hyphens, and numbers to differentiate between disciplines and types of drawings (print divisions). For example, architectural drawings are identified by the letter "A" and are sequentially numbered. Other drawing types include structural (S), plumbing (P), electrical (E), mechanical (M), site/civil (C), telecommunications (T), and other specialty systems, such as food service and laundry, if applicable. **See Figure 3-1.**

Architectural drawings (A) depict the layout of a building and the various spaces used for mechanical utilities and occupancy. Typically, architectural drawings first show large areas of space with minimal detail and then the subsequent drawings show increasing levels of detail. Because architectural drawings are the core set of drawings, other drawings are based on and coordinated with them.

Structural drawings (S) depict the various load bearing systems that physically support the facility. Structural drawings are based on the architectural drawings. They depict the spacing and distance between columns and beams.

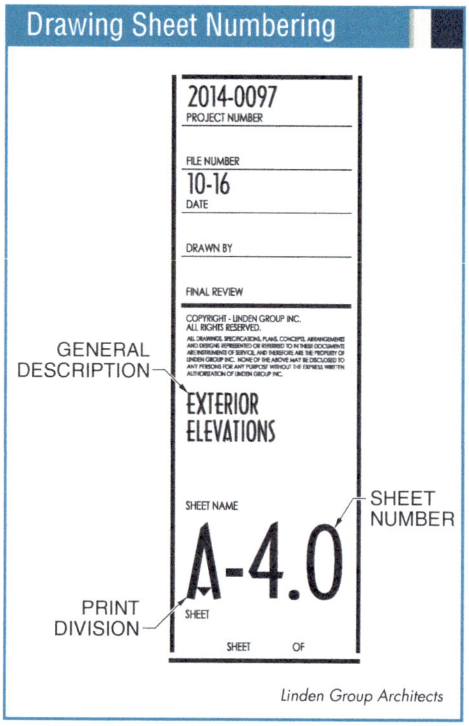

Figure 3-1. Construction drawings are organized by discipline and designated by both a letter representing a print division and a sheet number.

Plumbing drawings (P) illustrate detailed connections between fixtures and piping that are to be installed in a structure. Plumbing drawings sometimes must be referenced when installing electrical and VDV raceways and cabling. This is to ensure that there are no physical conflicts between systems.

Electrical drawings (E) depict electrical requirements, including power distribution, lighting, fire alarm, and any specialty wiring. Electrical drawings also depict work required by electrical trades, such as pathways and spaces to support specialty systems, which include telecommunications, security, access control, and audio/visual systems.

Mechanical drawings (M) depict the layout of mechanical, heating, ventilation, air conditioning, and sprinkler systems. Each type of system is identified by a specific letter and a sequential number that pertains to that particular system. For example, drawings for heating, ventilation, and air conditioning systems are identified by the letters "HVAC."

Site/civil drawings (C) depict the relationship between a structure, the building lot, and the services to support the facility. Facility services typically include sewer, sanitary, utilities (electric, gas, and water), roadways, curbing, and pedestrian walkways. The scope of site/civil drawings can also include depictions of landscaping, drainage systems, storm water retention facilities, parking, and lighting. The locations of utility lines such as electrical/telecommunications, water, and natural gas lines as part of outside plant (OSP) distribution systems on the property and the connection points to the utilities in the municipal-owned property (street) are also depicted. **See Figure 3-2.**

Telecommunications (Tel/Data) drawings (T) depict the requirements for VDV systems, including audio/visual (AV), access control, and security systems. Telecommunications drawings are outlined through floor plans, riser diagrams, site plans, and detail sheets.

Other specialty systems drawings can include fire alarm (FA), photovoltaic (PV), and distributed antenna systems (DAS).

DRAWING COMPONENTS

When referencing project drawings, electricians and installers must understand the individual drawing components prior to starting work on each project phase. Failure to understand drawing components may result in lost time and incorrect installations. Drawing components include cover sheets, title blocks, revision sections, plan views, elevations, schedules, detail drawings, diagrams, trade symbols, and abbreviations and acronyms.

Cover Sheets

The cover sheet is the first page in the contract set of drawings and provides basic information such as the project name, number, location, and the contact information for the architects and engineers that have designed the project. The cover sheet also provides an alphabetical and numerical listing of the drawings contained in the contract set. Additional items that may be included on a cover sheet are specific building code information, a map of the facility location, general notes, and written specifications. An architectural rendering of the structure may also be included on the cover sheet. **See Figure 3-3.**

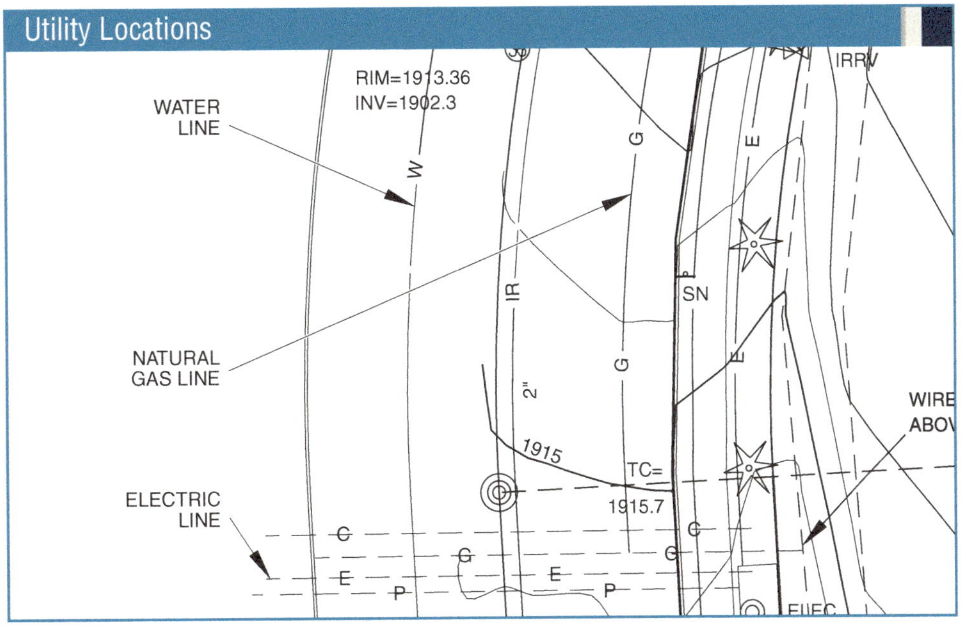

Figure 3-2. With site/civil drawings, existing utility locations are indicated with dashed or solid lines and letters noting the type of utility service.

Cover Sheets

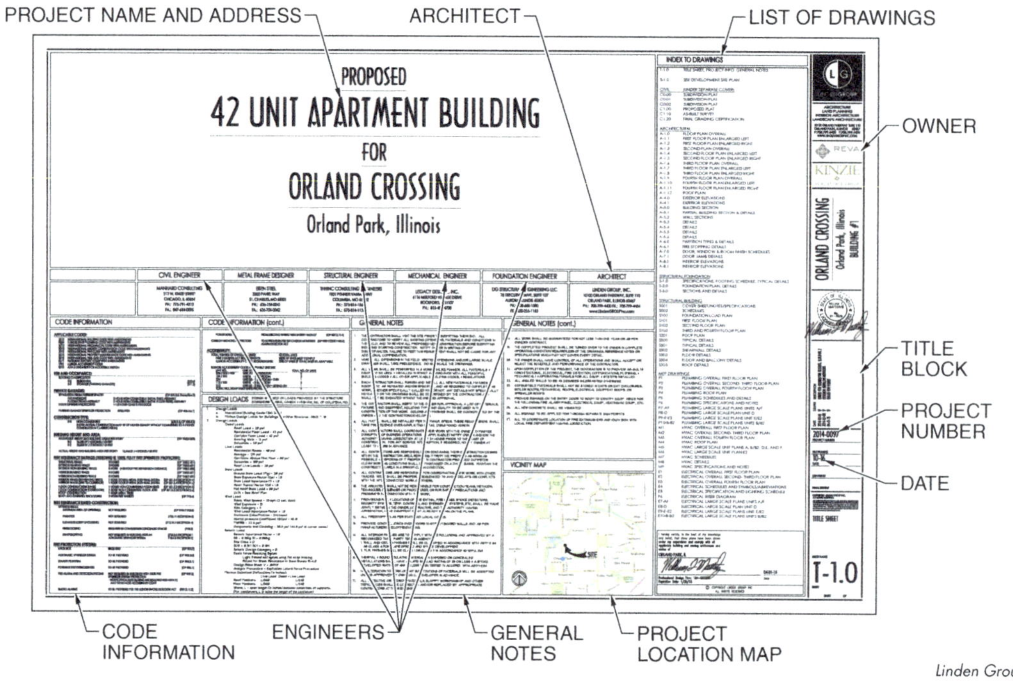

Linden Group Architects

Figure 3-3. The cover sheet in a set of construction drawings includes general project information such as the names of the consulting engineers, the project location, and applicable building codes.

Title Blocks

A *title block* is the area of a print that contains information about the contents of the print. On most prints, the title block is in the lower right corner or along the right margin on every drawing in the contract set. Information in the title block must be clearly understood before information is accessed from a print. The title block contains information such as the following:

- prefixed sheet number and letter depicting the discipline
- any revision made to the drawing
- subject or sheet contents
- project title and location
- name(s) of the architect and/or engineering firm, including office contact information
- state stamp identifying the architect or engineer as a licensed professional
- print division and print number
- date the print was drawn and the initials of the person who drew the print
- date the print was checked and the initials of the individual who checked the print
- drawing scale
- other job-related information, such as the project job number

The title block is the best place to begin reading a set of prints since it includes the project name and location, architect's name and office location, architect seal, plan completion date, drafter initials, sheet number and number of sheets in the set of prints, name of the sheet, and other information as determined by the architect. **See Figure 3-4.** The scale of the drawing (e.g., $\frac{3}{32}'' = 1'\text{-}0''$) or the scale for specific sections of the drawings is also listed on applicable pages, either as part of the title block or near the title block. The architect or engineer seal (professional stamp) and the signature of the party responsible for the design is required when obtaining permits and approvals to build.

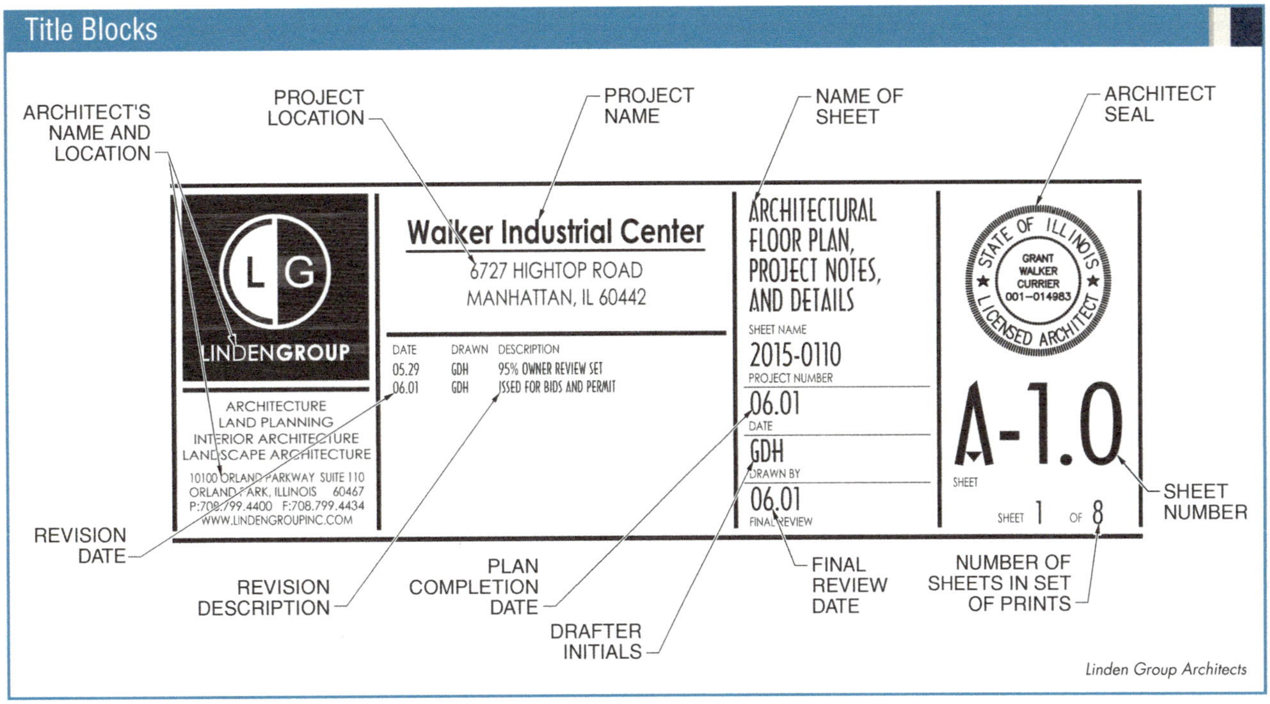

Figure 3-4. Title blocks provide basic information for the complete set of prints.

Revision Sections

The revision section provides the history of the drawing, from conceptual design to the final contract set. During a project, changes are identified and tracked through the revision column. Architects and engineers typically make several revisions to a set of drawings during the course of a project. Revisions are labeled sequentially as 1, 2, 3, etc. or as A, B, C, etc.

Depending on the size and scope of the changes, an entire drawing can be revised and the changes placed on a cloud-based digital storage system. With any change, the revision section (column) is updated according to the type of change. Changes noted in the revision section include the date and any pertinent change order identification. For drawings with minor revisions, the drawing is reissued with changes noted in a bubble or cloud outline. **See Figure 3-5.** Minor revisions may also indicate the most recent revision number inside of a triangle attached to the bubble outline. For example, a minor revision may be labeled as 1.1.

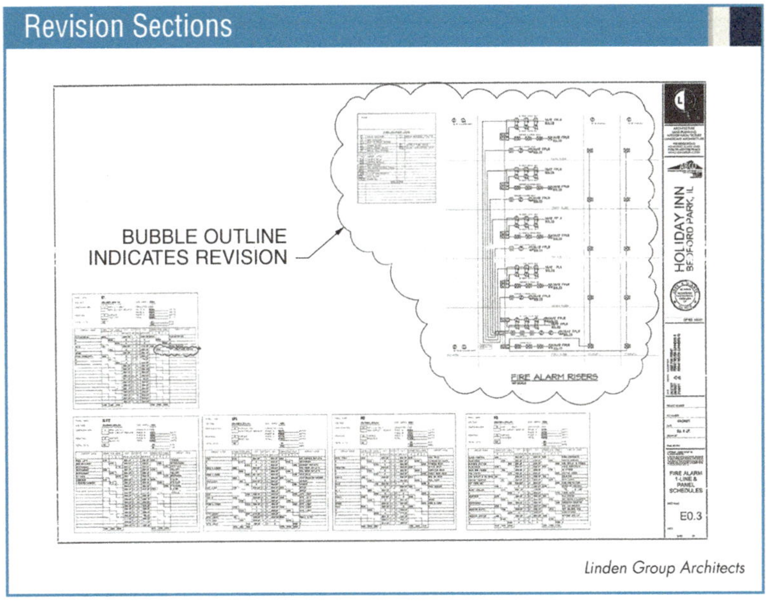

Figure 3-5. When there are minor changes, a revised drawing is reissued with the changes noted in a bubble or cloud outline.

TECH FACT

Other than the sheet number, title blocks are the same for each sheet in a set of prints.

Plan Views

A plan view shows a complete view of the floor plan. The most common type of plan view is the architectural floor plan, which shows doors, windows, walls, and partitions. Architectural plan views provide accurate, detailed dimensions and show both exterior and interior measurements of a space. Other types of plan views include reflected ceiling plans and partial plan views, in which a specific area is enlarged to provide more detail than that shown in the entire plan view. The plan view is the starting point from which an architect references other drawings for additional information. **See Figure 3-6.**

Elevations

An *elevation* is an orthographic view of a vertical surface without allowance for perspective. The two types of elevation drawings are interior and exterior elevations.

Exterior elevations depict the views of the exterior sides of a building to scale. Height measurements are provided from floor to floor of the interior and for exterior facade features. Symbols and conventional representations are used to identify door and window types. Notes and graphic symbols are used to identify the types of finish materials.

Plan Views

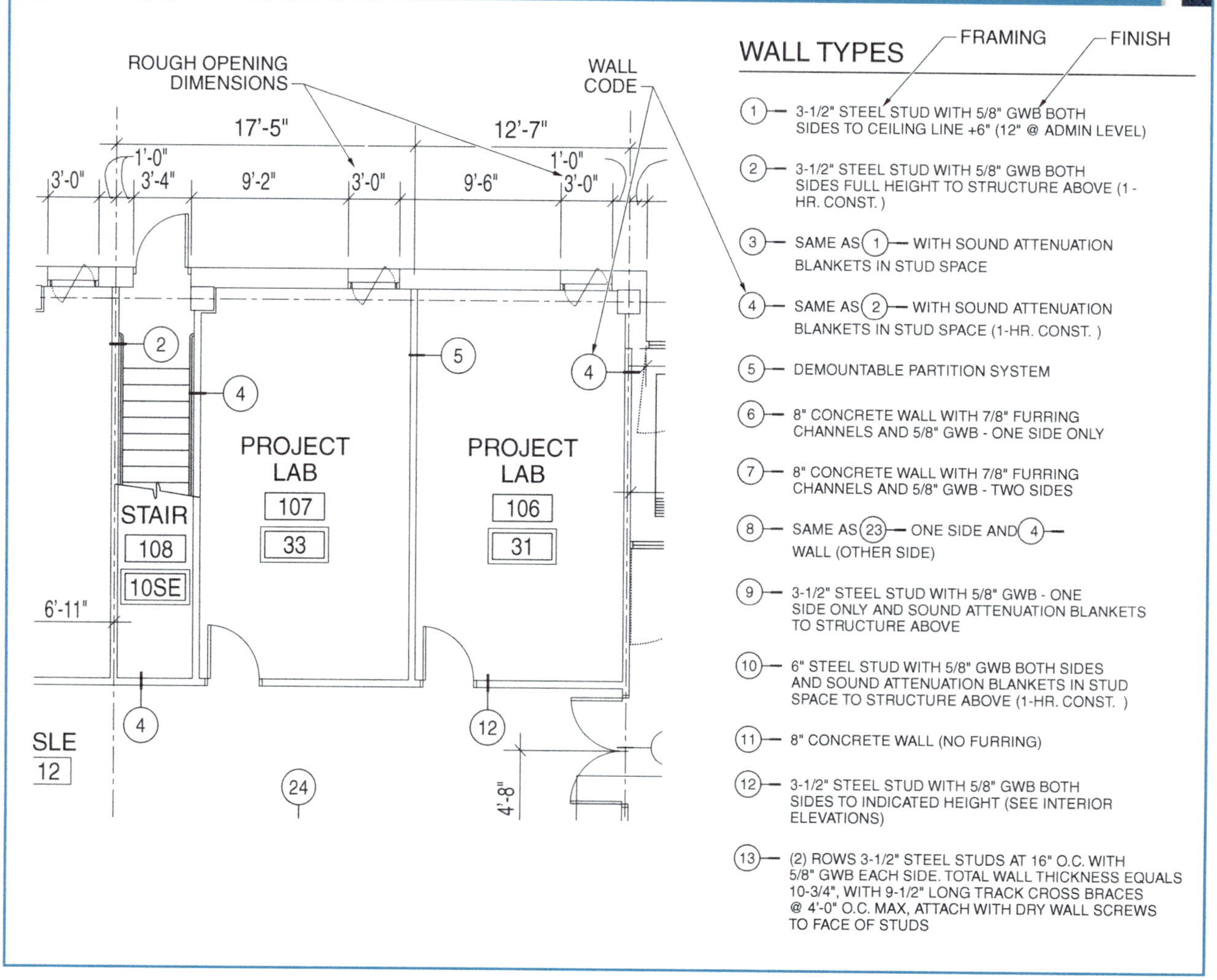

Figure 3-6. Plan views and architectural notes describe wall locations, framing and support members, and finish materials.

Interior elevations are similar to exterior elevations except that they depict features of interior spaces. Interior elevation views are used to identify dimension features not shown on plans or sections, such as wall finishes, wallcoverings, wainscoting, and crown molding. Interior elevation views also depict other interior elements, including cabinets, display boards, signs, and grab bars. The accessibility guidelines created from the Americans with Disabilities Act (ADA) are used for height measurements of restroom fixtures, grab bars, handrails, and ramps. Interior elevations are referenced on the floor plan with a symbol that references the drawing and applicable detail section. **See Figure 3-7.**

Schedules

A *schedule* is a table or detailed list on a print that provides information about specific building components and equipment such as doors, windows, and mechanical equipment. Since too much information can clutter a drawing, architects have designed schedules to organize repetitive information and minimize the amount of printed information on a drawing. However, schedules are not limited to architectural drawings. Schedules are used within any trade and often provide notes and detailed information for specific equipment and components.

TECH FACT

The Americans with Disabilities Act (ADA) is a civil rights law effective since 1990.

Interior Elevations

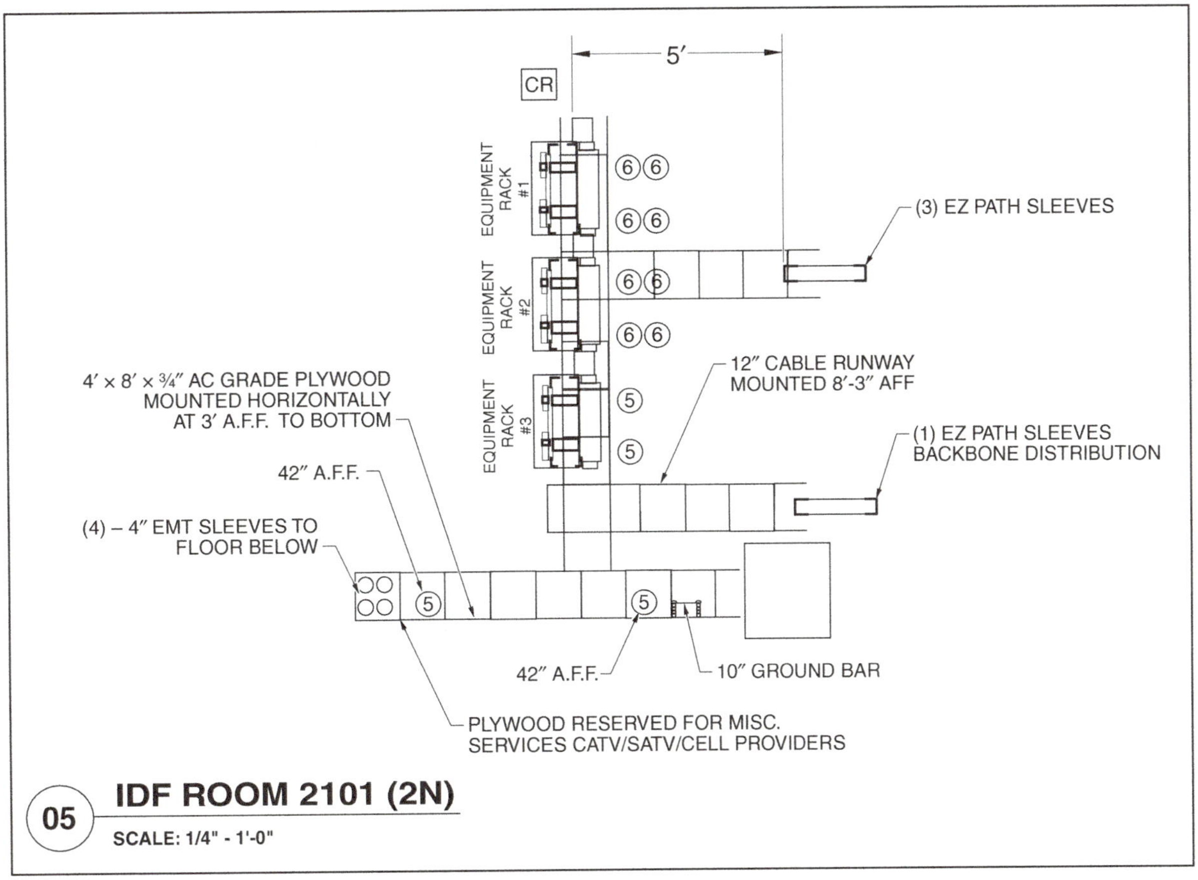

Figure 3-7. Interior elevation views are used to identify dimension features not shown on plans or sections.

Detail Drawings

A *detail drawing* is a drawing that shows a small part of a plan, elevation, or section view at an enlarged scale. Detail drawings are used to provide information not clearly shown in other parts of the drawings. Detail views can be located on plan and elevation drawings or grouped together on separate drawing sheets. Details are labeled and referenced similarly to wall sections. Detail tags have expanded detail that shows the specific features at an enlarged scale. **See Figure 3-8.**

> **TECH FACT**
>
> Detail drawings are typically used to show detail on VDV components such as outlets, patch panels, and rack systems.

Diagrams

Diagrams are simplified graphic representations that visually clarify systems or processes on the construction drawings. A diagram illustrates how various components of a system are configured and provides for coordination. For example, a riser diagram provides conceptual routing of a primary fiber-optic backbone. A riser diagram is intended to convey an overall project scheme, rather than a detailed specification. A riser diagram is typically not to scale but provides logical information that, when used in conjunction with floor plans, elevations, detail drawings, and a written specification, can assist in installation. **See Figure 3-9.**

Detail Drawings

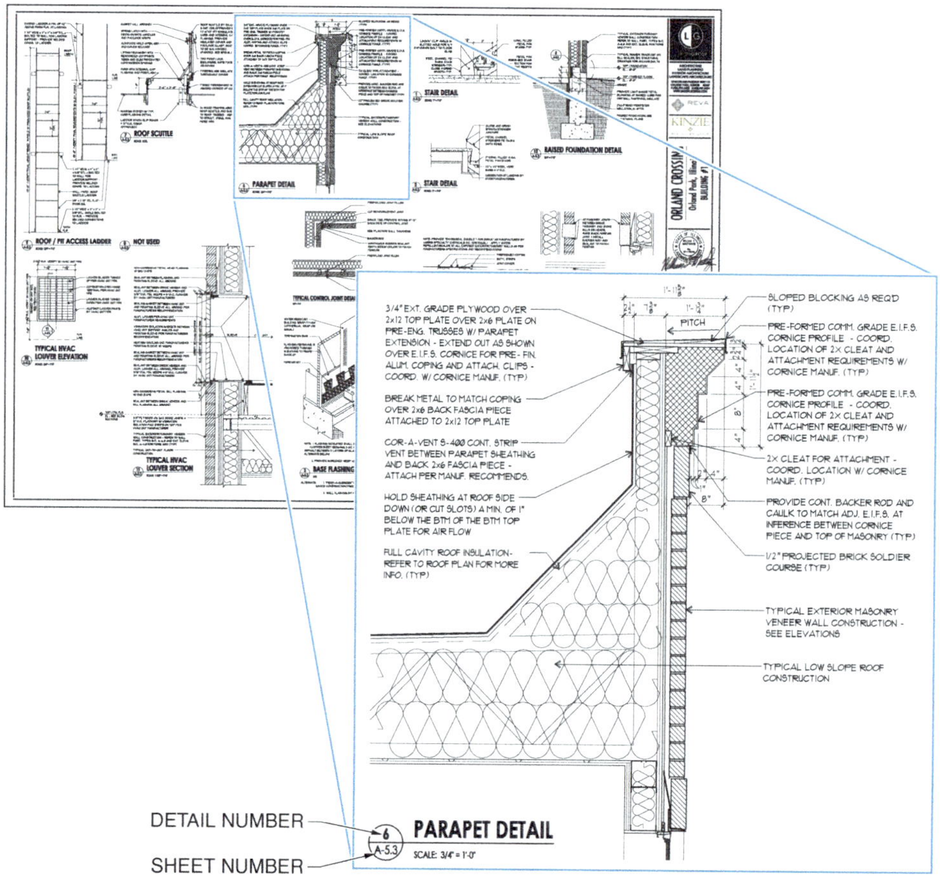

Linden Group Architects

Figure 3-8. Detail drawings are used to show specific features at an enlarged scale.

Diagrams

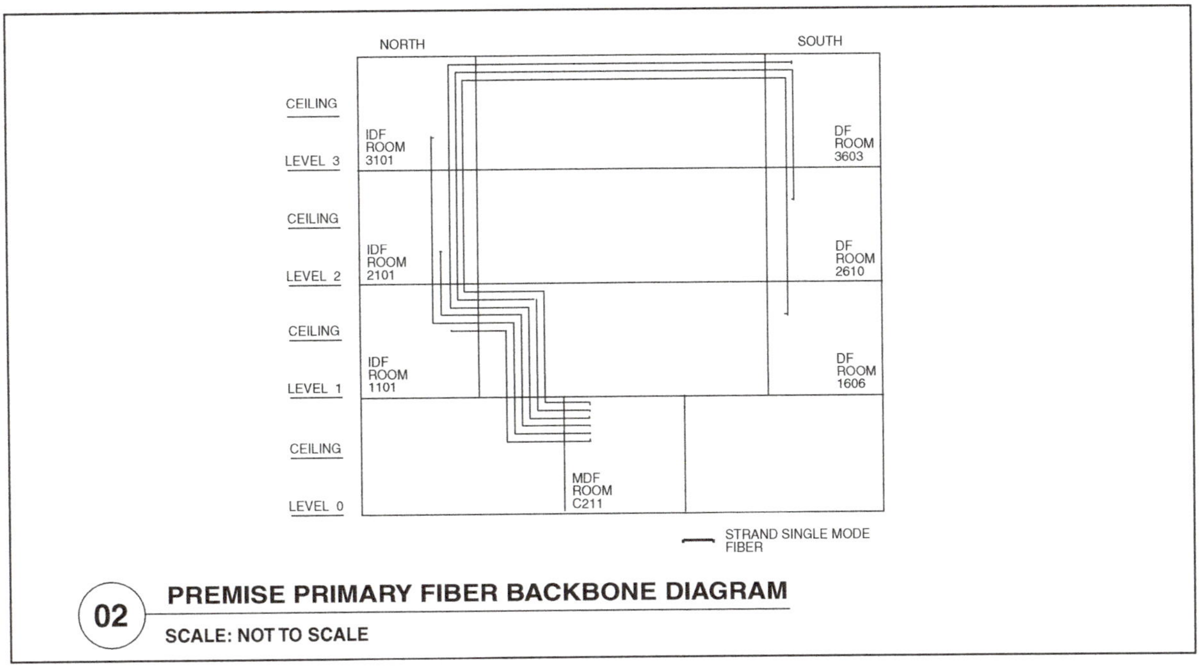

Figure 3-9. A riser diagram is used to provide conceptual routing of a primary fiber-optic backbone.

Trade Symbols

There are different trade symbols specific to each trade and type of system. In certain situations, the relationship of symbols overlaps between trades. For example, an electrical engineer may provide an open triangle for a tel/data outlet and a telecommunications engineer may provide a half-shaded triangle to depict the same component. It is required during the design and construction phase that the design and installation specialists coordinate symbols to minimize confusion on outlet heights and types. **See Appendix.**

Abbreviations and Acronyms

An *abbreviation* is a letter or series of letters or words denoting a complete word. Abbreviations are developed by standards organizations to provide simplified instructions to installers. An *acronym* is a word formed from the first letter of each of the successive parts or words of a compound term. For example, the acronym DAS refers to "distributed antenna system." Abbreviations and acronyms are used to save design and installation professionals time and to conserve space and make drawings less cluttered. **See Appendix.**

DRAWING SCALES

An architect's scale is used by designers and architects to produce and measure construction drawings. An *architect's scale* is a triangular scale, 1′ in length, with six graduated surfaces. Two different scales are marked on five of the surfaces, and inches and fractions of an inch are marked on the other surface. The relative size of a building in comparison to the objects shown on the drawing requires a measurement of fractions of an inch equaling feet. For example, a typical residential floor plan could be drawn at a scale of ¼″ that equals 1′-0″, or 1:48 actual size. *Note:* Some architects' scale types are flat and include two scales on each side.

An architect's scale has 11 different scale sizes ranging from full (actual) scale to 3⁄32″ = 1′-0″. For larger commercial projects, a scale of 1⁄8″ = 1′-0″, or 1:96 actual size, may be required in order to provide a complete plan on a single sheet. At 1:96 scale, a designer may require additional views, such as enlarged plans, to convey the necessary information. As a drawing object requires more details, it is often necessary to draw portions of the work at a larger scale.

For large-scale project drawings, such as those used for the construction of roads and bridges, an engineer's scale is used. An *engineer's scale* is a measuring and scaling tool that is graduated in the decimal system. The unit size of the engineer's scale is smaller than that of the architect's scale and is read in inches that represent lengths in feet. The scales on an engineer's scale include 1″ = 10′-0″, 1″ = 20′-0″, 1″ = 30′-0″, 1″ = 40′-0″, 1″ = 50′-0″, and 1″ = 60′-0″. **See Figure 3-10.**

Drawing Scales

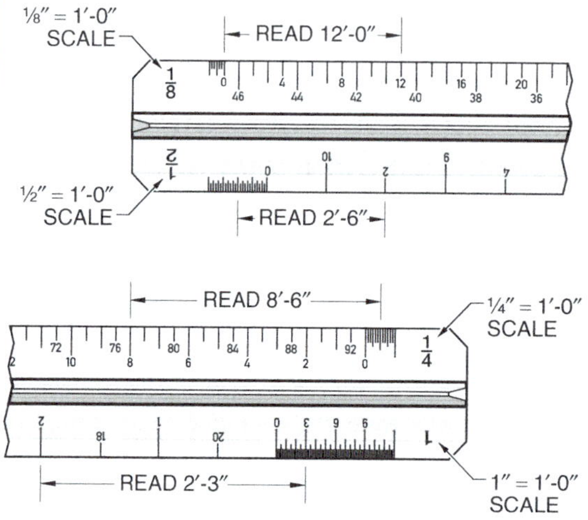

Scale	Size
3″ = 1′-0″	1⁄4 actual size
1½″ = 1′-0″	1⁄8 actual size
1″ = 1′-0″	1⁄12 actual size
3⁄4″ = 1′-0″	1⁄16 actual size
1⁄2″ = 1′-0″	1⁄24 actual size
3⁄8″ = 1′-0″	1⁄32 actual size
1⁄4″ = 1′-0″	1⁄48 actual size
3⁄16″ = 1′-0″	1⁄64 actual size
1⁄8″ = 1′-0″	1⁄96 actual size
3⁄32″ = 1′-0″	1⁄128 actual size

ARCHITECT'S

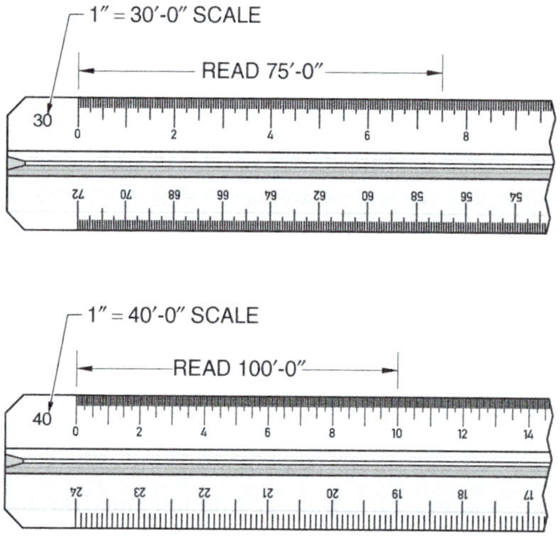

Scale	Size
1″ = 10′-0″	1⁄120 actual size
1″ = 20′-0″	1⁄240 actual size
1″ = 30′-0″	1⁄360 actual size
1″ = 40′-0″	1⁄480 actual size
1″ = 50′-0″	1⁄600 actual size
1″ = 60′-0″	1⁄720 actual size

ENGINEER'S

Figure 3-10. An architect's scale is used by designers and architects to produce and measure construction drawings while an engineer's scale is used for large-scale drawings.

ELECTRONIC DRAWINGS AND BUILDING INFORMATION MODELING (BIM)

Electronic drawings were first used in the construction industry in the 1970s with the introduction of computers and networking. In addition to architectural drawings, electronic drawings are used in manufacturing industries and provide two- and three-dimensional computer-aided design (CAD) models. The use of electronic drawings improved efficiency and accuracy and replaced hand-drawn blueprints, saving architects, engineers, and drafters time and costs on production, reproduction associated with design, and modifications.

Advancements in computer technology have led software manufacturers to develop an integrated concept called building information modeling (BIM) that is used for the design, construction, operation, and maintenance of buildings and infrastructures. *Building information modeling (BIM)* is the process of digitally representing the physical and functional properties of a facility. **See Figure 3-11.** In addition to the three dimensions of height, length, and width, BIM incorporates the added features of creating cost estimates and construction schedules based on object attributes. The model can be shared with other design disciplines in order to minimize information loss, which is a concern in the coordination of document preparation.

Object attributes are created in a virtual model. Object attributes include walls, doors, windows, equipment, and furnishings. The designer places object attributes within the geometry of the model. Object attributes contain information about the manufacturer, the model number, system requirements, cost, and other pertinent data as determined by the designer. Object attribute information is stored in the model database for easy access by others. Any conflict in the coordination of the design can be easily identified and corrected.

Building Information Modeling (BIM)

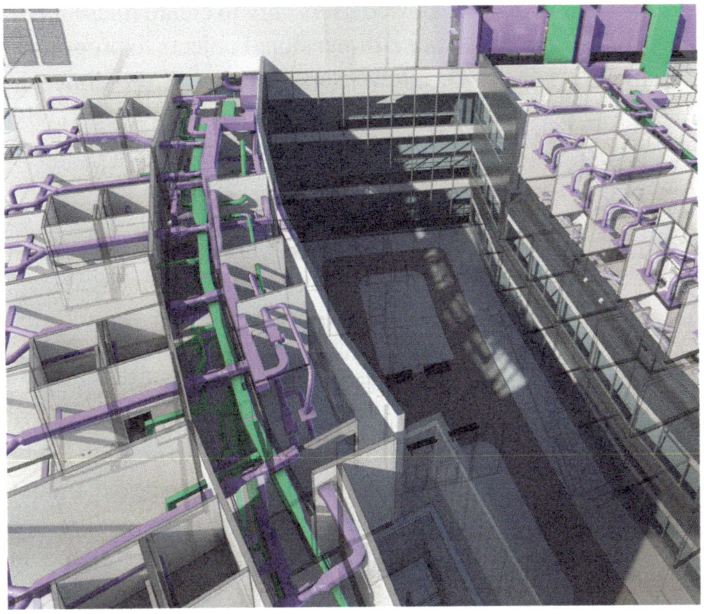

Autodesk, Inc.

Figure 3-11. BIM software allows for the use of color to differentiate between supply and return air ductwork in an HVAC system.

BIM is used not only to produce construction drawings for construction managers, but also to identify and maintain both products and systems for facility managers. BIM is implemented during the lifecycle of the facility, from conception to decommission. Common electronic drawings programs that use BIM software are AutoCAD®, Revit®, Bluebeam®, PlanGrid®, and Fieldwire®.

AutoCAD

AutoCAD is a commonly used computer-aided design (CAD) software program for designing construction projects. *Computer-aided design (CAD)* is the design and drafting of a component, member, or structure using a computer. CAD increases productivity by producing drawings with consistency that can be easily reproduced. With the implementation of modern design programs and fast personal computers, CAD has become more cost effective for producing construction documents. CAD programs such as AutoCAD have evolved from two-dimensional, vector-based software programs that allowed users only to create lines and basic, one-dimensional objects to programs that allow users to create three-dimensional building models that more accurately depict interior spaces, structural members, and mechanical systems.

Software manufacturers have created specialized CAD programs for architectural, structural, civil, mechanical, and electrical system design. In addition to the production of construction drawings for bidding and construction, CAD software can also run simulations for lighting design and structural analysis. Three- and four-dimensional modeling software applies colors, textures, and materials to surfaces to produce visually accurate views and renderings. **See Figure 3-12.**

Revit

Revit is a software program built for BIM with tools for architectural design, MEP, structural engineering, building details, and construction. Revit is designed to help engineers, architects, and designers design, simulate, visualize, and collaborate to capitalize on the advantages of the interconnected data within a BIM model.

BIM programs such as Revit are centered on models composed of different objects. When one piece of data changes in one view, it is updated in all other views automatically by Revit because each view is displaying the same data. Objects can also be related to other objects in the model. An advantage of Revit is that it can be used by designers of different systems in a work-sharing format in which all designers or other contributors can share and save their work to the same project.

Computer-Aided Design (CAD)

Autodesk, Inc.

Figure 3-12. CAD software allows users to create accurate views and renderings.

TECH FACT

Site visits are necessary in order to verify and inspect work progress and add any proposed changes to software programs. Any decisions regarding proposed changes will have an effect on the project.

Bluebeam Revu

Bluebeam Revu is a software program used by architects, engineers, contractors, estimators, and project superintendents that allows for project collaboration and efficiency. **See Figure 3-13.** Bluebeam Revu provides tools for estimating, performing submittal review, logging RFIs, and tracking punch lists and project closeout. Bluebeam software operates on a platform that converts hard-copy floor plans to reviewable three-dimensional drawings and scopes that can be marked and edited through electronic devices such as a tablet computer. The conversion platform reduces the time required for each project discipline to review the drawings, allowing project deadlines to be maintained.

Bluebeam® Revu® Software

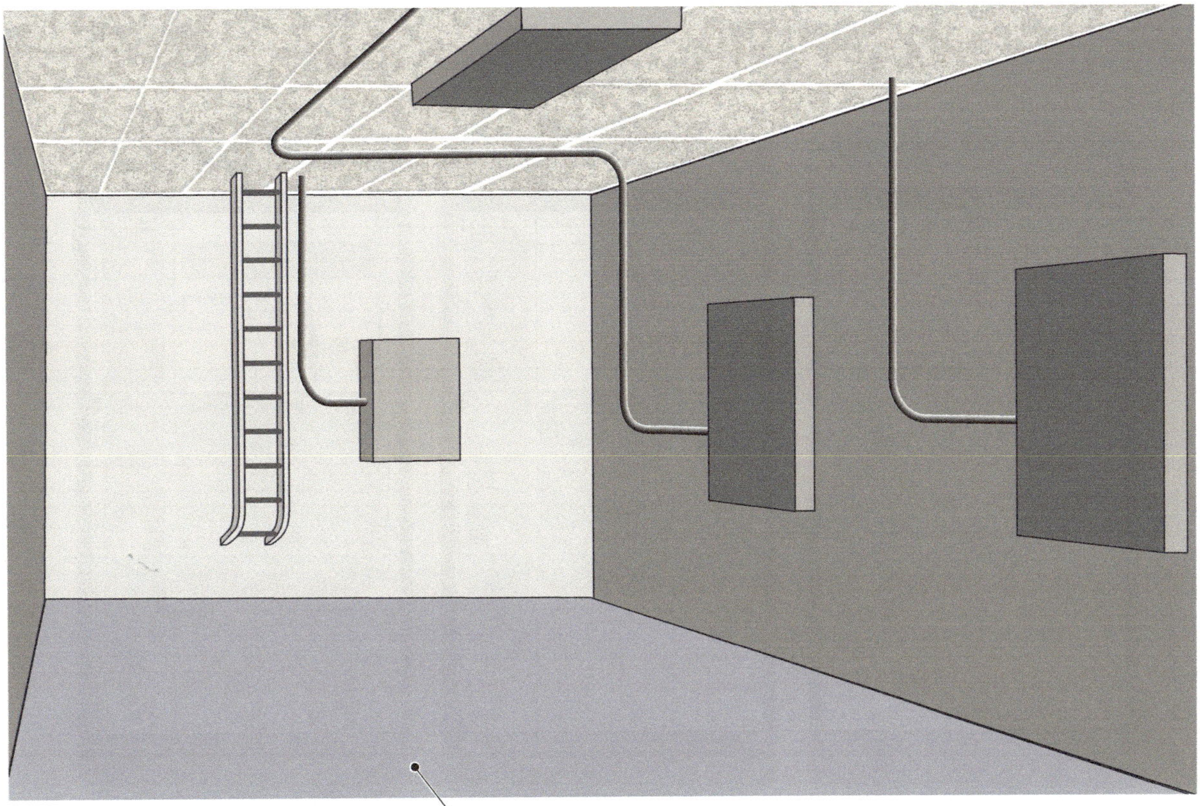

Figure 3-13. Bluebeam Revu software allows designers and installers to create and convert drawings to reviewable three-dimensional drawings.

PlanGrid

PlanGrid is a software platform used by installers to manage documents and plans, issue tasks to tradesworkers, generate field and daily reports, and track as-built documentation during the construction process. Other features include a safety and quality control module, as well as RFI and submittal tracking. PlanGrid software also provides installers with annotation tools to note drawings with cloud and arrow shapes, insert text, highlight sections of documents, track issues or changes, insert hyperlinks and measurements, and upload photos and videos to collaborate with each project discipline and track the progress of a project. **See Figure 3-14.**

Fieldwire

Fieldwire is a software platform that provides architects, engineers, contractors, subcontractors, installers, and supervisors with the necessary tools to coordinate, track, and manage construction projects. Fieldwire provides job-site supervisors with features for scheduling, task management, real-time document management, and communication with other construction professionals. In addition to the project documentation, Fieldwire also tracks tradesworkers, inspection, and punch lists. Fieldwire can be used on smartphones, tablets, or personal computers and is easily accessible from job sites.

PlanGrid® Software

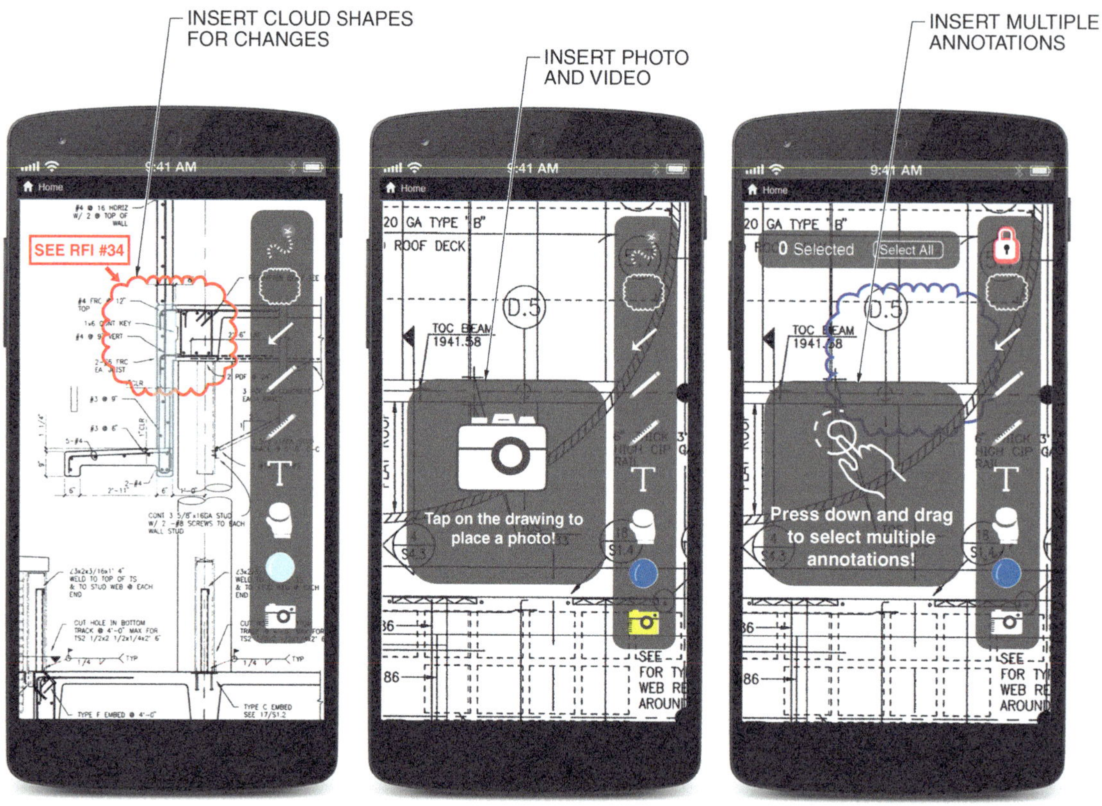

Figure 3-14. PlanGrid software has annotation tools that allow multiple users to insert project changes, technical support documents, photos, and videos.

SUMMARY

Systems drawings are used for all facets of a construction project. An understanding of drawings and their components is required for architects, engineers, contractors, and installers working on a project. Drawings can be in hard-copy or electronic formats; although, due to ease of use, most drawings are in electronic format. Various software platforms are used to provide architects, engineers, contractors, subcontractors, installers, and supervisors with the necessary tools to coordinate, track, and manage construction projects.

CHAPTER 3

Systems Drawings Review

Name ______________________________ Date ______________

True-False

T F **1.** Electrical drawings (E) support specialty systems such as telecommunications, security, access control, and audio/visual.

T F **2.** For drawings with minor revisions, a drawing is reissued with changes noted in a bubble or cloud outline.

T F **3.** Schedules are limited only to architectural drawings.

T F **4.** An architect's scale is a measuring and scaling tool that is graduated in the decimal system and is used for large-scale projects.

T F **5.** BIM is implemented only during the conception phase of a project.

Multiple Choice

______________ **6.** Most architects follow the ___ standard on drawing organization.
A. AIA
B. ASTM
C. CSI MasterFormat
D. UL

______________ **7.** ___ drawings depict the layout of sprinkler systems.
A. Architectural
B. Civil
C. Mechanical
D. Telecommunications

______________ **8.** The accessibility guidelines created from ___ are used to set the height measurements of restroom fixtures, grab bars, handrails, and ramps.
A. the ADA
B. the AIA standard
C. CSI MasterFormat
D. OSHA

______________ **9.** An architect's scale has ___ different scale sizes.
A. 6
B. 9
C. 11
D. 12

______________ **10.** ___ drawings are the core set of drawings that other drawings are based on and coordinated with.

A. Architectural
B. Civil
C. Mechanical
D. Telecommunications

Completion

______________ **11.** A(n) ___ is the area of a print that contains information about the contents of the print.

______________ **12.** The ___ section provides the history of the drawing, from conceptual design to the final contract set.

______________ **13.** A completed set of drawings is referred to as the "___" drawings, which includes the modifications completed during the design phase.

______________ **14.** ___ drawings depict the layout of a building and the various spaces used for mechanical utilities and occupancy.

______________ **15.** A(n) ___ is a drawing that shows a small part of a plan, elevation, or section view at an enlarged scale.

______________ **16.** A(n) ___ is a triangular scale, 1′ in length, with six graduated surfaces.

______________ **17.** ___ is the process of digitally representing the physical and functional properties of a facility.

______________ **18.** ___ is the design and drafting of a component, member, or structure using a computer.

______________ **19.** For large-scale project drawings, such as those used for the construction of roads and bridges, a(n) ___ scale is used.

______________ **20.** A(n) ___ is intended to convey an overall project scheme, rather than a detailed specification.

Short Answer

21. What are six types of facility services?

22. List the items commonly included on a title block.

23. Explain the difference between an abbreviation and an acronym.

24. What are five types of object attributes?

25. What are five common types of electronic drawing programs that use BIM software?

CHAPTER

4

Systems Specifications

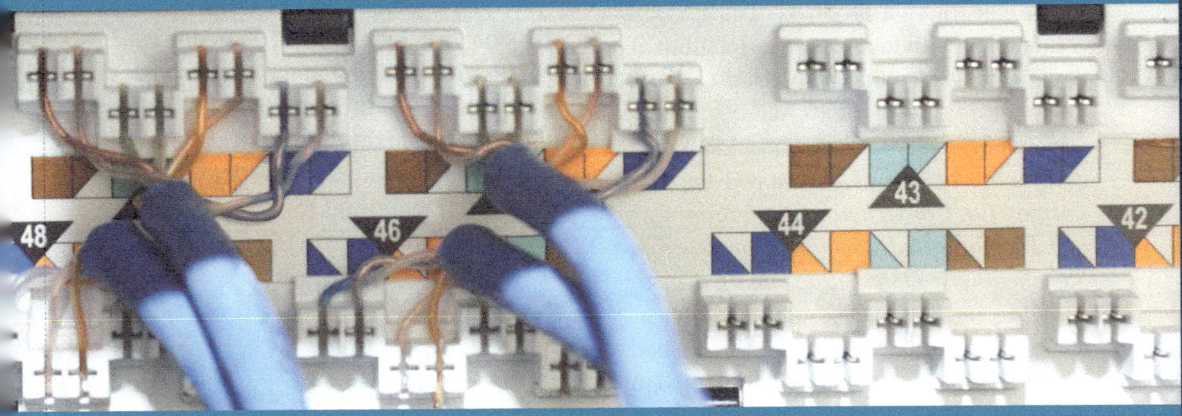

Specifications for VDV systems are developed in a variety of formats, depending on the scope and size of a project. Specifications are developed for every construction project and provide a written description of materials and workmanship that cannot be obtained from drawings. The specifications, accompanied by the corresponding drawing set, are the foundation of the contract between the owner and the contractor. Specifications are written and formatted with information based on the CSI Master-Format and can be performance based or descriptive.

Objectives

- Describe the format and organization of the CSI MasterFormat.
- Describe the development of specifications.
- Explain the differences between the various types of specifications.

Learner Resources

atplearningresources.com
access code 783821

CSI MASTERFORMAT

The Construction Specifications Institute (CSI) is a professional organization of individuals and construction-related industry trade groups that includes architects, engineers, constructors, specifiers, and suppliers of construction products. The CSI, in conjunction with the American Institute of Architects (AIA), the Associated General Contractors of America (AGC), the Associated Specialty Contractors (ASC), and other industry groups, developed the *CSI Format for Construction Specifications and The Uniform System for Construction Specifications, Data Filing, and Cost Accounting* for the concise classification of technical data, as previous processes had become obsolete or could not be adapted to meet the requirements of complex projects or emerging technologies. These efforts ultimately led to the MasterFormat.

The *MasterFormat* is a uniform system of numbers and titles for organizing information concerning construction requirements, products, and activities into a standard sequence. The MasterFormat is used for organizing information and cost data in project manuals, filing product information and other technical data, identifying drawing objects, and presenting construction market data. The MasterFormat is published jointly by the CSI and Construction Specifications Canada (CSC) with extensive public review and coordination with industry and trade organizations.

The MasterFormat is used for organizing information and cost data in project manuals.

There are two versions of the CSI MasterFormat in use by construction professionals: the 16-division version, seen in MasterFormat 95, and the 50-division version, first reorganized for MasterFormat 2004. The latest revised edition, MasterFormat 2018, which also has the same 50-division structure as the 2004 edition, allows for greater detail and expansion of construction standards from the 1995 edition. The 2018 version replaces all other MasterFormat versions; however, some municipalities continue to use previous versions.

MasterFormat 95

Because MasterFormat 95 has been in use as a construction standard for many years, some construction project specifications continue to be written, organized, and produced according to its 16-division structure. The MasterFormat 95 edition organizes related construction products and activities into 16 Level 1 titles, known as divisions. **See Figure 4-1.** The numbers and titles of the divisions are as follows:

- Division 1—General Requirements
- Division 2—Site Construction
- Division 3—Concrete
- Division 4—Masonry
- Division 5—Metals
- Division 6—Wood and Plastics
- Division 7—Thermal and Moisture Protection
- Division 8—Doors and Windows
- Division 9—Finishes
- Division 10—Specialties
- Division 11—Equipment
- Division 12—Furnishings
- Division 13—Special Construction
- Division 14—Conveying Systems
- Division 15—Mechanical
- Division 16—Electrical

The numbers and titles in MasterFormat 95 are also grouped under the following general headings:

- Introductory Information (00001 to 00099)
- Bidding Requirements (00100 to 00499)
- Contracting Requirements (00500 to 00999)
- Facilities and Spaces (no numbering)
- Systems and Assemblies (no numbering)
- Construction Products and Activities (Divisions 1 to 16)

Level 2 titles of the MasterFormat 95 edition identify clusters of products and activities that have an identifying characteristic in common. For example, the Level 2 notation, 06100 Rough Carpentry, includes products such as sill plates, floor joists, and studs, as well as activities such as methods of framing and blocking. Level 2 titles are the highest level used in titling and arranging units of construction information and are represented by the middle digit in a five-digit designation.

In the MasterFormat 95 edition, Level 3 notations include both numbers and titles. Level 3 numbers are represented by the last two digits of the five-digit designation. MasterFormat 95 suggests Level 4 titles but does not indicate numbers. Users create numbers by interpolating between assigned numbers when using Level 4 titles or creating new titles.

Not all divisions of MasterFormat 95 apply to construction projects. For example, Division 14—Conveying Systems typically does not apply to most construction projects and, consequently, is not included in construction project specifications. Instead, the architect can write the specifications using a condensed format and arrange the divisions to fit the scope of the building design.

MasterFormat 2018

The current edition of the MasterFormat is the 2018 version and contains 50 divisions that define broad areas of construction. **See Figure 4-2.** Assigned divisions are numbered 00 to 14, 21 to 23, 25 to 28, 31 to 35, 40 to 45, and 48. Divisions not assigned to specific, current construction areas, such as Divisions 15 to 20, allow for future expansion of the specification. The 2018 edition was revised from its original publication in 2004; however, the only difference is that in the 2018 edition, Division 40 is renamed Process Interconnections from Process Integration.

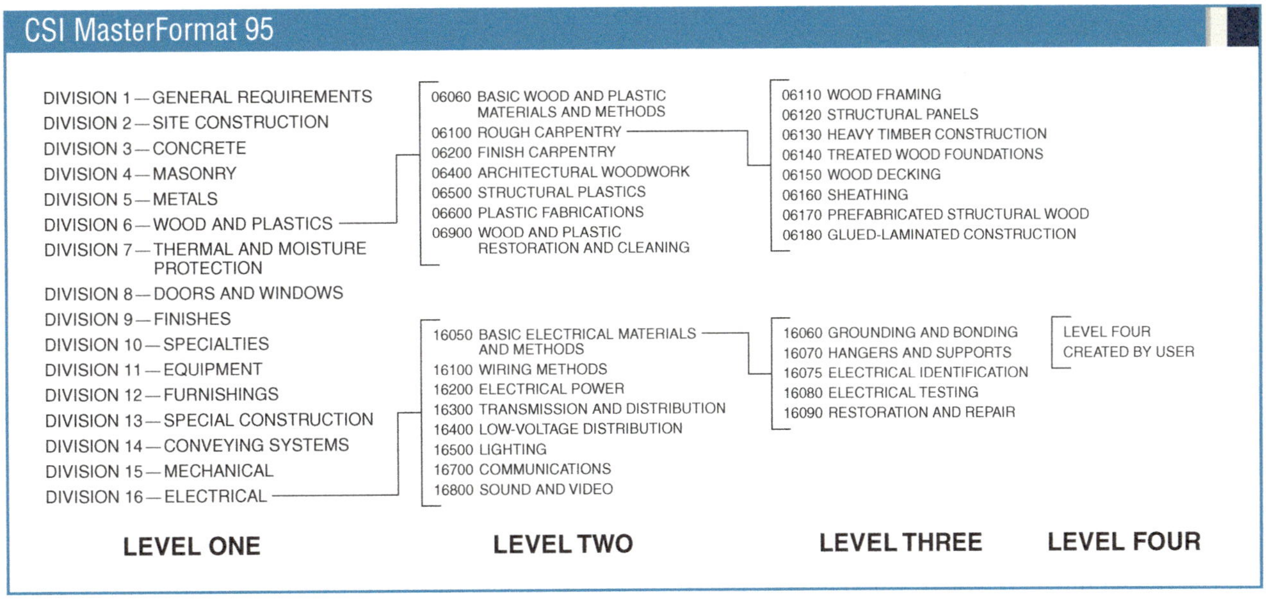

Figure 4-1. The CSI MasterFormat 95 is divided into 16 divisions, with each division providing specific information about various materials.

CSI MasterFormat 2018

Procurement and Contracting Requirements Group
- Division 00 — Procurement and Contracting Requirements

Specifications Group

General Requirements Subgroup
- Division 01 — General Requirements

Facility Construction Subgroup
- Division 02 — Existing Conditions
- Division 03 — Concrete
- Division 04 — Masonry
- Division 05 — Metals
- Division 06 — Woods, Plastics, and Composites
- Division 07 — Thermal and Moisture Protection
- Division 08 — Openings
- Division 09 — Finishes
- Division 10 — Specialties
- Division 11 — Equipment
- Division 12 — Furnishings
- Division 13 — Special Construction
- Division 14 — Conveying Equipment
- Divisions 15 to 19 — Unassigned

Facility Services Subgroup
- Division 20 — Unassigned
- Division 21 — Fire Suppression
- Division 22 — Plumbing
- Division 23 — Heating, Ventilating, and Air-Cond
- Division 24 — Unassigned
- Division 25 — Integrated Automation
- Division 26 — Electrical
- Division 27 — Communications
- Division 28 — Electronic Safety and Security
- Division 29 — Unassigned

Site and Infrastructure Subgroup
- Division 30 — Unassigned
- Division 31 — Earthwork
- Division 32 — Exterior Improvements
- Division 33 — Utilities
- Division 34 — Transportation
- Division 35 — Waterway and Marine Construction
- Divisions 36 to 39 — Unassigned

Process Equipment Subgroup
- Division 40 — Process Interconnections
- Division 41 — Material Processing and Handling Equipment
- Division 42 — Process Heating, Cooling, and Drying Equipment
- Division 43 — Process Gas and Liquid Handling, Purification, and Storage Equipment
- Division 44 — Pollution Control Equipment
- Division 45 — Industry-Specific Manufacturing Equipment
- Division 46 — Water and Wastewater Equipment
- Division 47 — Unassigned
- Division 48 — Electrical Power Generation
- Division 49 — Unassigned

DIVISIONS DEFINE BROAD AREAS OF CONSTRUCTION

26 00 00	**ELECTRICAL**
26 01 00	Operation and Maintenance of Electrical Systems
26 05 00	Common Work Results for Electrical
26 06 00	Schedules for Elelectrical
26 08 00	Commissioning of Electrical Systems
26 09 00	Instrumentation and Control for Electrical Systems
26 10 00	Medium-Voltage Electrical Distribution
26 20 00	Low-Voltage Electrical Transmission
26 30 00	Facility Electrical Power Generating and Storing Equipment
26 40 00	Electrical Protection
26 50 00	Lighting

NUMBERS AND TITLES DIVIDE DIVISIONS INTO MORE SPECIFIC AREAS

27 00 00	**COMMUNICATIONS**
27 05 00	Common Work Results for Communications
27 20 00	Data Communications
27 40 00	Audio-Video Communications
27 10 00	Structured Communications
27 30 00	Voice Communications
27 50 00	Distributed Communications and Monitoring Systems

Figure 4-2. The 50-division CSI MasterFormat 2018 provides room for inclusion of new construction materials and techniques.

Each division of the MasterFormat 2018 includes sets of numbers and titles that correspond to levels of detail. For example, Division 48 – Electrical Power Generation (48 00 00) contains multiple Level 2 numbers and titles, such as 48 01 00 Operation and Maintenance for Electrical Power Generation, 48 10 00 Electrical Power Generation Equipment, and 48 70 00 Electrical Power Generation Testing.

The use of a six-digit numbering system for the first three levels replaces the five-digit system used in MasterFormat 95. The six-digit system allows for more expansion per level of each division than the previous edition. Classification beyond Level 3 is possible using additional pairs of digits, or numbers or letters for Level 5 classification. Level 4 classifications use eight digits. **See Figure 4-3.**

For example, the classification for Plate Cell Solar Energy Collectors is at the fourth level of classification for 48 00 00 Electrical Power Generation (first level), 48 10 00 Electrical Power Generation Equipment (second level), 48 14 00 Solar Energy Electrical Power Generation Equipment (second level, a subset of 48 10 00), 48 14 13 Solar Energy Collectors (third level), and 48 14 13.16 Plate Cell Solar Energy Collectors (fourth level).

Generally, Level 4 classifications and above are used for specifications and project manuals, while Level 5 classifications are usually done by individual users for their internal use. The use of MasterFormat 2018 continues to increase as the old standard edition (MasterFormat 95) is phased out. Individuals referencing the specifications must be knowledgeable in the differences between the 16-division and 50-division versions of the MasterFormat to accurately read and interpret the specifications and locate necessary items. **See Figure 4-4.**

SPECIFICATIONS DEVELOPMENT

Almost every project requires written specifications to outline the scope of work, the materials and processes to be used, and the responsibilities of the individuals performing the work. A *specification* is written information included with a set of prints. The specifications can be either a separate document that is a section of a larger master document or listed on the cover sheet of the accompanied drawing set. When contractors bid on a project, they rely on the drawings and written specifications. While some specifications can be highly detailed, other forms of specifications are simple, broad statements that cover every aspect of a project. **See Figure 4-5.**

CSI MasterFormat 2018 Numbering System

DIVISION 48—ELECTRICAL POWER GENERATION

48 00 00 ELECTRICAL POWER GENERATION
- 48 01 00 Operation and Maintenance for Electrical Power Generation
- 48 05 00 Common Work Results for Electrical Power Generation
- 48 06 00 Schedules for Electrical Power Generation
- 48 08 00 Commissioning of Electrical Power Generation
- 48 09 00 Instrumentation and Control for Electrical Power Generation

48 10 00 ELECTRICAL POWER GENERATION EQUIPMENT
- 48 11 00 Fossil Fuel Plant Electrical Power Generation Equipment
- 48 12 00 Nuclear Fuel Plant Electrical Power Generation Equipment
- 48 13 00 Hydroelectric Plant Electrical Power Generation Equipment
- 48 14 00 Solar Energy Electrical Power Generation Equipment
  - 48 14 13 Solar Energy Collectors
    - 48 14 13.13 Amorphous Solar Energy Collectors
    - 48 14 13.16 Plate Cell Solar Energy Collectors
    - 48 14 13.19 Vacuum Tube Solar Energy Collectors
- 48 15 00 Wind Energy Electrical Power Generation Equipment
- 48 16 00 Geothermal Energy Electrical Power Generation Equipment
- 48 17 00 Electrochemical Energy Electrical Power Generation Equipment
- 48 18 00 Fuel Cell Electrical Power Generation Equipment
- 48 19 00 Electrical Power Control Equipment

48 70 00 ELECTRICAL POWER GENERATION TESTING
- 48 71 00 Electrical Power Generation Test Equipment

Figure 4-3. MasterFormat 2018 Level 4 classification uses eight digits: three pairs of digits followed by a decimal point and then another pair of digits.

Significant CSI MasterFormat Division Changes

MasterFormat 95	MasterFormat 2018
Division 2—Site Construction →	Division 02—Existing Conditions Division 31—Earthwork Division 32—Exterior Improvements Division 33—Utilities Division 35—Waterway and Marine Construction
Division 15—Mechanical →	Division 21—Fire Suppression Division 22—Plumbing Division 23—Heating, Ventilation, and Air-Conditioning (HVAC)
Division 16—Electrical →	Division 26—Electrical

Figure 4-4. Three MasterFormat 95 divisions have undergone significant changes in the conversion to newer versions of the MasterFormat, including several Level 2 and Level 3 titles in the area of electrical.

Simplified Specifications

1.2 SCOPE

A. Perform work and provide material and equipment as shown on Drawings and/or as specificed and/or indicated in this Specification(s). Completely coordinate work of this Section with work of other trades and provide a complete and fully functional installation.

B. Drawings and Specifications form complimentary requirements; provide work specified and not shown, and work shown and not specified as though explicitly required by both. Although work is not specifically shown or specificed, provide supplementary or miscellaneous items, appurtenances, devices and materials obviously necessary for a sound, secure and complete installation.

Figure 4-5. Some types of specifications can be simplified, broad statements that cover every aspect of a project.

Even the simplest construction job requires specifications. Specifications consist of short statements about the work to be done and the materials to be used. Specifications are included with both a working drawing and a written contract, which become the basis for agreement between the owner and the contractor. Complete specifications are written to provide detailed information about the project, such as additional details related to materials and methods that could not be shown on the prints or that require further description. Specifications also provide written descriptions of building components and procedures in conjunction with the prints.

Specifications supplement the set of prints with data that describe the construction job in a general manner. Specific building processes, such as steps taken related to sustainable construction techniques, are also included in the specifications. The set of prints and the specifications are intended to agree. In the event of any discrepancy or conflict between the prints and specifications for a specific project, the material in the specifications takes precedence over the material in the prints.

Specifications provide the general conditions of the contract and outline the responsibilities of the owner, architect, general contractor, and subcontractors. Specifications also include contractor guarantees. A *contractor guarantee* is a written assurance issued by a general contractor that furnished materials and equipment are of acceptable quality and workmanship as well as free from defects. With a contractor guarantee, the contractor promises to correct any defective materials or equipment for a specified period of time at no additional cost to the owner. The period of time for a contractor guarantee is typically one year from the date of completion. The general contractor is also responsible for scheduling and coordinating any corrective work required from a subcontractor.

The architect responsible for the specifications prepares the specification details in writing. The specifications define the architect's position in the project and the personal responsibility that must be assumed for inspection. The owner uses the specifications to verify that materials specifically desired in the building are provided and to obtain a specific and detailed overview of the finished building. The owner must review the various guarantees for future reference in case the structure or its equipment does not prove satisfactory.

The contractor and subcontractors use the specifications to obtain detailed information for estimating the costs of labor and materials when bidding competitively. Suppliers of building materials, such as roofing, lumber, and insulation, can determine the quality and types of materials to be used from the specifications. Equipment and fixture suppliers use the specifications to obtain detailed descriptions, with specific OEM catalog numbers or names, for items such as plumbing fixtures, HVAC equipment, and other similar items. Electricians comply with the requirements of the NEC® listed in the specifications when estimating costs for specific materials. **See Figure 4-6.**

The building department of the municipality in which the building is to be constructed uses the specifications and prints to determine their compliance with applicable building codes and zoning ordinances to meet structural, fire-resistance, environmental, and health standards. Financial agencies use the same information to appraise the building to determine its value. When government agencies, such as the Veterans Benefits Administration (VBA) or the Federal Housing Administration (FHA), provide any portion of the financing, they require a copy of the specifications for their approval. Certain sections of the specifications must be written on forms supplied by the individual agency when either VBA or FHA financing is used.

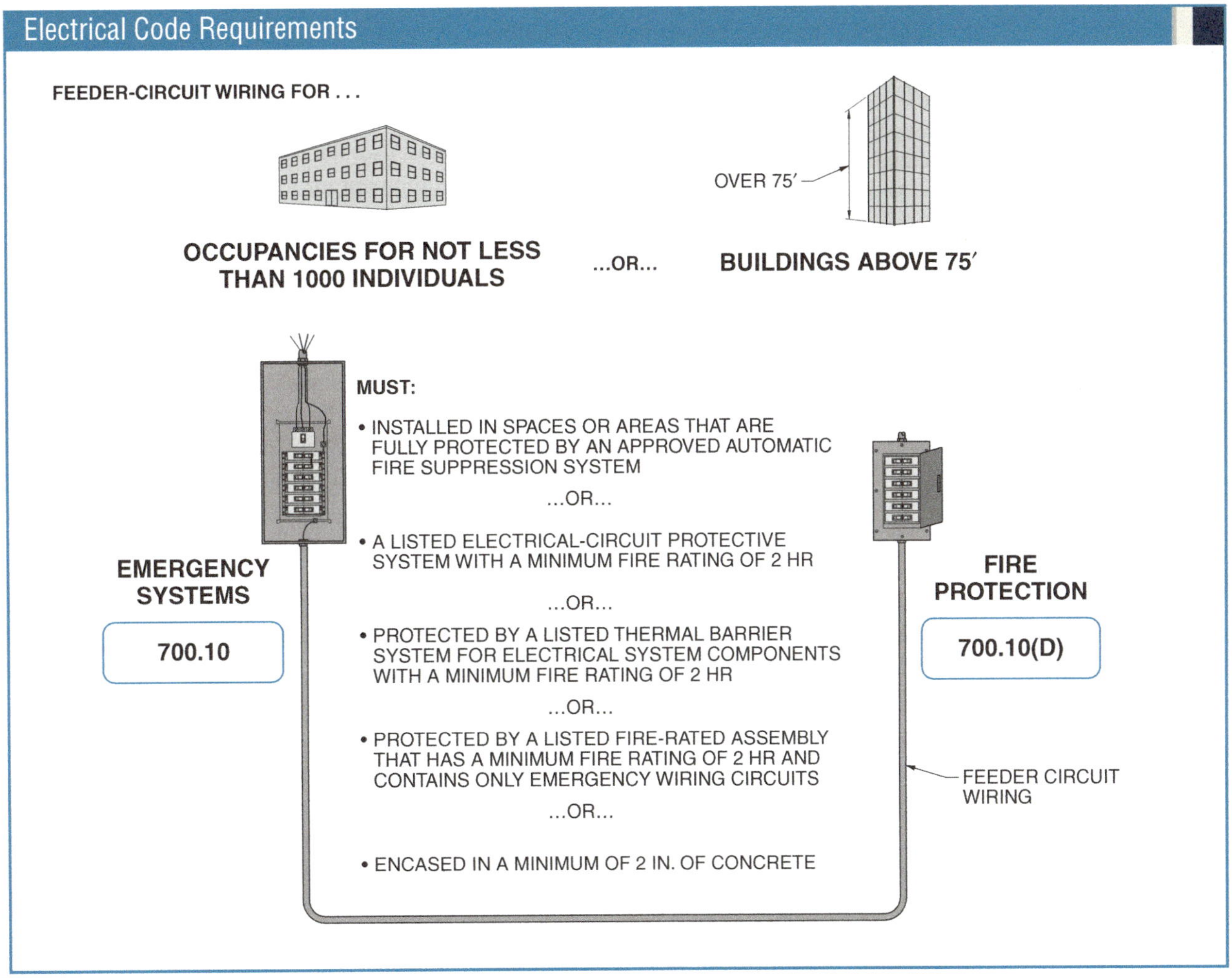

Figure 4-6. Many specifications and local building codes rely on information from the National Electrical Code® (NEC®).

Specifications serve, along with the contract agreement and the working drawings, as the legal basis for the transaction of erecting the building from start to finish. Since specifications are part of the legal contract, they may be used in court in the event of litigation. The potential for problems during a construction project is minimized when the working drawings are correct and the specifications are properly written.

Organizing Specifications

The specifications for a construction project typically begin with conditional statements, such as the party with the responsibility for examining the job site, the liability that can be assessed due to delays in the work, requirements for periodic cleaning of the construction area, and similar items. The remainder of the specifications is devoted to technical information concerning the divisions of work to be performed. For a specification, each technical division covering an area of work must contain the following four sections:

- Scope—information that is included in specific sections of work
- Materials—specific materials to be used in each section of work
- Applications—special instructions concerning material applications and usage
- Guarantee—a statement, which can be general, applying to each different trade group, and inserted in the general conditions, binding the contractor to a certain quality of work over a specified time

Depending on the type of project, specifications can be prepared with no detail or with a large amount of detail. Minimum specifications, such as for a minor remodeling project, may consist of a sketch showing the work to be done and a statement concerning the materials to be used on the project. For large-scale projects, certain specifications are prepared in a manner to provide general information concerning the construction and the features of a building without providing highly detailed procedures. For most projects, however, a full set of specifications covers every area of the building and the building site that is not shown in detail on the working drawings, including VDV systems.

Writing Specifications

The architect is responsible for writing the specifications so that they are consistent with the prints. Modern or revised developments in construction materials and methods must be reviewed so that the architect can recommend specific products to be used and building techniques to be followed. When preparing specifications, the architect includes the materials to be used and lists the equipment by catalog numbers. Alternative materials are also listed if the initial specified materials are not available. The process of writing specifications requires experience acquired from specifying and observing the performance of materials and equipment from previous projects.

Extensive information resources are available to architects from OEMs as well as other building information resources. One of the most useful tools for acquiring material and equipment information is Sweets™, which is an internet resource consisting of OEM information organized as a ready reference. Building product information on the Sweets website is organized according to the MasterFormat and provides technical information, such as size, color, and application, for each manufacturers' specific material. An advantage of having material and equipment information in an electronic format is that it facilitates quick searches of the most current information available.

Another example of reference material that an architect consults when writing the specifications are the various standards published by Underwriters Laboratories (UL), such as UL 263, *Standard for Fire Tests of Building Construction and Materials*. The UL 263 standard is a guide for determining the building assemblies that are suitable for various building applications based on life-safety requirements. UL also provides online resources such as product directories and application guides that help architects select products that meet regulatory requirements.

The American Institute of Architects (AIA) provides standard forms and overall directions for specification writing. Materials from UL and the AIA are particularly useful in stating the general conditions. The general requirements are stated at the beginning of the set of specifications, but they govern each separate area of construction. The general requirements also provide uniform legal language that protects the participants in a construction project, including the owner, architect, general contractor, subcontractors, suppliers, and related employees.

SPECIFICATIONS FORMAT

Specifications contain detailed written and graphic information contained in the project manual and may include notes that could pertain to instructions regarding owner and contractor responsibilities, product properties, or insurance requirements. For a small project, the technical specifications may be listed directly on the drawings. The most common format for specifications follows the MasterFormat and segments the specifications into three parts that collectively describe the design. A project specification is formatted into three distinct parts: Part 1 – General, Part 2 – Products, and Part 3 – Execution.

Part 1 – General

Part 1 – General, sometimes titled General Requirements, describes the intent of the project and other details about what the specifications are intended to cover. **See Figure 4-7.** Part 1 is a general description of the scope of work. *Note:* Depending on how the designer or architect writes the specifications, this part may be highly detailed or vague and general. Part 1 contains specific descriptions of the following items:

- General provisions—References to MasterFormat Division 01 – General Requirements and the "contract and general conditions" are typically mentioned and become applicable to the specification section. Other drawings and specifications related to the specified work may also be listed.
- Description of work—A summary of the scope of work specified is provided, as well as a reference to "furnish and install" items listed as part of the scope. The intent is to describe the product and installation requirements. References to owner standards and preferred manufacturer components, quality assurance, and labor and material warranties are also included. Older specification formats may include language that provides an alternative design meeting the requirements of the specification, which could provide cost savings and is often known as valued engineering.
- Definitions—The generic terms and definitions used in most specifications are provided. These generic terms and definitions may include the following:
  - "Contractor or Subcontractor" means the installer of the work specified under this section.
  - "Furnish" means to purchase and deliver to the project site, complete with all required apparatus.
  - "Install" means unload at the delivery point at the site and perform all work to establish secure mounting and operation at the proper location.
  - "Provide" means to "furnish and install."
  - "Work" means the labor, materials, equipment, apparatuses, and accessories required for a complete installation.
- Site inspection—The specifications direct bidders to the site and to review related construction documents to determine the conditions under which the work is to be performed and whether the bidders must provide the architect/owner any conditions, in writing, that may adversely affect the work of the specification section.
- Delivery, storage, and handling—The requirements for the delivery of materials to the project site are addressed. The requirement that materials be in new condition, in their original packaging, and with OEM labels and product identification is stated. With VDV specifications, specialty wire and cable may require preinstallation test data, which could become part of the project documentation requirement. Storage and protection of the materials

during the various phases of construction is also required be. Equipment delivery time-frame windows are also specified, as equipment typically cannot be delivered to a job site too early or too late.

- Submittals—Language pertaining to component product submittals is outlined to provide details on product construction, fire resistance, and the finished appearance of a component. A sampling of specific components may be requested to provide reference examples of finished products. Shop drawings prepared by the contractor require fabrication and installation details such as plans, elevations, and instructions for coordination between other trades.
- Obtaining information—The specifier must clearly state that the installation must adhere to the manufacturer's performance requirements. The specifications must reference industry-related documents, such as codes and performance standards. It is the contractor's responsibility to install the complete system per the referenced codes and standards even if they are not explicitly outlined in the specification or if the specification contradicts a code or standard.

Additional items may be outlined in Part 1 – Requirements. These include permitting, project coordination, project management, accurate record drawings, and identification and documentation of the specified system.

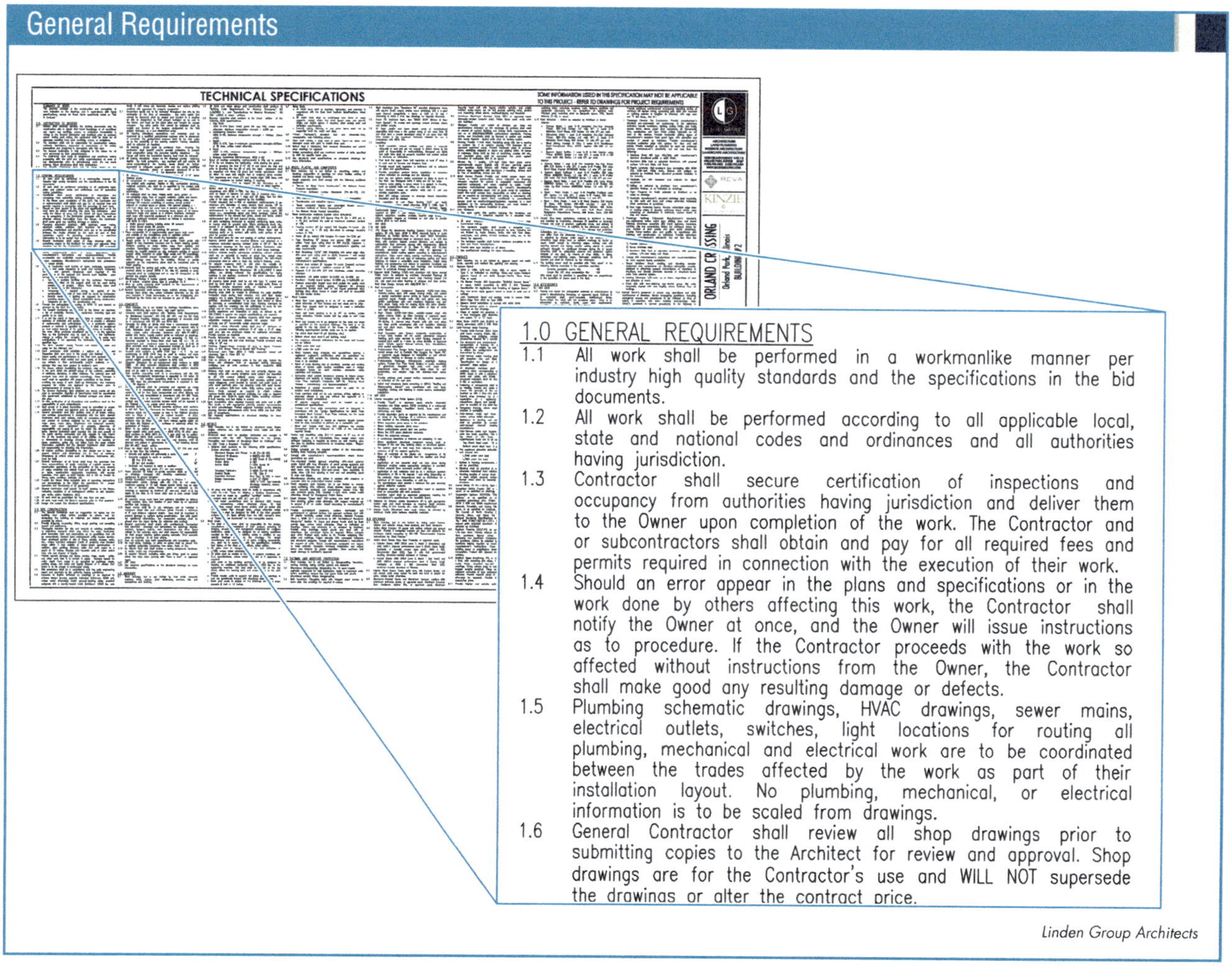

Figure 4-7. Part 1 – General Requirements describes the project intent and other details about what the specifications are intended to cover.

Part 2 – Products

Part 2 – Products provides a baseline of the acceptable OEMs and products that can be used on the project. This part also outlines the specific products that are to be installed and must include every assembly and component required for a complete installation. While specification formatting and types can vary, Part 2 usually contains specific descriptions of the following items:

- General—Wires and devices must comply with applicable standards and be UL listed. Components must be subject to approval, inspection, and testing by the architect, engineer, and owner at various stages of the installation. Manufacturer part numbers must be indicated throughout the section to establish quality and performance characteristics.
- Acceptable manufacturers—Acceptable OEMs of cable and related components by subsystem are listed as required, indicating OEM names, part numbers, and descriptions for each separate component. **See Figure 4-8.**
- Substitutions—Unless otherwise indicated, most specifications require bids to be submitted "as specified." Equal or equivalent products may be substituted if it is established that they are equal in quality, durability, appearance, and performance to the specified products. Substitutions can occur prior to the bid or through the submittal approval process prior to installation.
- Products—The specifier must provide a description and part number for each cable and component for each subsystem included in this section. The description and part number are to include following:
  - Horizontal cabling—category type for copper or fiber-optic cable, jacket color, quantities of each type, fire-resistance rating, pair count (for copper) or strand count (for fiber), and conductor sizes (for copper) or mode type (for fiber)
  - Riser cabling—category type for copper or fiber-optic cable, jacket color, quantities of each type, fire-resistance rating, pair count (for copper) or strand count (for fiber), and conductor sizes (for copper) or mode type (for fiber)
  - Grounding cable—conductor size and fire-resistance rating
  - Campus riser cabling—category type for copper or fiber-optic cable, fire-resistance rating, pair count (for copper) or strand count (for fiber), and conductor sizes (for copper) or mode type (for fiber)
  - Horizontal components—category type for copper or fiber-optic cable, termination hardware, cable management, connector types, faceplates, port quantities, and colors
  - Riser components—termination hardware, cable management, and connector types
  - Campus riser components—category type for copper or fiber-optic cable, termination hardware, cable management, and connector types
  - Horizontal connecting cords—category type for copper or fiber-optic cable, fire-resistance rating, pair count (for copper) or strand count (for fiber optic), lengths, and jacket color
  - Riser connecting cords—category type for copper or fiber-optic cable, fire-resistance rating, pair count (for copper) or strand count (for fiber-optic), lengths, and jacket color
  - Cable pathway supports—application, size, type, use, and accessory hardware
  - Cable pathway firestopping—horizontal application, size, type, and accessory hardware
  - Equipment room pathways—horizontal and vertical access, size, type, and accessory hardware
  - Equipment room racking—application, size, color, types, posts, and finishes
  - Equipment room cable management—application, size, color, type, and finish
  - Equipment room grounding—connector and termination type and size
  - Equipment room firestopping—horizontal and vertical applications, size, type, and accessories
- Equipment—The specifier must provide a description and part number for each specified piece of equipment. The description

must include information such as equipment type and port count and may include information on network hardware, access points, power distribution units, paging systems, and distributed antenna systems.

For example, VDV outlet details provide specific information for each type of VDV outlet, such as the exact termination scheme for the unshielded twisted-pair (UTP) cable to the jacks, the outlet-addressing format being used, the VDV outlet symbol, and the number and type of cables and jacks. **See Figure 4-9.** VDV outlet details also show cable part numbers, faceplates, and jacks. Additional information found in VDV outlet details includes information on the type of label used for the faceplate of a VDV outlet, such as size of the label, label font, and label font size.

Beast Cabling Systems, Inc.

Specialized VDV cable labeling equipment is often used to label high-cable-count projects, such as commercial or industrial installations.

Acceptable Original Equipment Manufacturers (OEMs)

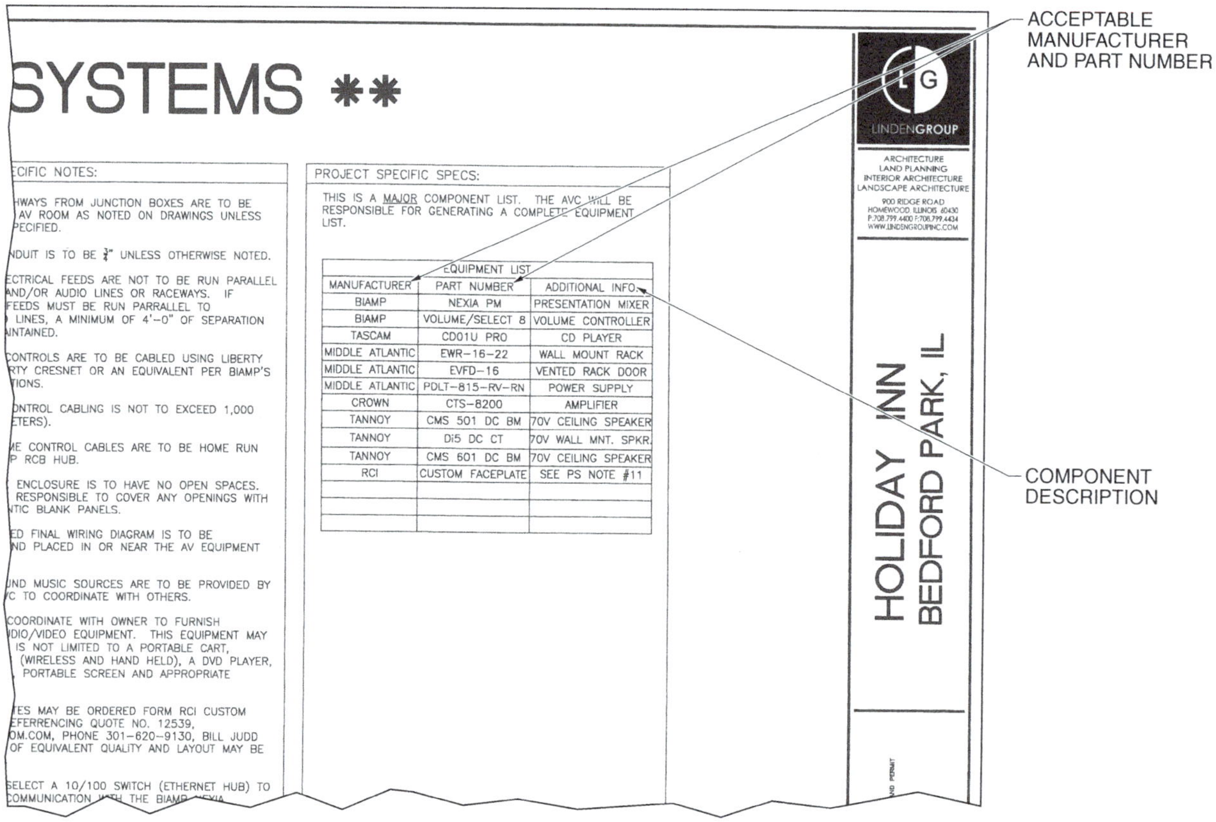

Linden Group Architects

Figure 4-8. Acceptable OEMs of cable and related components by subsystem are listed as required, indicating OEM names, part numbers, and descriptions for each separate component.

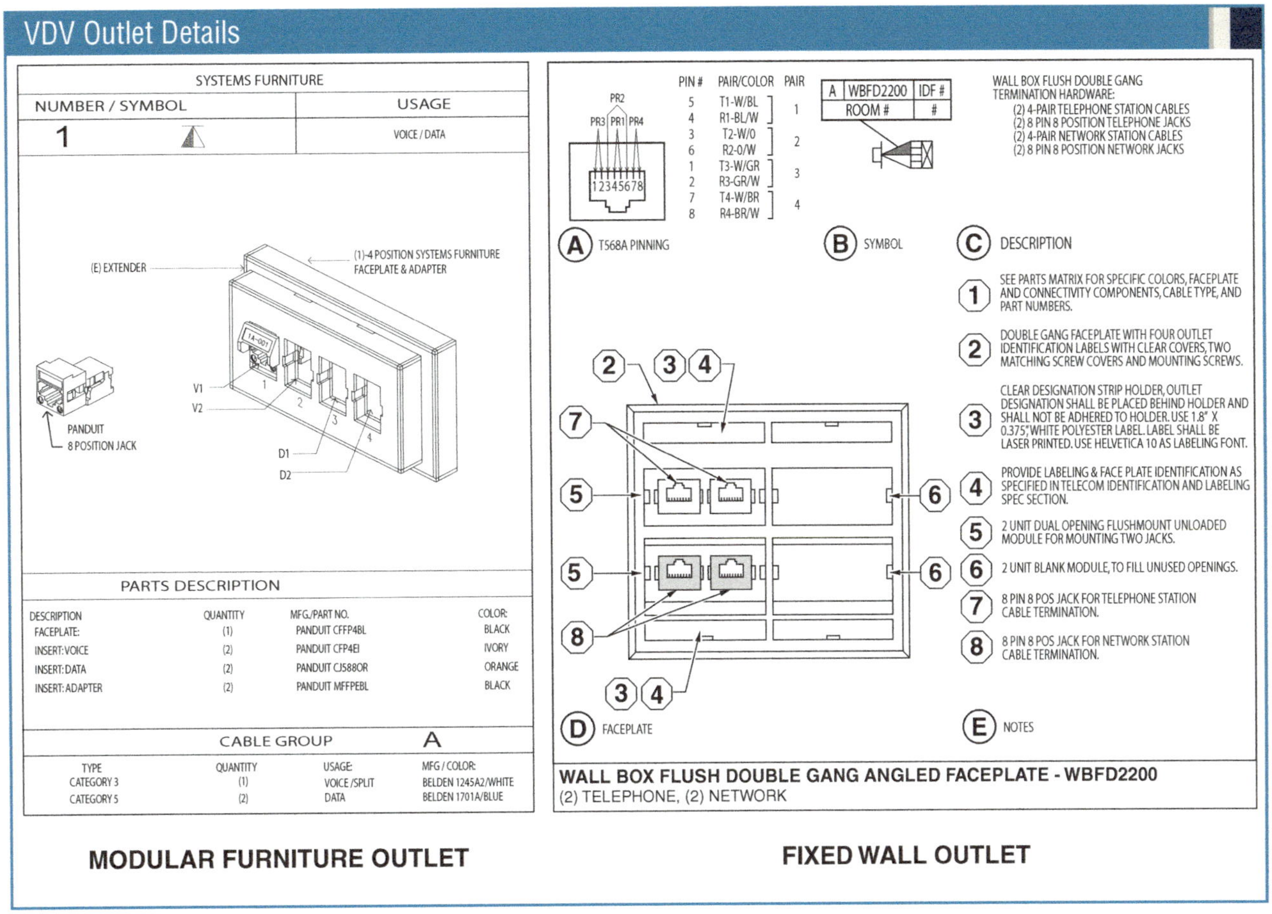

Figure 4-9. VDV outlet details provide information on the number and type of jacks, terminations, and labeling for each type of VDV outlet.

Part 3 – Execution

Part 3 – Execution typically provides details regarding installation, adjusting and cleaning, testing, manufacturer warranties, bid breakdown, and unit prices. The details that may be included in Part 3 – Execution are as follows:

- Installation—Detailed line-by-line descriptions of the component types by application are listed and may include cable type and termination hardware. A descriptive account of the installation also includes pathways, spaces, and associated components.
- Adjusting and cleaning—During the construction process, hardware and equipment can become contaminated with dust and dirt particles and require adjustments. The specifications may require that all equipment be cleaned and replaced as necessary at no additional cost to the owner.
- Testing—The specifications require each installed system to be tested. The specific test parameters for each system are referenced and outlined so that all testing be performed with a certain type of test equipment operated by qualified personnel. Test equipment calibration data may also be required, with the test results being a condition of project closeout. Any system that fails to meet specified test parameters is to be replaced by the contractor at no further cost to the owner. **See Figure 4-10.**
- Manufacturer warranties—Manufacturer warranties are commonly included in specifications and provide a level of confidence for the owner that the installed system performs at a high level for a specified minimum time. Most manufacturer warranties require periodic visual inspections

throughout the construction process and a thorough review of the project test data. The duration of a manufacturer warranty can range from 1 year to 25 years.

Certain manufacturer warranties may include an applications assurance warranty. An *applications assurance warranty* is a manufacturer warranty that ensures that standards-based technology designed to be used with a specific system is supported by the manufacturer. An applications assurance warranty provides a guarantee that the installation meets or exceeds the specification during the warranty period. Most manufacturer warranties also include materials and labor.

- Unit pricing—With most bid breakdowns, a unit-price schedule is often requested. A request can be either for common components that are typically added or deleted during a project or for a detailed summary of each system component broken down by materials and labor. Unit-price schedules are based on unit-price agreements. A *unit-price agreement* is a construction contract that is based on the cost of work or per item of material.
- Project pricing breakdown—Depending on the size, complexity, and type of project (industrial, commercial, residential, etc.), the project pricing breakdown often can be located at the end of Part 3 – Execution in the specifications. The project pricing breakdown can be either a lump-sum submission or an itemized description of each component or system, with distinctions made between labor and materials. **See Figure 4-11.**

Adjusting, Cleaning, and Testing Specifications

SHIRE B200
Lexington, MA

KlingStubbins
May 7
Issued for Construction

3.2 ADJUSTING AND CLEANING

A. Adjust hardware after installation for proper operation.

B. Clean and replace (if required) hardware and devices damaged during the installation at no additional cost to the owner.

3.3 TESTING

A. Furnish all test equipment and perform all work necessary to determine the system meets the requirements of this specification. Provide all test unit calibration data as part of the submittal process.

B. Test all copper cables for shorts, opens and inversions. Provide all test results as part of the close-out package.

C. Category 6 testing shall be performed with a Fluke DTX Class III tester in accordance with ANSI/EIA/TIA TSB -67.

D. Fiber optic testing shall be performed with a Fluke DTX Class III tester in accordance with ANSI/TIA/EIA-526-7: Optical Power Loss Measurements of Installed Single Mode fiber cable plant, Method A for Single Mode.

E. Technicians performing copper and fiber testing must be certified by Fluke to operate, analyze and interpret test data. Certificates must be submitted with manufacturer warranty certificates prior to start of project.

F. Test records consisting of power meter results must be legible and clearly identify the cable, fiber, test direction, test wavelength, date of test and name (s) of test technician.

G. No defective fiber strands will be accepted. In line splicing for repair, whether via fusion or mechanical methods, on newly installed optical fiber is not permitted. Cables containing un-repairable defective strands shall be replaced at the installer's expense.

H. Test all multi-pair UTP cable for continuity, shorts, opens and transposals with a continuity tester. Document all tests with date of test and name (s) of test technician.

I. Test all horizontal and vertical SATV/CATV cabling for continuity. Document all tests with date of test and name(s) of test technician.

Figure 4-10. Part 3 – Execution typically provides details regarding the adjusting, cleaning, and testing of products, hardware, devices, and equipment such as VDV systems.

Project Pricing Breakdowns

3.5 PROJECT PRICING BREAKDOWN

A. Provide the following breakdown for the various aspects of the project:

	Description	Material	Labor
1.	Horizontal Cabling & Connecting Hardware	________	________
2.	Premise Copper & Fiber Backbone Distribution	________	________
3.	MDF/IDF Equipment Room Fit Out	________	________
4.	Cable Tray, Conduit, Sleeves & Fire Stopping	________	________
5.	Campus Fiber Distribution (200 to 300)	________	________
6.	Guard Shack Fiber/Copper Campus and WAOs	________	________
7.	Copper and Fiber Optic Patch Cords	________	________
8.	Project Administration	________	________
9.	Taxes	________	
10.	Permits and Fees	________	
	Total Material	________	
	Total Labor	________	
			________
	PROJECT TOTAL	________________	

Figure 4-11. A project pricing breakdown is often an itemized breakdown of each system by description and distinguished between labor and materials.

SPECIFICATION TYPES

Different types of written specifications can be produced. Most specifications are written to closely follow the CSI MasterFormat, while other specifications are loosely based on the MasterFormat. With either type of specification, the goal is to provide a written narrative that can support the contract and complement the drawings.

A specifier must be accurate, brief, and concise with the information and instructions for the systems that they specify. Specifications should use simple sentences and words that are easily understood. Specifications should never be complex or require advanced training to comprehend. Specifications are typically based on proprietary systems, meaning that the specifier identifies products by OEM name, brand, model number, type, or some other distinctive identifier. The two methods used for specifying proprietary systems are the open and closed methods.

The open method is results-based, which describes the required system performance without mandating how system performances are to be accomplished. A specification in the open method describes a required product or system and also allows for suggestions of substituted products if the required product is not available. An advantage of the open method is that it typically results in lower costs. A disadvantage of the open method is that it can challenge inexperienced contractors. Another disadvantage is that substituted products may not match existing systems.

The closed method lists one or more products that are allowed as options, but it does not allow for the use of substitutions for the listed products, materials, and systems. The closed method not only describes required performance criteria but also the tools and technologies that must be used in the design of a product to meet the specifications. An advantage of the closed method is that the design can be completed in detail. A disadvantage of the closed method is that it limits supplier competition, which may result in increased costs or long-lead items since only certain suppliers' products and systems can be used on the project. Also, the number of bidders with the required experience and expertise to certify the system may be limited. Any corrections required for mistakes in a closed method specification are usually the specifying engineer's responsibility.

Performance-Based Specifications

A *performance-based specification* is a document that outlines the intended results of a project design. Performance-based specifications outline how the final installed system must operate. This specification type provides a description that is mostly based on a desired outcome, as well as accepted industry codes and standards. Performance-based specifications also provide information for the contractor to determine the best methods and materials to meet the final design. The advantage of performance-based specifications is that they provide the contractor flexibility and promote innovations. However, contractor flexibility also limits both the architect's and engineer's control over the project.

Descriptive Specifications

A *descriptive specification* is a type of performance specification that provides detailed requirements for material properties and workmanship, fabrication, and installation. Descriptive specifications are typically based on industry-accepted manufacturing, installation standards, and techniques. Descriptive specifications are never proprietary and require oversight by the engineer to verify that the completed installation meets the original intent of the design.

Specification Terminology

Regardless of the type of written specification, the expertise of the designer and the contractor determines the final installed product. During the process of determining specific products and techniques, all parties must understand the terms associated with final product selection. "As specified," "alternate," "option," and "substitution" are terms that have different meanings when used in product selection.

"As specified" suggests no differences or deviations from what has been stated. An *alternate product* is a product acceptable to the engineer or specifier for which the contractor submits alternate pricing. A *product option* is one of several listed products in the specification. Product options are also referred to simply as "options." A *product substitution* is a request to substitute one product for another product described in a specification. Product substitutions are also referred to simply as "substitutions." All types of product selection must be approved by the engineer and the owner.

SUMMARY

All construction projects, whether new or retrofit construction, require a written set of specifications that must be followed by the designers and contractors. Specifications can be written generally or highly detailed. Most specifications follow the CSI MasterFormat and are formatted in parts that cover general requirements, products, and execution. Specifications can also be performance based or descriptive.

CHAPTER 4

Systems Specifications

Review

Name ______________________________ Date ________________

True-False

T F 1. The MasterFormat is written and developed by Underwriter's Laboratories (UL).

T F 2. The current edition of the MasterFormat is from 2018 and contains 50 divisions that define broad areas of construction.

T F 3. Small-scale, simple construction jobs do not require specifications.

T F 4. In the event of any discrepancy or conflict between the prints and specifications for a specific project, the material in the specifications takes precedence over the material in the prints.

T F 5. A complete system must be installed per the referenced codes and standards even if they are not explicitly outlined in the specification or if the specification contradicts a code or standard.

Multiple Choice

______________ 6. With MasterFormat 95, Level ___ titles are the highest level used in titling and arranging units of construction information.

A. 1
B. 2
C. 3
D. 5

______________ 7. The MasterFormat 2018 edition was revised from its original publication in ___.

A. 1995
B. 1999
C. 2004
D. 2007

______________ 8. The period of time for a contractor guarantee is typically ___ from the date of completion.

A. six months
B. one year
C. two years
D. five years

__________ **9.** The architect ___ is responsible for writing the specifications so that they are consistent with the prints.
- A. architect
- B. authority having jurisdiction
- C. general contractor
- D. owner

__________ **10.** The MasterFormat is published jointly by the CSI and ___.
- A. American Institute of Architects (AIA)
- B. Associated Specialty Contractors (ASC)
- C. Construction Specifications Canada (CSC)
- D. Underwriter's Laboratories (UL)

Completion

__________ **11.** MasterFormat 95 is divided into ___ divisions.

__________ **12.** Specifications for VDV systems are covered in Division ___ of MasterFormat 2018.

__________ **13.** A(n) ___ is written information included with a set of prints.

__________ **14.** A(n) ___ is a written assurance issued by a general contractor that furnished materials and equipment are of acceptable quality and workmanship as well as free from defects.

__________ **15.** The ___ standard is a guide for determining which building assemblies are suitable for various building applications based on life safety requirements.

__________ **16.** A(n) ___ is a manufacturer warranty that ensures that standards-based technology designed to be used with a specific system is supported by the manufacturer.

__________ **17.** A(n) ___ is a construction contract that is based on the cost of work or per item of material.

__________ **18.** A(n) ___ specification is a document that outlines the intended results of a project design.

__________ **19.** A(n) ___ specification is a type of performance specification that provides detailed requirements for material properties and workmanship, fabrication, and installation.

__________ **20.** MasterFormat 2018 is divided into ___ divisions.

Short Answer

21. Briefly explain why Divisions 15 to 20 are not assigned to specific construction areas in the current version of the MasterFormat.

22. List and describe the four sections that must be covered in a specification's technical division for an area of work.

23. Describe the most common format and distinct parts of a project specification.

24. What is the goal of written specifications?

25. Explain the differences between the two methods used for specifying proprietary systems.

CHAPTER

5

Firestop Systems

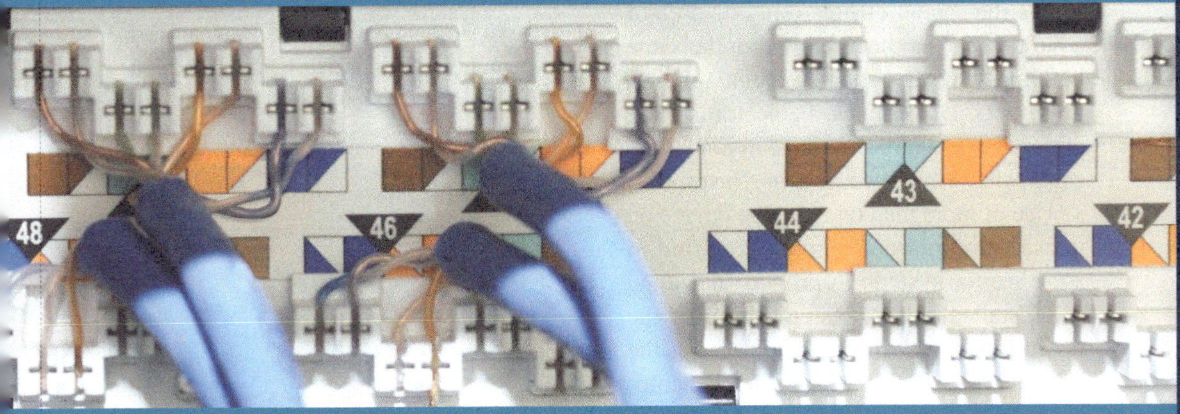

A major component in building construction is the fire protection system used to detect, suppress, and contain the spread of fire and smoke. Many building codes, insurance carriers, and end users have specific design requirements for fire protection of new and retrofit construction. Fireproofing and firestop systems exist to protect personnel from injury and property from damage caused by fire, smoke, and toxic gases. Depending on the size and type of construction, different types of systems from a variety of manufacturers are available for installation. Firestop system installations are documented for future reference and insurance requirements. Firestopping details are available on the project drawings and specifications.

Objectives

- Describe commercial building fire protection.
- Explain firestop system codes and standards.
- Describe types of industry-accepted fireproofing and firestop systems.
- Explain the various ratings used in firestop systems.
- Explain firestop system documentation.
- Explain firestop system drawings and specifications.

Learner Resources

atplearningresources.com
access code 783821

FIRESTOPS

A *firestop* is any material, device, or assembly of parts installed after a penetration of a fire-rated wall, ceiling, or floor area to prevent passage of flame, smoke, or gases through the rated barrier. A firestop resists the spread of fire from one area of a building to another area within the same building. Firestops may be plugs, pillows, grommets, slot frames, cable transit fire clamps, expandable foam, putty, sealant, caulk, sleeves, or electrical box inserts.

Use of more than one firestop device on the same barrier is commonly referred to as a "firestop system." A properly designed and installed firestop system maintains the fire rating of the barrier. A *fire rating* is the measure of resistance of a material or component to failure when exposed to fire and is expressed as the number of hours a material or component will retain its integrity.

Any amount of cables installed throughout a building is considered a fuel source for a fire. The materials used to insulate, route, and terminate cables and conductors, when properly designed and installed, can be considered deterrents to reduce the spread of smoke and fire. Sources of ignition can be the overheating of equipment, combustion of materials exposed to excessive heat, or a simple static electricity spark.

Fuel (combustible materials), heat (ignition source), and oxygen (air) must be present to start and sustain a fire. These three elements make up the model referred to as "the fire triangle." Further advancement of combustion and extinguishing technologies has revealed a fourth critical element to maintain a fire: the presence of a sustained chemical chain reaction. With the addition of this fourth element, the fire triangle is now commonly known as "the fire tetrahedron" (triangular pyramid). **See Figure 5-1.**

Fire-extinguishing agents remove one or more of the four elements in order to extinguish a fire. Removing any single element extinguishes the fire. For example, removing additional fuel, such as clearing away nearby combustibles, allows the fire to burn itself out. Cooling a fire with water or carbon dioxide removes the heat required to continue burning. Smothering a fire with fire blankets or dry sand also extinguishes a fire by removing oxygen. Extinguishing agents act by specifically disrupting the chemical reactions that sustain a fire. Because oxygen that flows through unsealed pathways and spaces provides fire and smoke a method of spreading throughout a building, firestop systems must be installed.

Figure 5-1. A fire requires specific elements to be present to burn; removing one or more extinguishes the fire.

COMMERCIAL BUILDING FIRE PROTECTION

Fire protection in buildings currently exists due to past tragedies where excessive loss of life and property damaged occurred. Stringent codes, standards, and regulations are often developed after a major fire event occurs. For example, on February 27, 1975, a fire at the New York Telephone Exchange in New York, NY, caused extensive property damage and eventual loss of life to first responders and other individuals exposed to the toxic smoke.

It was later determined that a spark in a cable vault in the basement of the building ignited the polypropylene (PP) insulation on the cables that routed horizontally and vertically throughout the building. The combination of the PP cable insulation and lack of firestop systems allowed a rapid spread of fire and smoke. The fire disrupted telephone service for over 200,000 customers for three weeks and caused close to $100 million in property damage (equivalent $460 million in 2018).

The fire fed vertically through slots and cable chases, feeding on almost 17,000 lb of toxic cable insulation. While no firefighters or other personnel were killed during the actual fire, many later developed terminal cancer due to exposure to the toxic smoke and fumes from materials in the cable insulation. Investigations of the fire lead to the change of cable insulation from PP to a less toxic polymeric material (Teflon®) that was also plenum rated. Also, various NFPA and UL codes and standards were developed to prevent similar incidents from occurring in the future.

A *plenum* is a building space that facilitates air circulation for HVAC systems by providing pathways for airflows typically greater than atmospheric pressure. **See Figure 5-2.** The spaces between a structural ceiling, dropped ceiling, or under a raised floor are typically considered plenums. *Note:* Some drop-ceiling designs create a tight seal that does not allow for airflow and therefore may not be considered a plenum air-handling space.

The three common elements related to commercial building fire protection are detection, suppression, and containment. When used together, detection, suppression, and containment can notify building occupants through audible and visual alarms, extinguish fires with overhead sprinkler systems, and limit the spread of fire by containment.

Detection occurs when the presence of smoke or flame is detected within a building, initiating fire alarms or other emergency systems that notify occupants of the situation and direct them to exit the premises. Suppression occurs when a suppression device, such as a sprinkler head, is initiated by heat or the fire alarm system and attempts to extinguish the fire. Fire containment occurs when barriers are designed and installed during the construction process to divide the building spaces into smaller units that can confine the fire to its point of origin, preventing its progression to other areas within the building. Containment is the final type of control method used if suppression fails to extinguish the fire. By partitioning buildings with rated walls, the architect and engineers can maintain the barrier rating of the compartments by utilizing engineered firestop systems.

Plenums include spaces under a raised floor.

FIRESTOP SYSTEMS CODES AND STANDARDS

There are several supplemental building codes that address specific building systems. The two building codes most commonly referenced for guidance with firestop systems for VDV cables are developed by the International Building Code® (IBC®) and the National Fire Protection Association (NFPA).

Plenum Spaces

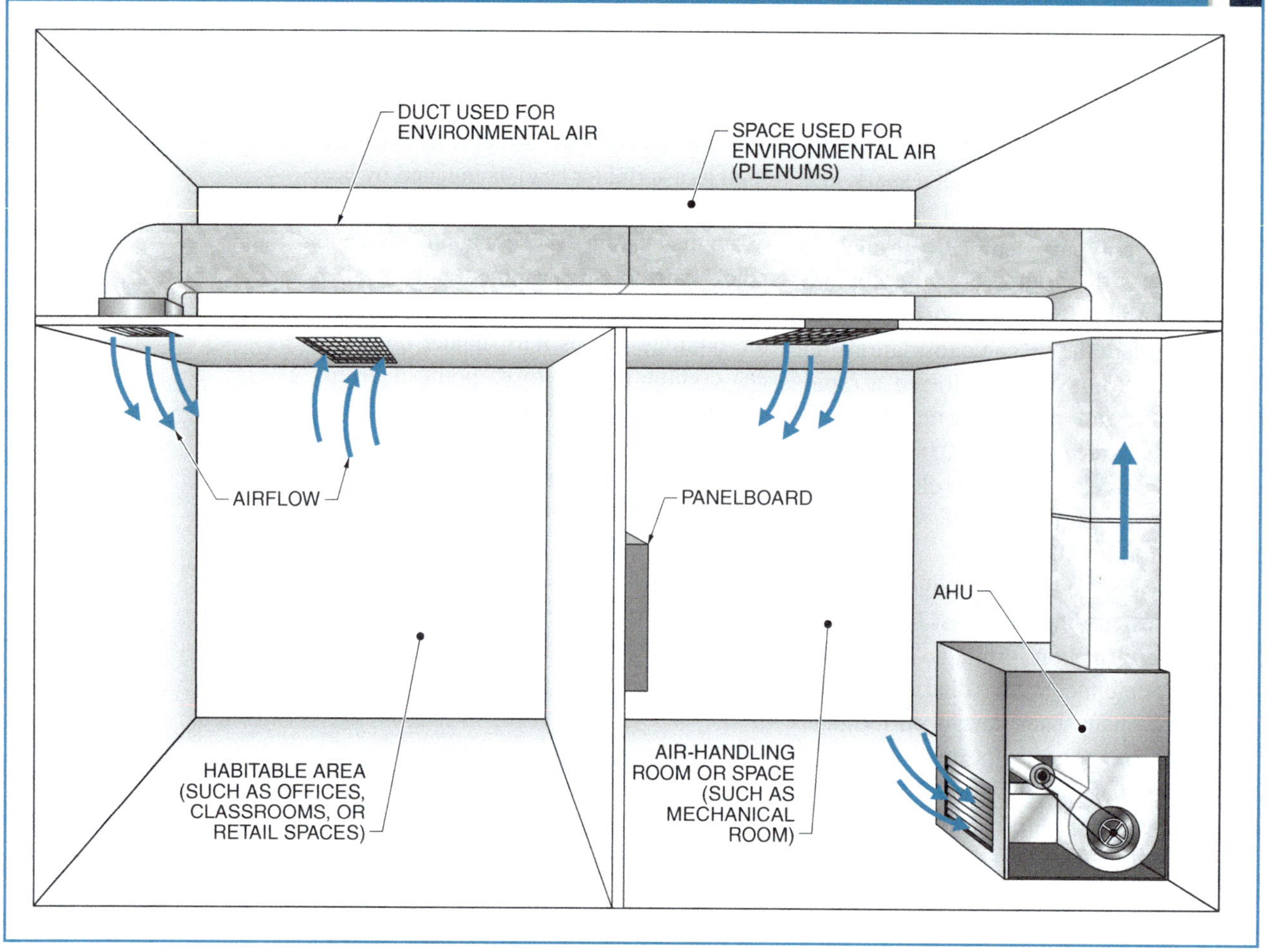

Figure 5-2. The spaces between a structural ceiling and dropped ceiling or the spaces under a raised floor are typically considered plenums.

International Building Code (IBC)

Local and state governments are required by law to enact and enforce a standard set of building codes. Drafting a building code can be expensive and time consuming, which most local municipalities cannot afford. It is easier for them to adopt a model building code and modify it to meet their specific needs.

A *model building code* is a complete set of requirements developed for the design and construction industries to provide a minimum standard for the health, safety, and welfare of building occupants. Model building codes are reviewed and revised by states and local municipalities to meet specific local and regional needs and then adopted by legislation into an enforceable code that is applicable to the municipality. Developing a model building code saves time and resources and provides continuity between codes of different states and municipalities.

The United States has been served by four major model codes. The *National Building Code* published in 1905 was the first model code. It was followed by three other model codes, each focusing on regional conditions. The 1927 *Uniform Building Code* (UBC) published by the International Conference of Building Officials was used primarily throughout the western part of the US. The southeastern areas of the US principally used the *Standard Building Code* (SBC) published in 1945 by the Southern Building Code Congress. In 1952, Building

Officials and Code Administrators International (BOCA) published the BOCA Basic Building Code, which was predominately used in the northeast area of the US.

In 1994, the three major code organizations merged, forming the International Code Council® (ICC®), with the purpose of establishing one international standard for building code construction. The resulting building code was to be adapted for regional and social conditions throughout the world. The newly formed ICC published its inaugural edition of the *International Building Code®* (IBC®) in 2000 and has since developed and published a library of international code manuals, with the latest version published in 2018.

As part of the IBC, the *International Fire Code®* (IFC®) was developed and published. While the IBC covers new construction, the IFC establishes fire prevention and safety standards for existing buildings. **See Figure 5-3.** The IFC establishes minimum requirements for fire prevention and fire protection systems. The IFC is not a federally mandated set of codes that apply universally across the US, nor is it a law that the federal government requires local governments to adopt. The IFC may be adopted voluntarily and individually by local, territorial, and state jurisdictions. However, the IFC is written such that it can be adopted "as is" and be recognized as a legally binding law.

As part of fire prevention, the IFC defines fire protection requirements related to penetrations and fire-rated barriers for various construction types as well as the following:

- the hazard of fire and explosion arising from the storage, handling, or use of structures, materials, or devices
- conditions hazardous to life, property, or public welfare in the occupancy of structures or premises
- fire hazards in the structure or on the premises, from occupancy or operation
- matters related to the construction, extension, repair, alteration, or removal of fire suppression or alarm systems
- conditions affecting the safety of firefighters and emergency responders during emergency operations

The IBC® and IFC®

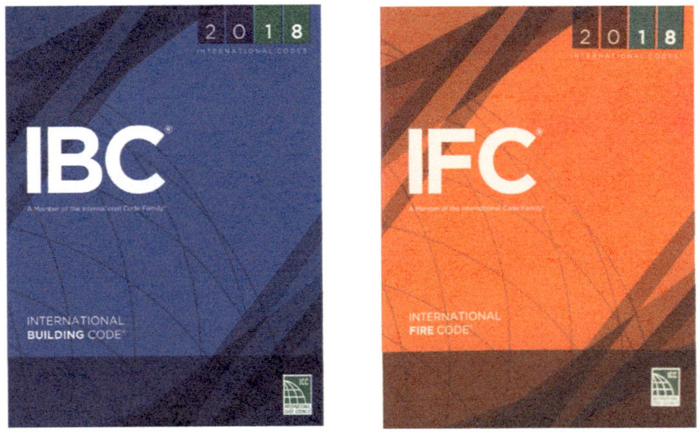

International Code Council, Inc.

Figure 5-3. The IBC is used for new construction while the IFC establishes fire prevention and safety standards for existing buildings.

National Fire Protection Association (NFPA)

The National Fire Protection Association (NFPA) is a national standards organization that provides guidance in assessing fire-related hazards. The NFPA publishes both the *National Electrical Code® (NEC®)*, which is one of the most widely adopted and recognized standards in the world, and the NFPA 70E® standard. *NFPA 70E®, Standard for Electrical Safety in the Workplace®*, is a standard that covers the electrical safety requirements of workers performing electrical work. **See Figure 5-4.** It covers the safety requirements for the installation of electrical equipment in residential, commercial, and industrial facilities.

The NEC® is focused on the practical safeguarding of individuals and property from the hazards of electricity. Along with electrical shock and burns, one of the main hazards of electricity is the ignition of fires. The NEC® specifies requirements for installation and use of fire protection systems in the following articles, sections, and subsections:

- Article 300, General Requirements for Wiring Methods and Materials

  - 300.21, Spread of Fire or Products of Combustion
  - 300.22, Wiring in Ducts Not Used for Air Handling, Fabricated Ducts for Environmental Air, and Other Spaces for Environmental Air (Plenums)
  - 392.18(D), Cable Tray Installation – Through Partitions and Walls
- Article 645, Information Technology Equipment
  - 645.3(B), Other Articles – Wiring and Cabling in Other Spaces Used for Environmental Air (Plenums)
  - 645.3(F), Other Articles – Cable Routing Assemblies, Communications Wires, Cables, Raceways, and Equipment
  - 645.5(E), Supply Circuits and Interconnecting Cables – Under Raised Floors
- Article 646, Modular Data Centers
  - 646.3(B), Other Articles – Wiring and Cabling in Other Spaces Used for Environmental Air (Plenums)
  - 646.3(F), Other Articles – Cable Routing Assemblies and Communications Wires, Cables, Raceways, and Equipment
- Article 800, General Requirements for Communications Systems
- Article 805, Communications Circuits
- Article 810, Radio and Television Equipment
- Article 820, Community Antenna Television and Radio Distribution Systems
- Article 830, Network-Powered Broadband Communications Systems
- Article 840, Premises-Powered Broadband Communications Systems

Compliance with the NEC® and the installation of engineered firestop systems decrease the spread of fire. However, the installation of engineered firestop systems alone does not make penetrations airtight. Both the IBC and NEC® agree that smoke barriers must be constructed to limit leakage. A *smoke barrier* is a continuous membrane, either vertical or horizontal, such as a wall, floor, or ceiling assembly that is designed and constructed to restrict the movement of smoke.

NFPA 70E®

Figure 5-4. NFPA 70E®, *Standard for Electrical Safety in the Workplace®*, is a standard that covers the electrical safety requirements of workers performing electrical work.

Approved Firestop Systems

Approved firestop systems are classified by independent testing and standards agencies such as Underwriters Laboratories (UL) and ASTM International (ASTM). Independent testing agencies such as UL test and certify the products of manufacturers and suppliers. A UL label or marking certifies that a product has been tested and meets UL standards. **See Figure 5-5.** ASTM International publishes technical standards for fire-resistant materials based on laboratory tests. These standards are typically cited in technical specifications and project manuals as a measure of performance or quality.

Testing agencies perform extensive testing of firestop products in simulations of actual field conditions. The testing agencies then document the results to gauge product performance. The analysis developed helps provide engineered solutions through a variety of industry standards, such as ANSI/UL 1479, *Standard for Fire Tests of Penetration Firestops*, and ASTM E814, *Standard Test Method for Fire Tests of Penetration Firestop Systems*.

A *through-penetration firestop system* is a firestop system that protects penetrations that pass completely through a barrier such as a wall or a floor. Both sides of the penetration are typically sealed with a firestop system. Additional standards that are often referenced in construction specifications for through-penetration firestop systems include ASTM E119, *Standard Test Methods for Fire Tests of Building Construction and Materials*, and ASTM E84, *Standard Test Method for Surface Burning Characteristics of Building Materials*.

UL listings for firestop systems use an alphanumeric code to designate, organize, and identify UL-approved firestop assemblies. The first letter of the listing designation identifies the type of penetrated structure, which includes the following:

C – both floor and wall penetrations

F – floor penetrations only

W – wall penetrations only

The second letter of the UL designation identifies construction type. These second letters include the following:

A – concrete floors with thickness of 5.0″ or less

B – concrete floors with thickness greater than 5.0″

C – framed floors

D – deck construction

J – concrete or masonry walls with thickness of 8.0″ or less

K – concrete or masonry walls with thickness greater than 8.0″

L – framed walls

M – bulkheads

N – composite panel walls

Finally, the numeric code in a UL listing identifies the common electrical and data communications penetrants. The numeric code ranges identify the following:

0000–0999 – no penetrant

1000–1999 – metallic pipe, conduit, or EMT

2000–2999 – plastic pipe, conduit, or tubing

3000–3999 – electrical cables

4000–4999 – cable trays with electric cables

5000–5999 – insulated pipes

6000–6999 – bus ducts

7000–7999 – HVAC ducts

8000–8999 – mixed multiple penetrations

9000–9999 – reserved for future use

A factor that must be considered when designing and selecting a firestop system is whether the installation will be permanent or retrofittable. The type of installation determines the type of product required. Examples of permanent through penetrations include conduits, bus ducts, and power cables. Permanent firestop sealing products are required for these types of penetrations because they provide a tamperproof seal and require little or no maintenance.

Retrofittable, nonpermanent products are typically specified by designers and installers for VDV applications. Retrofittable, nonpermanent products provide the ability move, add, and/or change VDV system cables and raceways with no disruption to the integrity of the barrier that they penetrate. Examples of retrofittable, nonpermanent products include sleeves, flashing, plugs, and cable transit clamps.

UL Labels and Markings

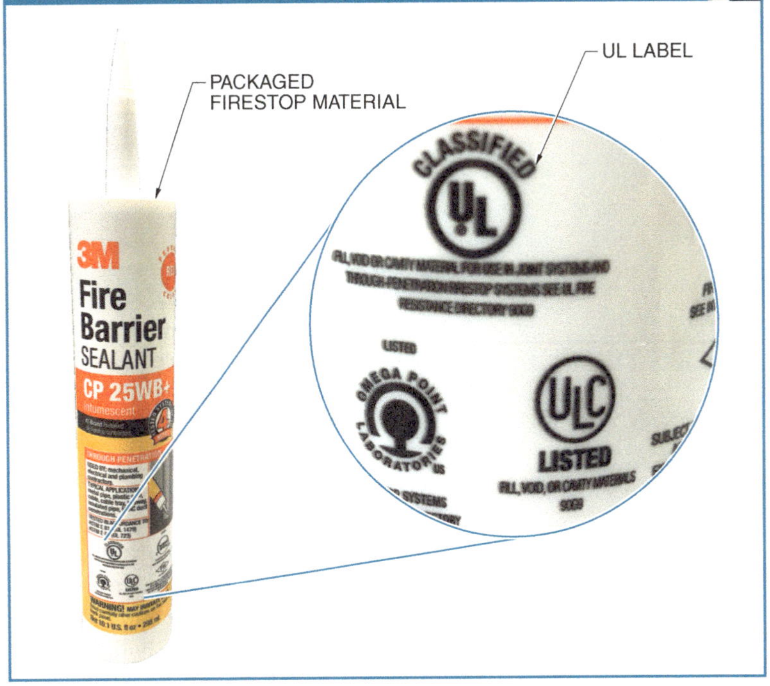

Figure 5-5. A UL label or marking certifies that a product has been tested and meets UL standards.

Additional ratings developed by UL for establishing the design of firestop systems include the following:

- Flame (F)—The flame (F) rating is expressed in hours and indicates the amount of time a barrier can withstand direct exposure to fire before it is consumed by flame through an opening.
- Temperature (T)—The temperature (T) rating is expressed in hours and indicates the amount of time that a barrier can be exposed to a temperature on the nonfire side of the wall that does not exceed 325°F.
- Smoke (L)—The smoke (L) rating is expressed in cubic feet per minute (cu ft/min) and indicates the amount of air/smoke that can leak through a penetration. *Note:* Smoke leakage requirements for cable bundles penetrating walls and floors require a low rating number such as 1 or 2, which indicates the minimum amount of smoke acceptable when leakage occurs. The lowest smoke rating that can be achieved is less than 1 but not zero. Tests to determine the rating must be conducted by a certified testing agency and require measuring the size of the through-penetration opening and the system to be installed within and around the opening. L ratings are tested with a differential pressure of 0.30 inches water column (75 Pa) at 75°F and at 400°F. OEMs of firestop systems must be consulted to determine the actual leakage values.

TECH FACT

For most normal construction, it is impractical to attempt to seal cables airtight because the cables would require an individual seal within and around each cable.

A *barrier* is a physical impediment that is intended to block passage. In the construction industry, barriers are erected to separate different parts of a building. The most common construction materials used in the construction of barriers are metal and wood framing, gypsum wallboard, poured concrete, and hollow concrete blocks. When used together, these materials can establish a range of hourly ratings.

Barriers fire ratings are determined by local building codes. Barriers are rated by their ability to withstand heat in 1-hour increments. Because the core of a building usually contains elevator shafts, non-egress stairwells, electric equipment rooms/closets, VDV equipment rooms, and mechanical rooms, the barriers that protect the core typically have higher ratings from 2 hours to 4 hours, depending on the size of the building and local building code requirements. Common materials used in the construction of barriers and the hourly ratings of each are as follows:

- ⅝″ thick gypsum wallboard – ½ hour
- poured concrete – 4 hours
- hollow concrete block – 2 hours

Note: Although these ratings are typical, they may be different depending on the project specifications. The project drawings must be referenced for the actual ratings. For example, the rating of gypsum wallboard installations can increase up to 2 hours as layers of wallboard are added to the installation.

FIREPROOFING AND FIRESTOP SYSTEMS

A variety of fireproofing and firestop materials are applied throughout a building. Specifications and the appropriate local fire protection codes contain information regarding the required fireproofing and firestops in a structure.

Fireproofing Systems

Fireproofing is the process of applying protective material to a structural or finish member to increase the fire resistance of the member and protect it from failure in the event of excessive heat or fire. Fireproofing systems are installed in most commercial and industrial buildings. Fireproofing systems used in commercial and industrial building construction can be cementitious or endothermic/intumescent. **See Figure 5-6.**

The most common material in commercial construction that requires direct application of fireproofing is structural steel. In a building fire, structural steel loses strength as the temperature increases. Systems consisting of high-strength, gypsum-based wet mix in the form of a slurry are applied by being sprayed directly onto the steel framing and decking. The bonding strength per square foot and application amounts can vary based on the height and design standards of the building. The information on bonding strength and application amounts is used to provide a fire resistance rating.

Cementitious Fireproofing. A *cementitious material* is a material that can harden like cement. Cementitious materials are used in some types of fireproofing systems. Cementitious fireproofing materials include mixtures that are wet mixed and pumped as a slurry. Cementitious fireproofing materials completely encase the structural or finish member to prevent direct contact with flames and inhibit the spread of heat to the member. A variety of cementitious fireproofing materials are available, each with different material bases and fireproofing qualities. For example, a cementitious fireproofing mixture may be a gypsum or portland cement base with vermiculite.

Cementitious fireproofing materials are typically sprayed onto a structural or finish member. The base surface must be clean and free of any material that would prevent proper adhesion. OEM information related to the proper mixture and application of the fireproofing material must be followed carefully to achieve the required fire rating.

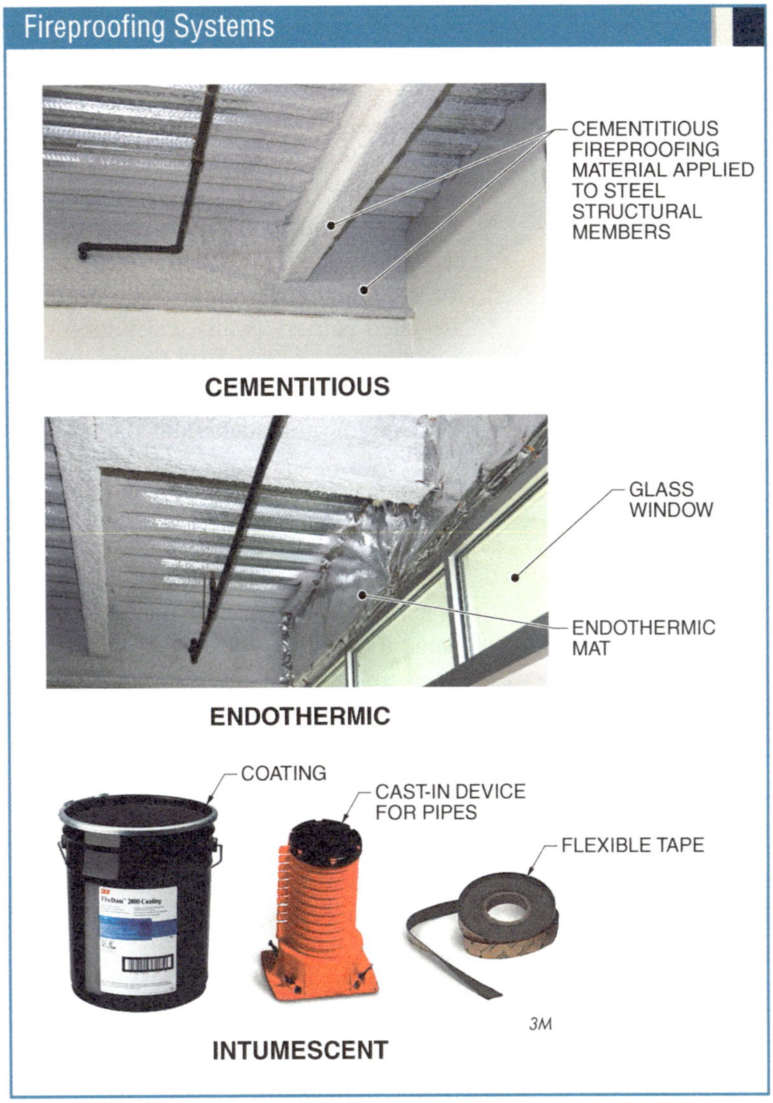

Figure 5-6. Fireproofing systems are used in most commercial and industrial buildings and can be cementitious or endothermic/intumescent.

Endothermic/Intumescent Fireproofing. An *endothermic fireproofing material* is a fireproofing material that releases water vapor when exposed to high temperatures. Endothermic fireproofing materials cool the outer surface of the member that they are applied to, inhibiting the spread of fire in a building. For some applications, endothermic mats or tape are wrapped around the outside diameter of electrical raceways and cable trays containing circuit conductors or placed against barriers where spray-type fireproofing materials are not suitable.

An *intumescent fireproofing material* is a fireproofing material that expands rapidly when exposed to high temperatures, closing voids left by burning or melting construction materials. Intumescent fireproofing materials are commonly applied to structural steel members, such as open web steel joists, in the form of a sprayable coating, cast-in devices, or flexible tape. Intumescent fireproofing may also be designed to provide a certain amount of sound insulation. In some applications, a base coat may be required for proper adhesion and coverage. Fire ratings of up to 3 hours can be achieved with the proper application of intumescent fireproofing materials.

Firestop Systems

Firestopping is the process of applying and installing a material or member that seals open construction to inhibit the spread of fire, smoke, and fumes in a structure. A variety of products for firestop systems designed for commercial structures are available that have been tested and proven as reliable methods for maintaining through-penetrations of fire-rated barriers (walls and floors). The installation of an approved firestop system must include the following:

- backing/forming materials
- fill, void, or cavity materials
- firestop devices

Backing/forming materials are essential to the performance of firestop systems. They increase fire resistance and act as a depth gauge for sealant products. They can also act as a form for liquid components. The most common backing material is mineral wool. However, other types of heat-resistant firestop materials in the form of panels are available as well. Mineral wool is the most commonly used backing material in most firestop systems for its low cost and ease of installation. **See Figure 5-7.**

Mineral Wool

Figure 5-7. Mineral wool is the most commonly used backing material in firestop systems because of its low cost and ease of installation.

Fill, void, or cavity materials are categorized as permanent or retrofittable. Permanent products are products that cannot be reused and are typically installed with electrical conduits and cables. Retrofittable products are reusable products that can be removed and are typically installed with VDV system cables.

Firestop devices are categorized as passive or intumescent. Passive firestop products do not react to heat and have limited use with firestopping. Passive firestop products are typically used for use with noncombustible applications. Intumescent firestop products expand when exposed to high heat or fire and can compensate for the burning or melting of combustibles. Intumescent firestop products have a broader product and application range. In addition to backing material, other firestop systems include pillows, foam blocks, caulk, putty, sleeves/flashing, plugs, cable transit fire clamps, and compartmentation. **See Figure 5-8.**

Firestop Pillows. Firestop pillows combine an insulating core with intumescent coating material and are protected by an outer polymeric bag. Firestop pillows are considered retrofittable and may require other products such as mesh or fireproofing sheet material to maintain a tight compression fit. *Note:* Firestop pillows are also sometimes referred to as firestop seal bags.

Foam Blocks. Foam blocks are used as an insulative foam with intumescent components. Foam blocks are permanent and are used for specific applications, such as fire walls, cable trays, and fire doors. Foam blocks can be cut to size for specific application requirements and have temperature ratings up to 1400°F.

Caulk/Putty. Fire-rated caulk and putty are nonhardening and may be applied as a firestop material around conduit and other piping where it penetrates walls and floors. Caulk and putty create airtight seals to inhibit the spread of fire, smoke, and fumes in a structure. Some types of firestop caulk expand when heated to ensure an airtight seal. Firestop caulk is available in tubes or bulk shapes for application with a putty knife. Firestop putty can be applied by hand and pressed into place. Firestop caulk and putty are considered permanent and should not be reused once exposed to smoke or heat. Caulk and putty are typically preferred by cable installers rather than other types of firestopping due to low cost and ease of installation.

Sleeves/Flashing. Sleeves or sheet metal flashing may also be used as firestopping for wall and floor penetrations. Cast-in-place sleeves are attached to the formwork before concrete is placed to provide an opening for pipes, boxes, or other pathways. Some sleeves are intumescent devices and do not require firestop foam or caulk. Other sleeves are filled with foam or caulk to create an airtight seal after pipes have been positioned. Expandable firestop sleeves may also be fitted onto piping and conduit and slid into position at wall and floor penetrations.

In addition to firestop systems, smoke detectors are installed in plenum spaces to help prevent the spread of smoke and fire in commercial structures.

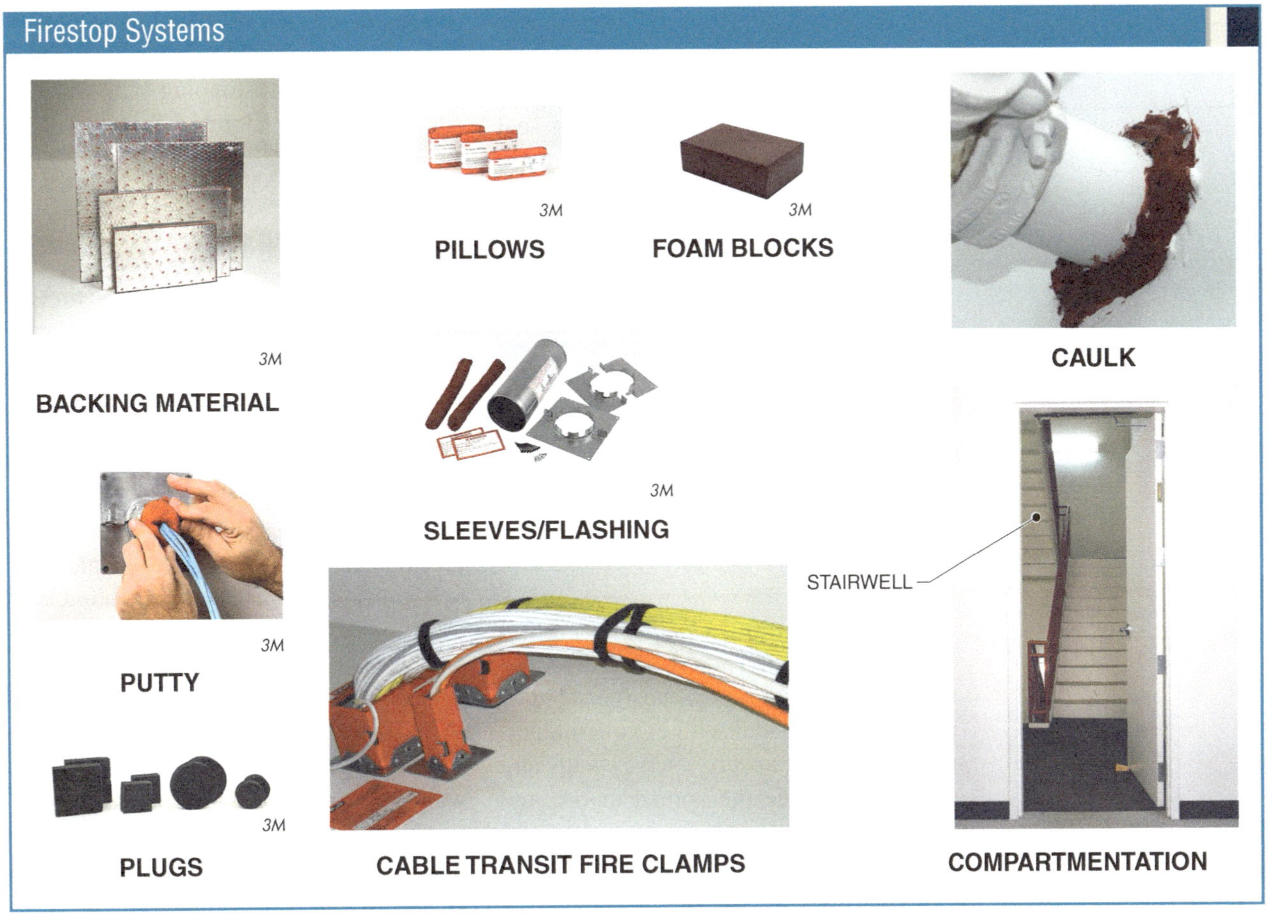

Figure 5-8. Firestop systems include backing material, pillows, foam blocks, caulk, putty, sleeves/flashing, plugs, cable transit fire clamps, and compartmentation.

Sheet metal flashing may also be used as a firestop material. Thin-gauge sheet metal is cut and fastened in place around the diameter of pipes and conduit as needed to stop airflow. Caulk or putty may also be applied to the exposed edges of the sheet metal to ensure an airtight seal around the flashing.

Plugs. Firestop plugs are an easily installed component that creates a seal against fire and smoke. Plugs are used for both temporary and permanent installations for blank openings and cable penetrations. Firestop plugs are commonly sized to fit 2″ to 4″ openings (round, square, or rectangular), such as the openings for conduit stubs, sleeves, pass-through devices, and pipes.

Cable Transit Fire Clamps. Cable transit fire clamps are raceway devices that are designed to allow cables to pass through fire-rated walls without the installation of additional firestop materials. Cable transit fire clamps are equipped with a fire-sealing system that adjusts to the number of cables installed and expands to provide a seal when exposed to heat. Cable transit fire clamps typically provide more resistance to smoke seepage than standard sleeves, putties, or caulks.

TECH FACT

Every trade on a construction project is responsible for verifying that the firestop rating of the materials used for the barriers their specific system penetrates meets the required specification.

Compartmentation. *Compartmentation* is the deliberate division of a building into sections or units. Properly designed buildings are constructed with compartmentation to provide barriers, such as walls and floors, to impede the spread of fire and smoke. Each section or unit (compartment) has a fire rating measured in the amount of time a wall or floor can withstand the spread of fire. Typically, the core of a building, such as where stairwells, elevator shafts, utilities, and equipment rooms are located, have higher fire ratings than other areas of a building. The higher fire ratings for the core areas also provide protection from the spread of fire vertically through the building.

TECH FACT

Not every firestop product works with every type of application. Firestop products must be engineered for each specific application. There is no standard for the color-coding of firestop products. OEMs manufacture firestop products in various colors for the purpose of distinguishing between their different types of products or systems.

Alternative firestop systems such as expandable foam, fire barrier strips, and pass-through devices for metal pipes are acceptable firestop system components for unused openings. These products are also acceptable for firestopping with through penetrations. **See Figure 5-9.**

FIRESTOP SYSTEM MANUFACTURERS

A variety of firestop OEMs provide firestop systems and products. In some cases, certain firestop systems from a specific OEM can be interchanged with components from different OEMs to provide protection. However, most firestop systems are proprietary and are not interchangeable with other OEM systems. The design and installation of firestop systems requires proper planning by the installer, such as coordination with other trades. Before installation can begin, approval of the system is also required either from the architect(s) and engineer(s) that designed and specified the system or from the AHJ. OEMs of these systems assist installers with submittals for proper approval.

FIRESTOP SYSTEM DOCUMENTATION

Documentation of an installed firestop system by the installer and/or contractor is critical from the submittal approval process to the end, as well as to the as-built documentation submission provided to the end user/owner. By documenting the products used through submittals and as-builts, in addition to photographs and video, installers and contractors can control liability in the event of a fire or tampering by others of an installed system.

As another form of documentation, most firestop system OEMs suggest the installation of a permanent label at each penetration indicating the date, installer name, and a unique identifier to the penetration (such as a product number). **See Figure 5-10.** Sometimes, a photograph and time stamp are used to ensure that the installation was done properly. Facility managers maintain these records for insurance purposes and future moves, adds, or changes (MACs). Some firestop OEMs also have created software that provides the documentation required by the specifications, end user, and property insurance carrier requirements.

Alternative Firestop Systems

Figure 5-9. Alternative firestop systems such as expandable foam, fire barrier strips, and pass-through devices for metal pipes are acceptable firestop system components for unused openings.

FIRESTOP SYSTEM DRAWINGS AND SPECIFICATIONS

Floor plans and firestop detail drawings depict horizontal and vertical VDV pathways routing through a building. VDV cable pathways penetrate floors and ceilings. Installers must reference architectural drawings to determine the rating of the barriers through which pathways penetrate. They must also reference specifications to determine the correct firestop application and location.

Firestop details provide a section view of the fire-rated wall or floor and the raceway penetration. **See Figure 5-11.** Firestop details also provide exact information for each firestopping situation. That information includes the type of floor or wall, dimensions for locating the opening, the maximum permissible diameter of the opening, the type of raceway penetrating the wall or floor, and the type of firestop material used and how to apply/install it.

Firestop Documentation Labels

Figure 5-10. As a form of documentation, most firestop system OEMs suggest the installation of a permanent label at each penetration indicating the date and installer name and with a unique identifier at the penetration (such as a product number).

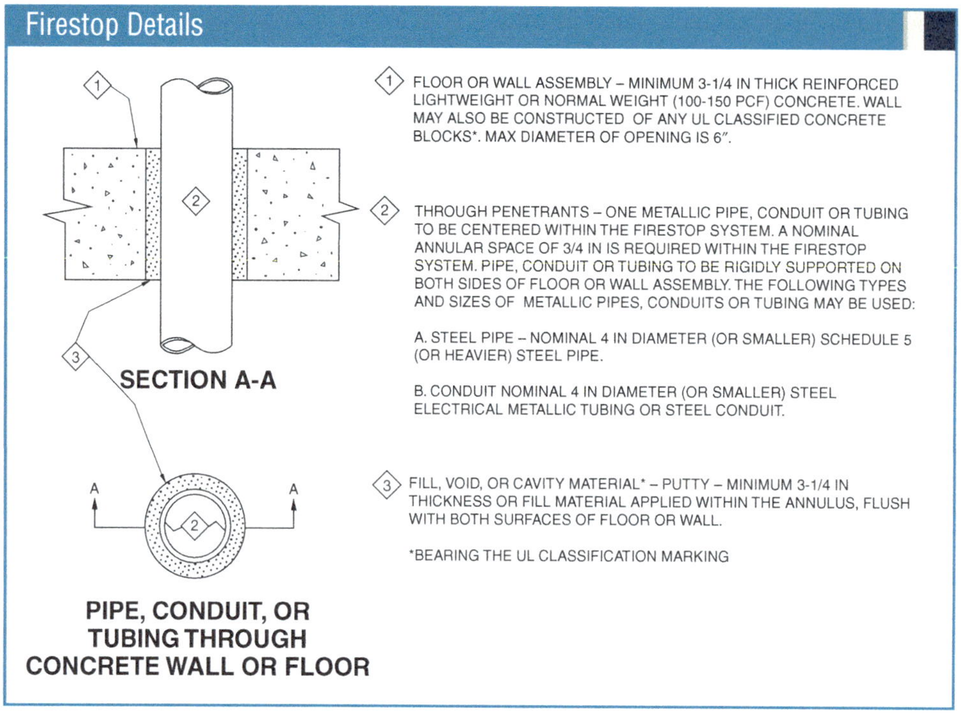

Figure 5-11. Firestop details provide a section view of the fire-rated wall or floor and the raceway penetration.

SUMMARY

VDV technicians must be familiar with each aspect of commercial installations, including the specifications and installation of firestop systems. Firestop systems must adhere to IBC and NFPA standards. Various OEMs provide fireproofing and firestop systems, which are specified in the project drawings and specifications. Installed fireproofing and firestop systems must be documented by the installer and/or contractor for reference in the event of future MACs and for insurance purposes.

CHAPTER 5

Firestop Systems

Review

Name ______________________________ Date ______________

True-False

T F 1. Many building codes, insurance carriers, and end users have specific requirements when designing fire protection for new and retrofit construction.

T F 2. Any amount of cables installed throughout a building is not considered a fuel source for a fire.

T F 3. Stringent codes, standards, and regulations are typically developed by fire engineers at annual code meetings.

T F 4. Containment of a fire is the final type of control method used if suppression fails to extinguish the fire.

T F 5. The IFC is a federally mandated set of codes that apply universally across the US and is part of a law that the federal government requires local governments to adopt.

T F 6. Installation of engineered firestop systems alone can make penetrations airtight.

T F 7. The lowest smoke (L) rating in a firestop system that can be achieved is less than 1 but not zero.

T F 8. The most common material in commercial construction that requires direct application of fireproofing is structural steel.

T F 9. Cementitious fireproofing materials are typically pressed and molded onto a structural or finish member.

T F 10. Firestop plugs are commonly sized to fit 2″ to 4″ openings.

Multiple Choice

______________ **11.** The ___ establishes fire prevention and safety standards for existing buildings.
- A. BBC
- B. IBC
- C. IFC
- D. UBC

______________ **12.** The ___ is focused on the practical safeguarding of individuals and property from the hazards of electricity.
- A. IBC
- B. IFC
- C. NEC®
- D. NFPA

______________ **13.** A(n) ___ fireproofing material is a fireproofing material that releases water vapor when exposed to high temperatures.
- A. cementitious
- B. endothermic
- C. intumescent
- D. vermiculite

______________ **14.** Foam blocks can be cut to size for specific application requirements and have temperature ratings up to ___ °F.
- A. 1400
- B. 1600
- C. 2000
- D. 2500

______________ **15.** Fire ratings of up to ___ hours can be achieved with proper application of intumescent fireproofing materials.
- A. 3
- B. 6
- C. 10
- D. 12

Completion

______________ **16.** A(n) ___ is material, device, or assembly of parts installed after a penetration of a fire-rated wall, ceiling, or floor area to prevent passage of flame, smoke, or gases through the rated barrier.

______________ **17.** A(n) ___ is a complete set of requirements developed for the design and construction industries to provide a minimum standard for the health, safety, and welfare of building occupants.

______________ **18.** A(n) ___ is a continuous membrane, either vertical or horizontal, such as a wall, floor, or ceiling assembly that is designed and constructed to restrict the movement of smoke.

______________ **19.** A(n) ___ is a physical impediment that is intended to block passage.

____________________ **20.** A(n) ___ is the measure of resistance of a material or component to failure when exposed to fire and is expressed as the number of hours a material or component will retain its integrity.

____________________ **21.** Firestop pillows are also referred to as ___.

____________________ **22.** ___ is the deliberate division of a building into sections or units.

____________________ **23.** A(n) ___ is a building space that facilitates air circulation for HVAC systems by providing pathways for airflows typically greater than atmospheric pressure.

____________________ **24.** The three common elements related to commercial building fire protection are detection, suppression, and ___.

____________________ **25.** The UL designation for a smoke rating is the letter ___.

Short Answer

26. Briefly explain the differences between detection, suppression, and containment.

27. List the designations of the first two letters in the UL-approved firestop assembly identification system.

28. List three examples of permanent through-penetrations.

29. Explain the difference between fireproofing and firestopping.

30. Explain why firestop caulk and putty is preferred by installers rather than other types of methods.

CHAPTER

6

Systems Estimating

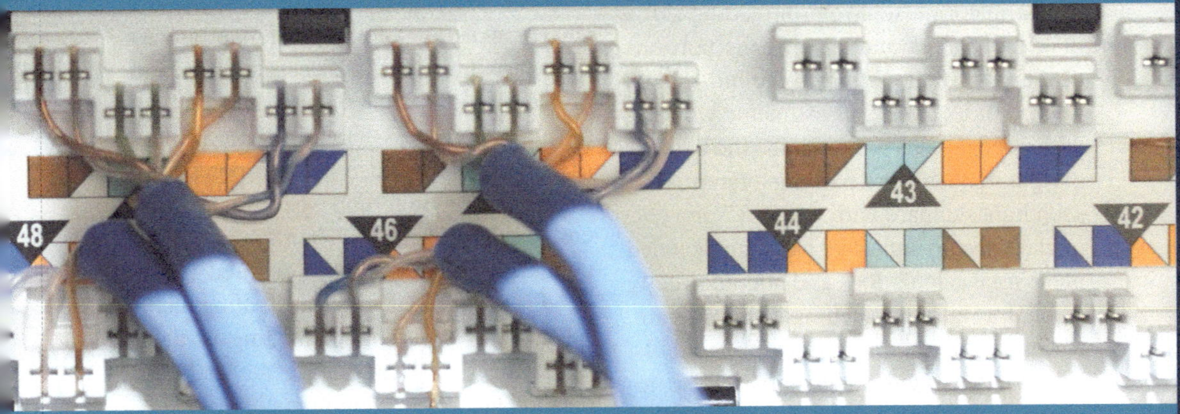

VDV systems estimating requires an experienced estimator. An estimate is typically based on detailed drawings and specifications or a scope document summarizing the project. Every estimator uses a specific process when developing an estimate. An estimator provides an educated, complete evaluation of a project based on calculations of labor units, material quantities, costs, site surveys, and other related aspects that encompass the project and the realistic costs for completion. An estimator's accuracy is just one of several factors that determine whether a project is successful and profitable or fails to cover the cost of completion.

Other factors that affect the estimating process include the specific type of installation and the GC's ability to complete projects of various sizes and complexities. GCs must cover their costs and maintain profitability while assuming a substantial amount of risk each time they enter into a construction contract. While the estimate is the basis of the GC's cost, other factors such as poor oversight and lack of productivity can have a negative outcome. It is not always the lowest or highest bidder but usually the GC with a record of being able to complete the work according to the contract documents within a specified time frame that is awarded the project.

Objectives

- Describe the three estimating methods.
- Describe how technology is used to formulate an estimate.
- Describe the responsibilities of an estimator.
- Identify the various components of an estimate.
- Describe how labor units are determined and applied to an estimate.
- Describe how a site survey is used with systems estimating.

Learner Resources

atplearningresources.com
access code 783821

ESTIMATING PRACTICES

Estimating is the process of determining labor and material costs required to complete a project. An *estimator* is an individual who performs quantity takeoff and pricing to determine the cost of a construction project. An estimator must have printreading skills in order to mark and take off the portions of the plans that pertain to the bid-related work. For example, an estimator working with a set of plans to determine the number of light standards in a parking lot may mark each light standard with a red "X" or a colored number to indicate it has been included in the total calculations and not incorrectly labeled as an item such as a sewer drain. Accurately interpreting drawing symbols, abbreviations, and dimensions is also required for quantity takeoff and estimating.

While an accurate estimate can help obtain jobs and increase profitability, a poor estimate can negatively affect a business in terms of lost jobs and profits. Estimating methods tools such as architect's and engineer's scales, digital plan measurement takeoff tools, and tally counters are still used to assist in gathering and quantifying the information on the drawings used to develop the estimate. **See Figure 6-1.** Prior to the development of digital (electronic) estimating systems, estimating was performed manually and extended on a ledger sheet.

A *ledger sheet* is a grid on a sheet of paper consisting of rows and columns into which item descriptions, locations, code numbers, quantities, and costs are entered. As prints and specifications are marked and quantities determined, the estimator enters these quantities on a ledger sheet. **See Figure 6-2.** Estimators should use standard categories on ledger sheets to ensure consistency and to present the information in a format familiar to other professionals involved with the construction process. An estimator's skill performing calculations reduces the amount of time required for the final tabulation of material quantities. A well-designed ledger sheet can be cross-referenced to other portions of the bid and used in other estimating calculations.

Estimating methods can be categorized into three types, and each must be developed and adapted for the contractor and markets served. The estimator must have knowledge of the factors that have an impact on the project, such as building codes, labor rates, material costs, operating costs, other overheads such as general and administrative (G&A) costs, profit margins, project schedules, and project continuity that may otherwise affect the normal progress of the work. Regardless of the method applied, estimators commonly mark up drawings and specifications to ensure that the necessary items are properly counted and no items are counted twice. **See Figure 6-3.** The three estimating methods are the traditional estimating method, preliminary cost estimating method, and detailed takeoff estimating method.

Traditional Estimating Method

The first estimating method is a traditional method in which the costs are developed for the estimate based on the experience of the GC, verbal instructions, or visual inspection, but lacks any clear written information. The *traditional estimating method* is an estimating method in which material quantities are calculated and entered onto ledger sheets for pricing. The quantities are multiplied by the appropriate unit costs to determine total material costs. The material costs are then added to determine the final material cost. Material quantities may also be used as the multiplier when unit labor costs are used to calculate total labor costs.

The traditional estimating method is typically used by GCs for small-scale projects, often only requiring a few days to complete, and consists of minor applications. Many project owners and GCs classify this work as "service" or "maintenance," where the project costs are typically less than the costs of more complex projects due to the small size and scope. The traditional estimating method is considered to be risky because of an increased chance of losing money on jobs due to the lack of written detail and other unknown variables.

The traditional estimating method can also be performed using the total area (in sq ft) of a project and percentage estimates. For total area estimates, some GCs prepare a cost per square foot of a building estimate based on their experience with projects of similar size and scope. For this approach to be successful, the estimator must be experienced and have a systemic approach to adjust for varying conditions.

The percentage estimate is commonly used for medium- to large-scale projects. A *percentage estimate* is a type of general-knowledge estimate formulated by using data from similar projects with known completion costs and calculating the completion cost against the entire cost of the finished building. For example, if data from past projects with similar requirements indicate that the typical building costs $20.00 per square foot to construct, and the installed network systems averaged 10% of the cost with little variation, then it can be reasonably estimated that the project cost will be close to $2.00 per square foot. Precautions must be taken to ensure that no special equipment or system is to be installed that could significantly increase the cost. Contingencies, allowances, and assumptions should be clearly stated when dealing with percentage estimates.

Contingencies are often identified for unknown factors, such as with retrofit projects where existing conditions often cannot be verified until after the work commences. For example, if asbestos were discovered during a retrofit, it may require that individuals receive training on how to work in such environments or that individuals trained in proper asbestos removal be hired. This would increase costs to the owner for the labor and proper disposal of the asbestos needed to complete the work. In some instances, there may be indications that additional, unidentified work or equipment may be required in the future. A contractor can provide a list of detailed allowances and requirements that assists the owner with the project scope to avoid costly change orders. Many contractors have standard practices built into their proposals to cover such circumstances.

Estimating Tools

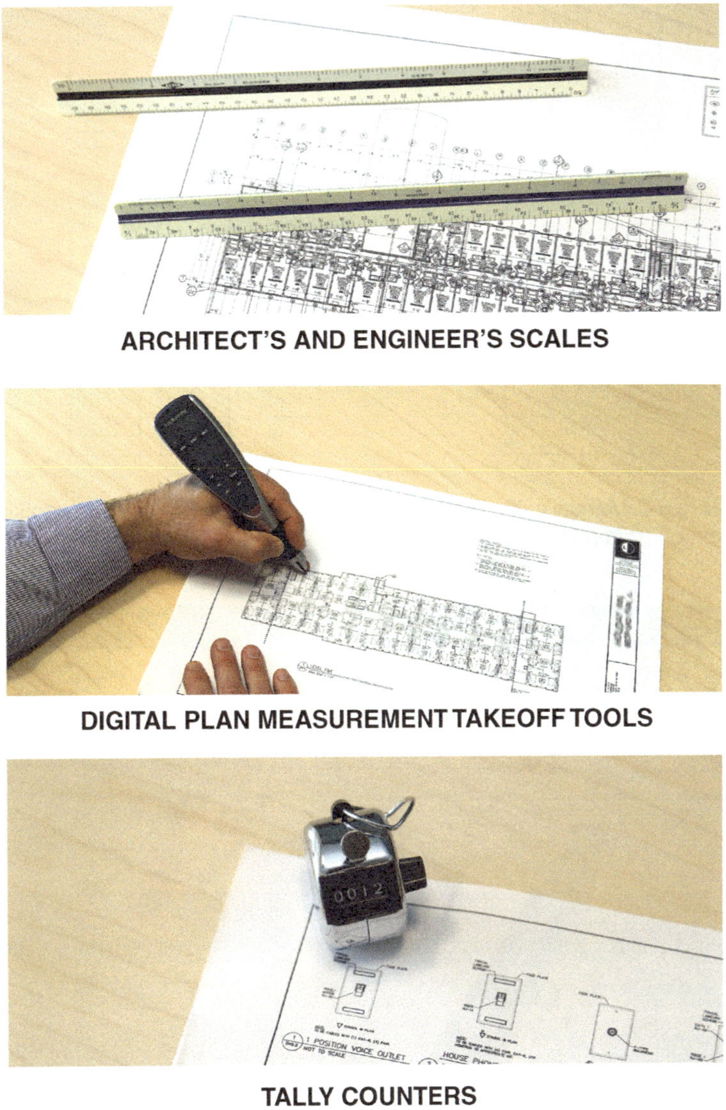

Figure 6-1. Estimating tools such as architect's and engineer's scales, digital plan measurement takeoff tools, and tally counters are used to gather and quantify the information on the drawings to develop the estimate.

Preliminary Cost Estimating Method

The second estimating method can be classified as a "preliminary cost estimate." A preliminary cost estimate is typically based on the schematic or design development drawings, sketches, and general specifications. A preliminary cost estimate is often used during the negotiation stage of a project or when a design-assist (DA) procurement method is to be utilized.

Ledger Sheets

QUANTITY SHEET

Sheet No. 1

Project: Airport Site

Date:

Estimator: RFK

Checked:

No.	Average Elevation ft	Average Cut ft/sq ft	Quantity Cu ft	Average Fill ft/sq ft	Quantity Cu ft	Total Cut Cu ft	Total Fill Cu ft	Difference Cu ft
	0′ grade level = 162′							
1	169.22	7.22	72,200					
2	171.30	9.30	93,000					
3	168.50	6.50	65,000					
4	164.55	2.55	25,500					
5	161.40			0.60	6,000			
6	157.87			4.13	41,300			
31	172.17	10.17	101,700					
32	166.60	4.60	46,000					
33	160.40			1.60	16,000			
34	154.00			8.00	80,000			
35	146.65			15.35	153,500			
36	139.00			23.00	230,000		2,022000	978

Figure 6-2. Manual estimating methods use ledger sheets to track quantities taken off from prints.

Drawing Interpretation

Figure 6-3. Estimators commonly mark up drawings and specifications to ensure that all necessary items are counted and that no items are counted twice.

A *design-assist (DA) procurement method* is a construction project procurement method in which one or more subcontractors are retained, prior to the completion of design, to assist the architect, engineer, or construction manager in the development of the design and construction documents. Since the construction documents are not complete in detail, however, the estimator's experience and knowledge alone should be accurate enough to provide an acceptable estimate for negotiations.

The disadvantage of using a DA procurement method for estimating is the tendency of estimators to overestimate certain aspects of a project. To avoid overestimating items on a project estimate, an estimator must be as clear and concise as possible with the estimate and base it on accurate, up-to-date information and facts. This allows the estimate to be properly revised as it evolves. If information is limited or incomplete, notes, drawings, and other documentation must be presented to indicate the estimator's reasons and conclusions for developing the final estimate.

Detailed Takeoff Estimating Method

The third and most accurate estimating method is a detailed takeoff estimate, which is based on a detailed material takeoff. A *detailed takeoff estimate* is an estimate in which each individual construction component is accounted for, or taken off. For example, a detailed takeoff estimate for an equipment room may include individual entries and calculations for the square feet of removable floor tiles, cable trays, and cable racks; subsurface preparation; construction labor, including rough wiring; cable and equipment installation; firestop system installation; and final equipment testing costs. **See Figure 6-4.** When an estimate is developed for a project, detailed takeoff estimates should only be prepared for work to be performed by the contractor. Subcontractors should prepare detailed takeoff estimates for work that they are hired to perform.

A detailed material takeoff requires a review of the symbols on drawings that depict certain aspects of the work. The symbols represent accurate data and known elements of the project. The materials are quantified by segregating drawing numbers on material takeoff sheets. The takeoff determines the amount and types of materials and equipment required to complete the work and, upon award of the project, is generated as a bill of materials. For example, for a VDV symbol that represents an outlet, it is required that each symbol is represented as it relates to other components that are associated with the symbol, such as faceplates, wires, and connectors. Failure to accurately account for the symbols depicted in the drawings can have a negative impact on the estimate, as it can cause material and labor miscalculations.

The next step when compiling a detailed cost estimate is to calculate the quantity of materials. Productivity factors, such as how the project labor force is comprised, the construction methods utilized, supervision, and job-site working conditions are applied to the takeoff and used to determine the overall labor cost to complete the estimate.

ESTIMATE FORMATS

Estimate formats can be manual or digital (electronic). Most final estimates developed for commercial and industrial projects are digital. However, manual estimating is performed on the job site at the tradesworker level.

Manual Estimates

Manual estimates are developed for small projects, such as minor residential projects or projects where only a section of a building will be retrofit. Manual estimates require handwritten information to be recorded into a ledger sheet or a similar document by the estimator or the contractor's representative for later input into an electronic system by the project manager. Manual estimating is also performed for large projects by tradesworkers on the job site for input into the final digital estimate.

Digital Estimates

With the development of software programs, most contracting firms use digital technology to provide accurate and timely estimates. In some cases, an estimate can be built using an electronic spreadsheet that can be edited with construction estimating software. Some construction estimating software also tracks labor and material costs and provides comparisons in the estimate for similar materials and associated labor. Digital estimating programs often use spreadsheets to manage information.

Detailed Takeoff Estimates

System Roughwiring/Roughwiring Attachment

#	Description	Qty	Manufacturer	ModelNumber	UnitCost	ExtCost	CostMarkedUp	ExtCostMU	LaborUnit	ExtLabor		LABOR
1	3 Pair Cat6A Plenum Cable (Yellow)	206	Mohawk 4 Pair	M58648	$ 480.00	$ 98,880.00	$ 552.00	$ 113,712.00	7.00	$ 1,442.00		213.00
2	4 Pair Cat6A Plenum Cable (Yellow)	198	Mohawk 4 Pair	M58648	$ 480.00	$ 95,040.00	$ 552.00	$ 109,296.00	1.00	$ 198.00		199.00
3	Cable Lace & Dress @ Termination Field	2260	labor	Installation Labor	$ 0.20	$ 452.00	$ 0.23	$ 519.80	0.10	$ 226.00		2260.10
4	Hardware For 4 Port Faceplate	1147	Hardware	H103	$ 6.00	$ 6,882.00	$ 6.90	$ 7,914.30	0.00	$ -		1147.00
5	Firestop Sealant 18.5cu in	15	Nelson Fire Stop	AA529	$ 21.15	$ 317.25	$ 24.32	$ 364.84	0.15	$ 2.25		15.15
6	4-Pair Cat6A Plenum Cable (White)	8	Mohawk 4 Pair	M58648	$ 480.00	$ 3,840.00	$ 552.00	$ 4,416.00	1.00	$ 8.00		9.00
7	12 Strand Fiber Plenum SM Armored	1.7	Corning Fiber	012E88-31131-A3	$ 1,508.41	$ 2,564.30	$ 1,734.67	$ 2,948.94	20.00	$ 34.00		21.70
8	50 Pair Cat3 Plenum Voice Trunk	1.7	Mohawk Multipair	XXXX	$ 1,325.30	$ 2,253.01	$ 1,524.10	$ 2,590.96	14.00	$ 23.80		15.70
9	#6 Green Ground Wire	0.5	LCN	#6 Green Ground Wire	$ 486.48	$ 243.24	$ 559.45	$ 279.73	12.00	$ 6.00		12.50
10	Genspeed Cat 5e	2	General Cable 4 pair	7131282	$ 155.00	$ 310.00	$ 178.25	$ 356.50	3.00	$ 6.00		5.00
11	RG 11 CMP	0	Belden Coax	89292	$ 1,025.00	$ -	$ 1,178.75	$ -	1.00	$ -		1.00

Figure 6-4. A detailed takeoff estimate is an estimate in which each individual construction component is accounted for, or taken off.

A *spreadsheet* is a computer program that uses cells organized into rows and columns to perform various calculations when formulas and numbers are entered. The foundation of most estimating programs is a computer spreadsheet. Spreadsheets are designed with a standardized row and column format for the entry of material quantities and descriptions, labor costs and material prices, and other items required by the estimator. Since the quality of the final bid is determined by the accuracy of the information entered in the spreadsheet, an estimator must take care to verify that the spreadsheet is performing the proper calculation for the proper range of cells. Estimators can use software or digital templates with preformatted cell calculations that can be copied into an existing spreadsheet. **See Figure 6-5.**

Spreadsheets

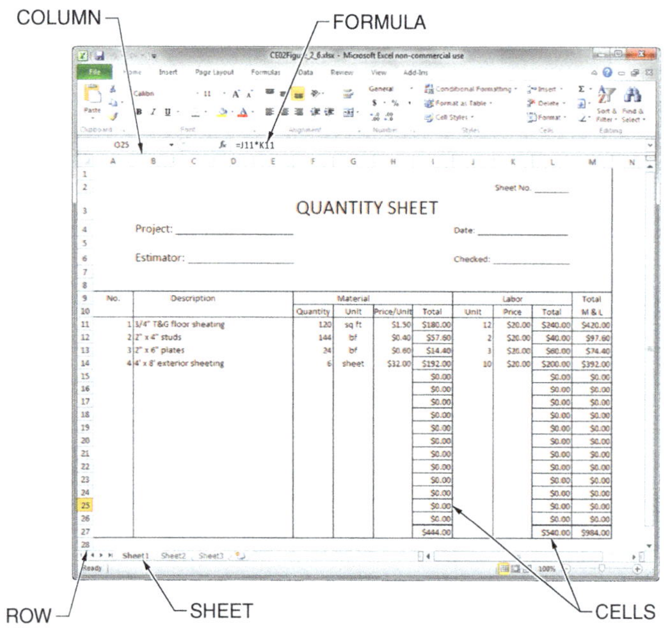

Figure 6-5. Digital estimating programs include spreadsheets that can be used to manage information.

A *digital estimating method* is an estimating method in which quantity takeoff and pricing calculation is accomplished using electronic spreadsheets and integrated database systems. Digital estimating methods have improved the estimating process by reducing the time to build an estimate, reducing errors, and improving the efficiency of the entire estimating process.

Digital estimating methods include computerized spreadsheet and estimating programs. Computerized spreadsheets have streamlined the calculation process by allowing estimators to enter quantities, costs, and standard calculations into individual cells. Estimating programs allow material pricing, labor costs, and quantity calculations to be determined digitally by using standard materials, crew cost blending, and/or assembly information that can either be entered manually or retrieved from database history files.

The advantages of digital estimating methods are faster estimates, easier last-minute changes to the bid, the ability to share estimating components for a large structure with a team of estimators, and assistance in reducing mathematical errors or missed bid items. In addition, digital estimating methods enable standardization throughout a large contracting company; integration with other company departments such as accounting, scheduling, and job costing operations; possible integration with subcontractor and supplier bid information; and flexibility of final bid reporting.

Digital Estimating Software. Digital estimating software also allows estimators to easily import and utilize information from previous bids in new bids. Some digital estimating programs allow easy conversion from US customary units to metric units. Estimating programs range from spreadsheet programs that are developed and customized by an estimator or construction estimating department, to programs with ledger sheet formats, to more sophisticated estimating programs that integrate the quantity takeoff process with design documents and scheduling operations through the use of a digitizer, digital construction documents, and quantity and pricing calculations.

Advanced features of digital estimating software include database interactivity and the use of standard assemblies and preset formulas for various estimating activities, as well as a digitizer to calculate quantities. Digital estimating software also includes the ability to interface with CAD drawings, visual modeling systems (such as BIM), or detail drawings that were scanned into a computer. **See Figure 6-6.** Digital estimating software has the ability to compile a variety of final reports and forms.

Most contracting companies see the value of digital estimating software and use this technology as a tool to increase the efficiency and accuracy of their estimating methods. Some software systems allow for the electronic floor plans to be uploaded to the system so that most of the counting and measuring can be performed electronically. While electronic estimates are considered more accurate than manual estimates, an estimate is only as accurate as the skill of the estimator who inputs, reviews, and interprets cost data such as materials, labor units, and quantities.

Since some items on a drawing include multiple parts and components associated with a single symbol, such as a VDV symbol, the time savings from using digital estimating software occurs when the associated components are extended and subsequently included in the digital estimate. For example, a single VDV symbol, when properly counted and applied to an estimate, typically also accounts for a faceplate, wires, connectors, identification (labeling) methods, and system testing. These additional VDV components and processes are included in the digital estimate. Prior to the use of digital estimating, quantifying the additional components was manually performed, which required additional time to include in the estimate and often resulted in errors.

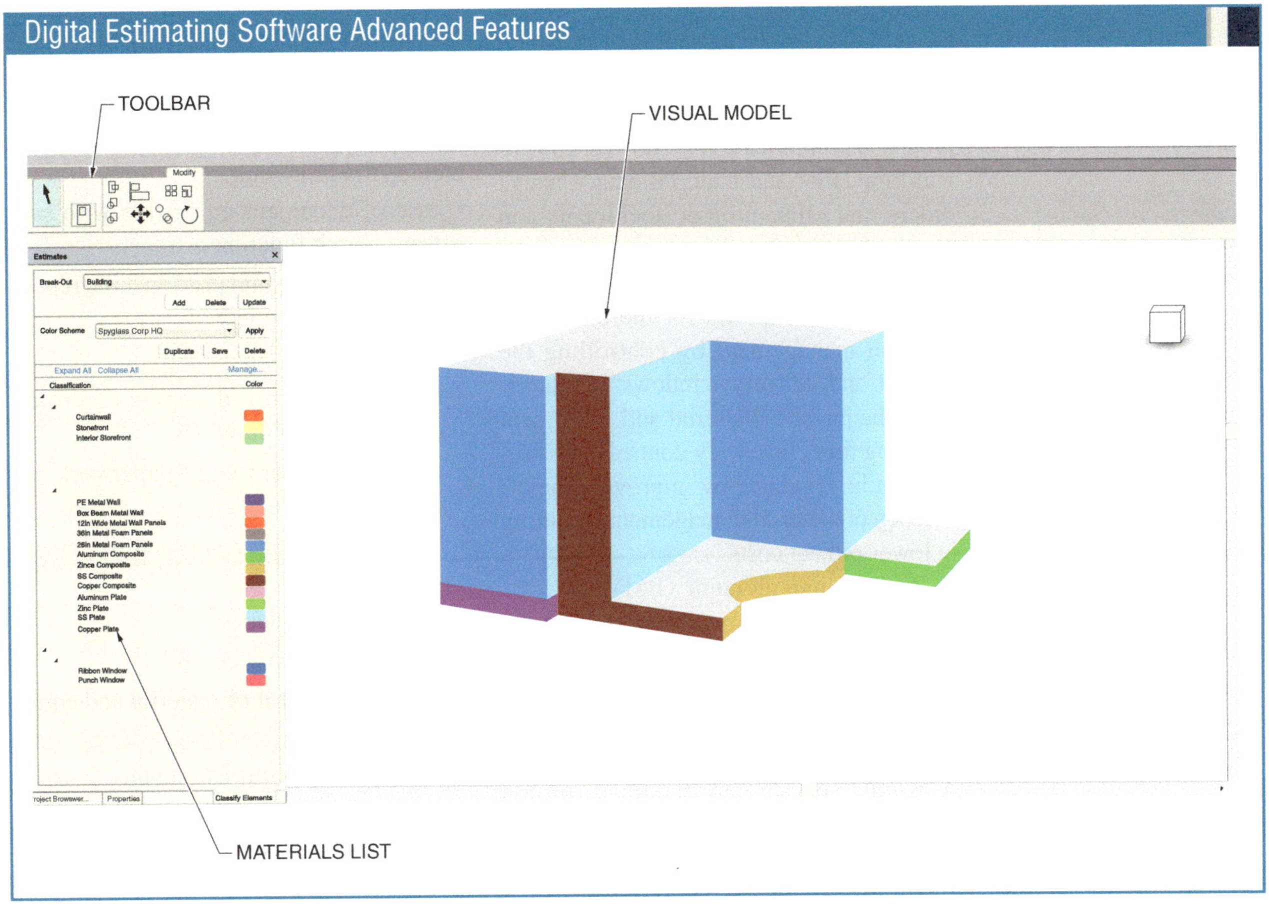

Figure 6-6. Advanced features of digital estimating software include the ability to interface with visual modeling systems, such as BIM.

Another advantage of using digital estimating software is more efficient use of the estimator's time. The ability to provide detailed bills of materials and quantify the materials by type, phase, or system allows the estimator and purchaser more flexibility in formulating costs and procuring materials from several vendors. Having the ability to export the data in a variety of electronic formats and transmit it to vendors provides an advantage for both the estimator and vendors. Digital estimating software can also be integrated with other construction management software tools to provide supervisors and field personnel real-time data for any phase of the project.

Project Management (PM) Software. Project managers can integrate project management (PM) software into an estimating system that provides the ability to track the project by building, floor, job site, or construction phase. The advantage of integrating PM software into an estimating system is that it allows for coordination with field supervisors concerning the work breakdowns, schedules, scope changes, project communications, and tracking of real-time labor and material data. Change orders and other project documents can also be uploaded and become part of the electronic project file.

PM software allows for more efficient material management by controlling the amount and delivery windows of materials to the job site. Material and equipment handling time, inventory control, and storage can be regulated by improving the efficiency of material management, resulting in lower project costs.

Project management (PM) software also has the capability of connecting with a pricing service. A pricing service, as applied to construction projects, is an online method to establish the best price for materials and services from several distributors. A pricing service can maximize profits while considering consumer and market costs for materials.

The use of a pricing service is usually accomplished through a subscription service with material distributors, which provides accurate and current pricing on most of the equipment and materials listed in an estimate. Most estimating and PM software systems can access pricing services. A commercial business subscribes to a pricing service to ensure that they have the best information available for the markets they serve, as material prices can vary by regions and change by the day. For example, the cost of copper cabling and conductors can be managed by tracking spot prices of raw copper. **See Figure 6-7.** *Note:* A pricing service database must be maintained with part numbers and prices that are verified on a regular schedule to ensure that they are accurate and current.

Modern estimating methods utilize the most modern technology available. For example, electronic files, such as PDF or AutoCAD files, can be utilized for the estimating process and referenced by installers on the job site through electronic devices such as smartphones, tablets, and laptop computers. In addition to ease of access, advantages of estimating and PM software include the following:

- ease of document navigation with software search function
- ability to create and print specific reports related to the project
- remote access
- backed-up/archived copies of project documents and drawings
- reduction in hard-copy paperwork and clutter
- improved efficiency and control of project variables
- higher bid frequency
- efficient inventory control
- improved control of material and equipment storage
- management of project finances such as costs and billing
- reduced overhead costs
- ability to manage cash flow
- higher profit margins

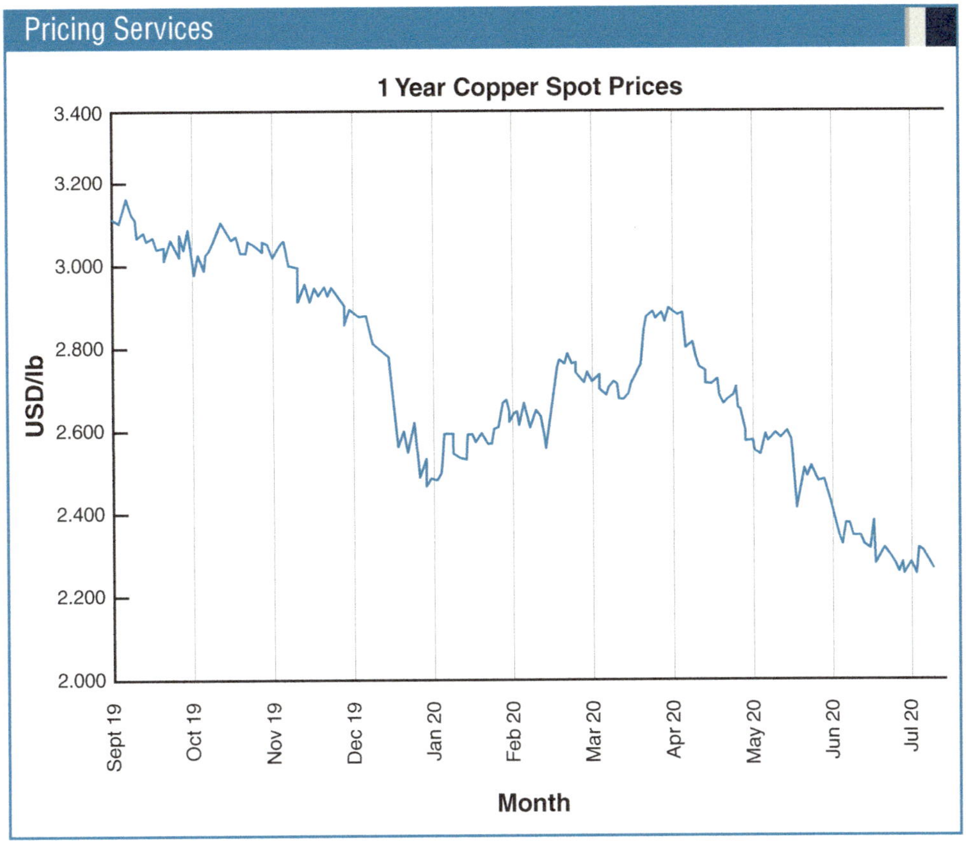

Figure 6-7. Project management (PM) software can connect with a pricing service to help manage the procurement of certain goods and services.

ESTIMATOR RESPONSIBILITIES

Estimators and contractors base many of their calculations on historical data by reviewing detailed job cost tracking data from previously completed projects. The method of preparing an estimate requires experience and knowledge gained through formal education, field experience, on-the-job training (OJT), or a combination of all three. An estimator's skills should include the following:

- field experience in the trade discipline for which the estimate is developed
- knowledge of trade-related acronyms and abbreviations
- good judgment with the ability to qualify and justify components of the estimate
- ability to perform research of labor units and material procurement
- ability to use drawings and specifications to visualize the construction process and finished product
- excellent mathematical skills, since mathematical mistakes, such as a sum not properly calculated, can be carried through an entire estimate, resulting in inaccuracies

An estimator must understand the type of building construction and how project schedules for other trades can affect an estimate. There are several considerations concerning the type of building construction when preparing an estimate. These considerations include whether the facility is residential, commercial, or industrial and whether it is an existing building, a historic structure, or new construction. **See Figure 6-8.** An estimator must prepare a detailed material takeoff, with labor units and material costs applied, to ensure the development of an accurate estimate that will minimize the chance of underestimating these items on a project.

Figure 6-8. Building construction types include residential, commercial, or industrial and whether the building is an existing building, a historic structure, or new construction.

In addition to labor rates and material costs, estimators must also have knowledge of other factors that affect their projections. These factors include local building codes, labor productivity, operating costs, G&A overhead, profit margins, and any job conditions that might adversely affect job continuity and the normal progress of work.

ESTIMATE COMPONENTS

There are four basic components of an estimate. The first component is "time," which refers to the amount, in man-hours, that a various task takes to be completed. Time also equates to labor units. A labor unit can be defined as a target used by an estimator to apply to a specific task. Typically, one man-hour is equal to one labor unit. **See Appendix.** Even though there are 60 minutes in one labor unit, only a specific percentage of the unit is considered productive time and must be applied to the estimate as such. Items such as material procurement, review of specifications and drawings, work breaks, and job-site cleanup are not considered productive time. **See Figure 6-9.**

The second component of an estimate is "material." The material component of an estimate includes the material, devices, and equipment required for project installation as outlined in the specifications and shown on the drawings. This component should also include any material that may not be specified in the specifications or on the drawings but are necessary for a complete and thorough installation.

The third component of an estimate is referred to as the "intangible items." An intangible item is usually associated with the costs of items that are required to complete a project. Intangible items typically vary by project but include permits, fees, bonding requirements, commissioning, warranties, and any other items that must be included to cover the cost of the installation.

The fourth component of an estimate is referred to as the "level of difficulty." The level of difficulty for a project affects the labor units and can vary from project to project. Estimators must study the varying levels of difficulty and factor in unproductive time into the labor units. Levels of difficulty for estimating are typically considered normal, moderate, and high. **See Figure 6-10.** For example, a multistory construction project has a higher level of difficulty than a single-story construction project due to the number of man-hours and equipment required to move personnel and materials up vertically from the ground as the building is constructed.

Labor Unit Analysis		
Task	**Amount***	**Amount†**
Installation/construction	69	5.5
Construction layout	6	0.5
Equipment mobilization/transport	5	0.4
Punch list development	5	0.4
Job-site personnel work breaks	3.5	0.3
Miscellaneous work	2.5	0.2
Drawing and specification review	2.5	0.2
Material procurement	2.5	0.2
Material and equipment receiving/storage	2	0.15
Job-site cleanup	2	0.15

* in %
† in hours per 8 hour shift

Figure 6-9. Even though there are 60 minutes in one labor unit, only a specific percentage of the unit is considered productive time and must be applied to the estimate as such.

Estimating Level of Difficulty			
Premise/Condition Type	**Normal**	**Moderate**	**High**
Residential	X		
Commercial (single-story)	X		
Commercial (multi-story)		X	
Institutional			X
Industrial		X	
Special construction			X
Normal environmental conditions	X		
Extreme environmental conditions			X
Normal schedule	X		
Compressed schedule		X	

Figure 6-10. Estimating levels of difficulty are considered normal, moderate, and high.

Another factor that must be considered when calculating level of difficulty is environmental conditions, such as temperature and humidity extremes that can adversely affect the efficiency of workers. If workers become fatigued or uncomfortable due to excessive heat, cold, or humidity, the level of difficulty is not the same as it is under normal conditions. For example, in cold conditions, workers must perform tasks while wearing gloves, reducing their sense of touch and ability to manipulate hand tools.

Another factor affecting level of difficulty is job site access. Some job sites are easier to access than others. For example, a job site in an open field is easier to access than a job site in an urban location. Also, some OEMs have guidelines for installing materials in temperature-controlled environments for the optimal performance of their product, which can affect the level of difficulty.

Productivity Estimates

Productivity estimates can also be affected if a project has a compressed or accelerated construction schedule. For example, when tradesworkers are required to work over 40 hours per week (overtime), there is an increased chance of productivity losses, inefficiency, and increased labor costs over the duration of a project. Overtime can be corrected by adding more workforce to a project. However, overcrewing a job may not be productive due to the sequencing of work or overall progress of related work. In some cases, multiple shifts can be added to the work schedule, provided that the project is properly managed and activity of the effected trades is synchronized.

VDV System Materials and Labor Units

An estimate for VDV systems is broken down into four parts. These parts detail the quantity of material and labor units for each item as follows:

- Part 1—The first part includes a complete list of wires and cables, consisting of each wire/cable type and lengths required for the project, including backbone cabling, horizontal cabling, patch cords, etc.
- Part 2—The second part includes all termination components, which include connectors, jacks, faceplates, racking systems, raceway systems, equipment room hardware, etc.
- Part 3—The third part includes copper and fiber-optic link tests, including certification and qualification tests (copper systems) and light transmission, signal measurement, and visual fault tests (fiber-optic systems). Product warranties should also be included in this part.
- Part 4—The fourth part includes any special equipment, such as test instruments and certification equipment, termination tools, and cable routing equipment, specified in Parts 1 to 3.

Estimator Checklists

An estimator's attention to detail is critical, and most contractors have a standardized check list of questions for the estimator and management for when the estimator creates the estimate. Examples of questions that must be addressed prior to creating an estimate include the following:

- Have all bid documents, including drawings, specifications, bid instructions, and bid response requirements been thoroughly reviewed?
- What is the project scope and are there any special conditions that need to be factored into the estimate?
- Are there any addenda to the drawings and specifications and have all RFIs been answered?
- What is the timeline to complete the estimate?
- What type of floor plans are to be used as reference to estimate the job (electronic, hard-copy, or a combination of both)?
- What is the size of the project, in square feet?

Questions that must be addressed regarding the takeoff section of the estimate include the following:

- Has all major equipment been accounted for in the takeoff?
- Is there equipment provided by others to be installed? (This question is commonly found on labor-only estimates.)
- Is all VDV work included in the takeoff?
- What, if any, VDV work is excluded from the takeoff?
- Will conduits or pathways be provided and installed by other subcontractors?
- Will outside plant (OSP) work be required?
- What is the scope of work in the equipment rooms?
- Is any work, such as conduits/backboards/equipment outlets and cable runways, to be completed by others?
- Will coordination with local exchange carriers (LECs) be required prior to completing the installation of VDV equipment?
- Is the material and labor pricing in the database current?
- Are there to be labor increases during the time frame of the project, and have they been factored into the takeoff?

- Have all vendor material quotes been received and reviewed?
- Are vendor quotes within the range of historical pricing?
- Have any discrepancies in the range of historical pricing been addressed?

Questions that must be addressed regarding the management assessment section of the estimate include following:

- Has the biggest challenge of the job been identified and considered in the final estimate?
- Has the estimator had adequate time to prepare a thorough estimate?
- Have all takeoffs been verified?
- Has the detailed estimate been compared to a projected price based on square footage or preliminary cost projection?
- Has the estimate been compared to previous completed projects of similar size and scope with any discrepancies accounted for?
- Has the estimate been reviewed by a senior estimator?
- How does this specific project affect company cash flow?
- Are there additional upcoming jobs that are more desirable?
- What is the amount and size of other jobs that are currently in the bid stage?
- What are the overall current market conditions?

Site Visits and Surveys

A visit to the site to evaluate the job conditions and logistical complexities can be useful with the development of an estimate. A site visit provides an opportunity to gather information pertinent to formulating an estimate or proposal to satisfy the customer's requirements. A site visit and survey are necessary for projects that require a retrofit of existing structures. **See Figure 6-11.** Existing structures sometimes have complex construction that can affect the installation and construction of new floor plans and equipment.

Site Visits

Figure 6-11. A site visit and survey are necessary for projects that require a retrofit of existing structures.

There are several factors and questions to consider when conducting a site survey. These factors and questions include the following:

- What are the requirements and costs of municipal permits and associated legal documents?
- Does the site allow for the storage of delivered materials and equipment or must a storage container or a location to store materials and equipment off-site be provided?
- What costs are to be incurred by storing materials and equipment off-site?
- Are there special considerations concerning accessibility and working time frames with the site?
- Are there on-site refuse facilities or must the contractor provide their own refuse disposal?
- Are there training or specific safety and security policies that must be met in order to enter the job site and perform the work?

- Do job-site personnel have dedicated space to park their personal vehicles on site or must they park and be shuttled from a remote location?
- Is the project in an existing building or is it new construction?
  - For existing buildings, is any demolition required?
  - For existing buildings, is the construction wood frame or structural steel?
- What are the ceiling heights and their construction type?
  - hard surface
  - spline
  - lift-out tile
  - open
- What methods are to be used to support and attach pathways in the building?
- What type of firewalls and firestop systems exist in the building?
  - Are there any specific requirements or standards for breaching firewalls during installation?
  - What is the specified method for sealing firestop through penetrations?
- What type of construction is used for hard-wall offices and cubicles?
  - gypsum wallboard
  - concrete block
  - systems furniture
- How are the work area outlets to be mounted?
  - Are existing raceways a large-enough size or must new or additional raceways be installed?
- Is the building environmental air return a plenum environment, a PVC environment, or a combination of both?
- Is there sufficient space allocated to install the equipment, including racks, raceways, and cabinets, in the equipment room?
- Where is the location of the building entrance facility?
  - What type and cost of service is required to connect the entrance facility to the main equipment room?
- What miscellaneous material is required to complete the installation that may not have been included in the specifications, takeoff, or drawings?

Each project is unique when compared to other projects. An accurate site survey can result in a successful estimate and proposal and prevent the underbidding/overbidding of a project.

SUMMARY

Prior to bidding on any project, an estimate of project costs, along with the desired profit margin, must be developed. Although estimates can be prepared manually, most contractors use digital estimating software to complete their estimates due to the ease of use and accuracy. The estimator that prepares the estimate has several roles and must take many details into consideration when developing an estimate. Prior to beginning work on an estimate, a site survey is conducted to provide the estimator with a better understanding of job-site variables that may not be listed in the project specifications or drawings. Estimates are developed in the most accurate manner possible to ensure that the project is not underbid or overbid.

CHAPTER 6

Systems Estimating

Review

Name ______________________ Date ______________

True-False

T F **1.** Estimating methods are categorized into six types.

T F **2.** The traditional estimating method is typically used by general contractors (GCs) for small-scale projects.

T F **3.** When an estimate is developed for a project, detailed takeoff estimates should only be prepared for work to be performed by the contractor.

T F **4.** Manual estimating is typically only performed at the architect level.

T F **5.** A single VDV symbol, when properly counted and applied to an estimate, typically also accounts for a faceplate, wires, connectors, identification (labeling) methods, and system testing.

T F **6.** PM software does not have the capability of connecting with a pricing service.

T F **7.** Estimators are not required to have knowledge of local building codes.

T F **8.** Time equates to labor units.

T F **9.** Even though there are 60 minutes in one labor unit, only a specific percentage of the unit is considered productive time and must be applied to the estimate as such.

T F **10.** A site visit and survey are not necessary for projects that require a retrofit of existing structures.

Multiple Choice

________________ **11.** A ___ service can maximize profits while considering consumer and market costs for materials.
- A. design
- B. drafting
- C. pricing
- D. purchasing

________________ **12.** OJT is an acronym for ___.
- A. observing-the-job training
- B. on-the-job training
- C. open job testing
- D. over-the-job training

________________ **13.** A ___ estimate is a type of general knowledge estimate formulated by using data from similar project with known completion costs and calculating the completion cost against the entire cost of the finished building.
- A. design-assist
- B. detailed takeoff
- C. preliminary cost
- D. percentage

________________ **14.** ___ is the process of determining labor and material costs required to complete a project.
- A. Engineering
- B. Estimating
- C. Identifying
- D. Reviewing

________________ **15.** An estimate for VDV systems is broken down into ___ parts detailing the quantity of materials and labor units for each item.
- A. two
- B. three
- C. four
- D. six

Completion

________________ **16.** A(n) ___ is an individual who performs quantity takeoff and pricing to determine the cost of a construction project.

________________ **17.** A(n) ___ is a grid on a sheet of paper consisting of rows and columns into which item descriptions, locations, code numbers, quantities, and costs are entered.

________________ **18.** A(n) ___ estimate is an estimate in which each individual construction component is accounted for, or taken off.

________________ **19.** A(n) ___ is a computer program that uses cells organized into rows and columns to perform various calculations when formulas and numbers are entered.

_______________ **20.** A(n) ___ method is an estimating method in which quantity takeoff and pricing calculation is accomplished using electronic spreadsheets and integrated database systems.

_______________ **21.** A(n) ___ item is an item that varies by project, such as permits, fees, bonding requirements, commissioning, warranties, and any other items which must be included to cover the cost of the installation.

_______________ **22.** ___ surveys may also show utility services, landscaping, and other information needed to perform the work required before constructing the foundation.

_______________ **23.** A(n) ___ is an estimating method in which material quantities are calculated and entered onto ledger sheets for pricing.

_______________ **24.** A(n) ___ method is a construction project procurement method in which one or more subcontractors are retained, prior to the completion of design, to assist the architect, engineer, or construction manager in the development of the design and construction documents.

_______________ **25.** Contingencies, allowances, and assumptions should be clearly stated when dealing with ___ estimates.

Short Answer

26. Explain the risks of using the traditional estimating method.

27. What are four productivity factors that are applied to the labor portion of a takeoff?

28. List at least ten advantages of using estimating and project management (PM) software.

29. Briefly explain why contingencies need to be identified.

30. Explain how the level of difficulty for a project can be considered high as opposed to normal or moderate.

CHAPTER

7

Project Management

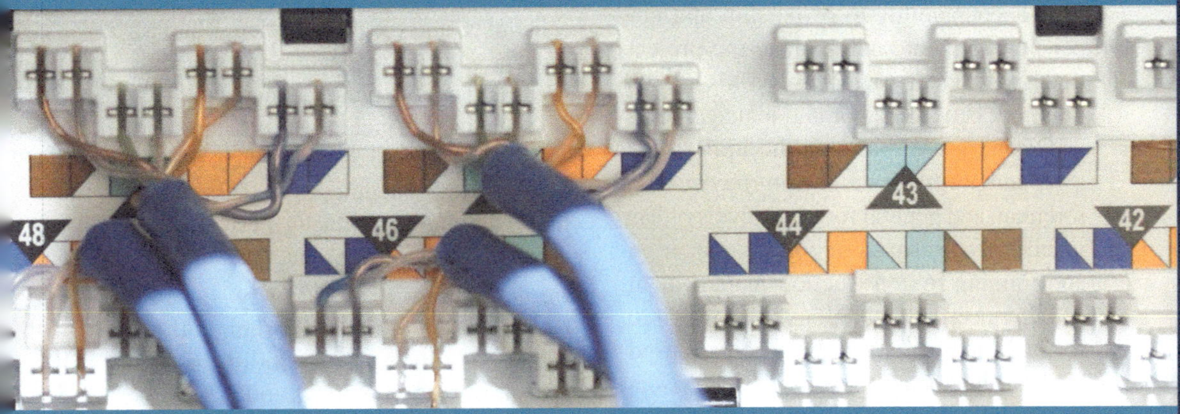

A project manager is responsible for successfully completing construction projects within the specified budget and time constraints. The project manager not only leads the project team but also oversees the responsibilities of the construction project team members and the methods, techniques, and practices employed to achieve the project goals. Project plans are the documents used throughout the project to ensure that the project is completed in a timely manner and within specifications. Project documentation provides a written and graphic account of the methods used to complete work, decisions that were made, and records compiled during a project.

Objectives

- Describe the skills and traits that a project manager must possess to successfully oversee a construction project.
- List the different members of a project team and their responsibilities.
- List the different project plan tools used by the project management team during the construction process.
- Describe the importance and purpose of documentation and recordkeeping during a construction project.

Learner Resources

atplearningresources.com
access code 783821

PROJECT MANAGERS

A *project manager (PM)* is the contractor's representative who oversees the overall planning and execution of a construction project. Most PMs demonstrate excellent leadership, provide project oversight, have accountability to ownership and upper management, and manage the project tradesworkers. In many cases, PMs are the only link between labor and management.

Often, a PM starts at the apprentice level and progresses to the journeyworker, foreman, and general foreman levels or has received formal postsecondary education and training in construction management and engineering technology. It is required for a PM to understand the processes used by the estimator and the job-site supervisors that implement the project plan. A PM must be actively involved in the project and bring a collaborative approach to the estimators, job-site supervisors, and tradesworkers on the project. A PM must be accessible and provide positive feedback to the team while also encouraging feedback from team members. **See Figure 7-1.** A successful PM must possess the following skills and traits:

- Organizational skills—A PM must lead by example and be organized. Organization of a project can be accomplished through effective planning and coordination by systematically arranging the resources required for a project, including job-site personnel, tools, materials, schedules, and the project plan. The PM must provide the leadership and structure to implement the plan. Elements of project organization include the planning and sequencing of work and how the work fits in the overall schedule, as well as assigning tasks and monitoring performance through detailed oversight.
- Implementation of authority—A PM must be able to implement their authority, as they have responsibility for the actions of the work crews and managers under their management. A PM must control all variables and understand the scope of the project, be an expert in their field, delegate orders, and confidently make decisions related to the project. Due to their position, they must often make decisions quickly, understanding that the results could seriously affect the project outcome.

Project Managers

Project Manager Skills and Traits

- Organizational skills
- Implementation of authority
- Leadership
- Communication skills

Figure 7-1. Project managers (PMs) must be accessible and provide positive feedback to the team while also encouraging feedback from team members.

- Leadership—A PM must be able to lead or have demonstrated leadership qualities in previous positions. The PM must have the confidence of their project workforce. The project workforce must be confident in the PM's leadership and know that although the PM's position is to supervise and monitor performance, the PM also cares about the workforce's safety and well-being. A successful leader regularly meets with the team and encourages shared effort toward achieving the team's goals. The PM must keep the job-site workforce informed, keep them properly equipped, provide the correct installation instructions, and maintain the flow of materials to the job site. A successful leader realizes that they are only as successful as their workforce and understands that journeyworkers take pride in their work, are expected to complete the job properly, consider reinstallation of work to be a failure, and do not waste time.
- Communication skills—An effective PM must have strong interpersonal relationship skills. Strong interpersonal relationship skills include the ability to listen, communicate effectively, solve problems, and relate to others on a personal and professional level. Communication can be in spoken or written format, or a combination of both. Regardless of the type of communication method used, it is important to verify that the communication has been received and understood by the individuals and work crews that it is directed to. An effective communicator must also be a strong collaborator, open to feedback from others, and trust opinions and input from team members.

The role of a PM and their site supervisors is to achieve a successful project plan, which means that they must not only have a clear understanding of the estimate and scope of work but also be able to manage and implement them both. Success is measured in profits; and profits are driven by the materials, labor, and ability of the workforce to perform per the plan. Regular scrutiny of materials and labor is required so that inconsistencies or mistakes can be identified and adjusted accordingly. If a surplus or shortage of materials are carried, they can positively or negatively affect the workforce's efficiency. Labor estimates can be affected by problems such as schedule delays and other subcontractor failures to complete work in sequence. When problems arise, supervisors must make critical decisions in adjusting the labor workforce and schedule.

The goal of the project team is to meet the requirements for contractual performance. Two aspects that are key to the success of a project include the schedule and the corrective actions taken during a project. The PM and site supervisors must be proactive with their counterparts at the GC level. This occurs in project and coordination meetings on site. The expertise that each party brings to the project provides for the sharing of ideas and problem-solving techniques, which can contribute to achieving substantial completion.

The development of a project plan begins with a project kick-off meeting and weekly project status meetings, along with constant coordination on the job site and reviewing the existing conditions. The lack or refusal of any project team member to coordinate, communicate, and/or collaborate with other project team members could result in failure of the project.

PROJECT TEAMS

A project team is composed of several individual staff members, each of whom have specific tasks and goals needed to complete a project. The manager of the project, usually the PM, must manage the manager-level staff members and trust them to manage the work crews that perform the specific tasks in their department or specific trade. Project team staff members typically consist of purchasing managers, safety and compliance managers, field superintendents, foremen, journeyworkers, and apprentices.

Purchasing Managers

The purchasing manager for a project is the staff member(s) responsible for coordinating the purchases of various products and services in the form of purchase orders (POs). POs are used by purchasing managers to order materials from suppliers. The conditions of the PO should be clearly stated and the agreement should include itemized quantities, unit prices, delivery and payment terms, applicable tax, and shipping and insurance costs. **See Figure 7-2.**

Purchase Orders

Sherwood Supply
305 Stony Ledge Road
Pendleton, New York

PURCHASE ORDER
467-02887

Customer ID:
MPO-19765

Ship To: M & N Electrical
325 Pinehurst Street
Amherst, New York
Attn: Albert Walls

Project: Erie Canal Western Terminus Museum
Sweeney Street, North Tonawanda, NY
Project No: 2010-876
Delivery: March 22

Quantity	Stock No.	Description	Unit Price	Line Total
38	PN-087	CAT 5e Connectors	$ 3.76	$ 142.88
28	SL-226	CAT 5e Single-gang faceplate	$ 2.23	$ 62.44
1	PP4-40120	1080′ SPOOL CAT 53 cable	$ 144.50	$ 144.50
42	PP3-30072	Cable tray section x 6 ft	$ 11.23	$ 471.66
				$ –
				$ –
				$ –
				$ –
				$ –
				$ –
				$ –
				$ –
				$ –
				$ –
			Subtotal	$ 821.48
			Sales Tax	$ 69.30
			Total Cost	$ 890.78

Payment Method:
☐ Due upon Delivery
☐ Cash Payment
☑ Credit to Account

Shipping Terms:
The purchaser shall review all items for completeness and content.
Any items delivered incorrectly or damaged must be noted upon delivery.

Received: Albert M. Walls Date: 3/22

Figure 7-2. A purchase order (PO) is a document with the details of a purchase that is often coded with a specific number.

When large quantities of materials or products are ordered or the long-term rental of equipment is required, suppliers may offer discounts, rebates, or lower unit costs. For construction projects, the purchasing manager is often a GC, contractor, or subcontractor. For example, VDV equipment is typically ordered by either the electrical contractor or the subcontractor hired to install the VDV system.

Purchasing managers solicit bids from distributors and OEMs of equipment and commodities used on a construction project. Purchasing managers base their purchases on the bill of materials (BOM) prepared by the estimator during the bid process, as well as any items noted during careful review of prints and drawings. Purchasing managers collaborate with the estimator during the bid process to determine the best pricing on equipment and commodities, so they may provide the most competitive price when submitting the bid. Upon beginning a project, the purchasing manager enters into agreements with the distributors to provide such equipment and commodities. They work with the project team on staging materials so as not to overburden the job with excess materials or too little materials to keep the team productive and the project moving ahead on schedule.

The purchasing manager typically specifies a window of time or a specific date for the delivery of products to a job site. For example, certain products require indoor storage and cannot be delivered to a job site until such accommodations can be provided, whether it is an enclosed building or storage unit designed to store such equipment. The delivery date of such products is specified in the PO, with a clause that allows the GC to refuse the delivery and withhold subsequent payment if equipment is delivered prior to the specified date in the PO. An experienced purchasing manager is an asset to the project team because they understand how to negotiate a lower cost and reduce overall risk by ensuring timely deliveries and management of relationships with suppliers.

Safety and Compliance Managers

The primary role of a safety and compliance manager on a project team is to ensure a safe work environment for the team and those working with them on the job site, including any outside visitors to the job site. A safety program is a requirement for all construction job sites. Safety and compliance managers provide plans to optimize protection of a company's personnel, property, and purposes.

It is the responsibility of the GC to implement and enforce a job-site safety program. The GC must prepare and implement the job-site safety program that best suits the project to minimize risks that may involve the loss of human resources, materials, and equipment. The implementation of a safety program is an ongoing process that may directly affect schedules and create additional costs. To enforce job-site safety, the GC should make periodic safety inspections of the work and job site on a recurring schedule. A *safety inspection* is the process of critically examining and evaluating the safety and occupational health hazards of a job site. **See Figure 7-3.**

Figure 7-3. To enforce job-site safety, the GC should make periodic safety inspections of the work and job site on a recurring schedule.

Safety and compliance managers provide training for new employees and hold regular safety meetings with all employees to review policies and raise awareness. In the event of accidents or near accidents, the safety and compliance manager works with safety agencies and employees in documenting the accidents and develops plans to prevent them in the future. Safety and compliance managers work with the GC's team and the other trades to ensure a safe worksite, based on the requirements of OSHA.

The *Occupational Safety and Health Administration (OSHA)* is a federal regulatory agency that develops and publishes safety and occupational health regulations and has legal authority to inspect workplaces, enforce regulations, and propose fines and penalties for violations. OSHA was established in 1971 as an agency under the US Department of Labor (DOL). OSHA workplace regulations are also required to be posted at all places of employment. **See Figure 7-4.** OSHA works to accomplish the following:

- establish and enforce safety standards
- conduct job-site inspections
- investigate and document employee complaints regarding work conditions
- provide technical assistance, consultation programs, and educational support

OSHA monitors and enforces safety regulations for all industries in the US. OSHA maintains separate departments within the agency that provide expertise and assistance for various industries. Two such departments related to the construction industry are the Office of Construction Services and the Office of Engineering Services.

The Office of Construction Services assists labor and management in establishing and facilitating communication and safety awareness. This department coordinates training and education for various construction agencies and special-interest groups. It monitors a database that evaluates construction industry events regarding job safety and health issues.

The Office of Engineering Services provides engineering expertise to those who investigate accidents that involve fatalities and catastrophes that receive high-level attention. Members of the department act as technical consultants and advisors regarding unusual occupational safety and health issues.

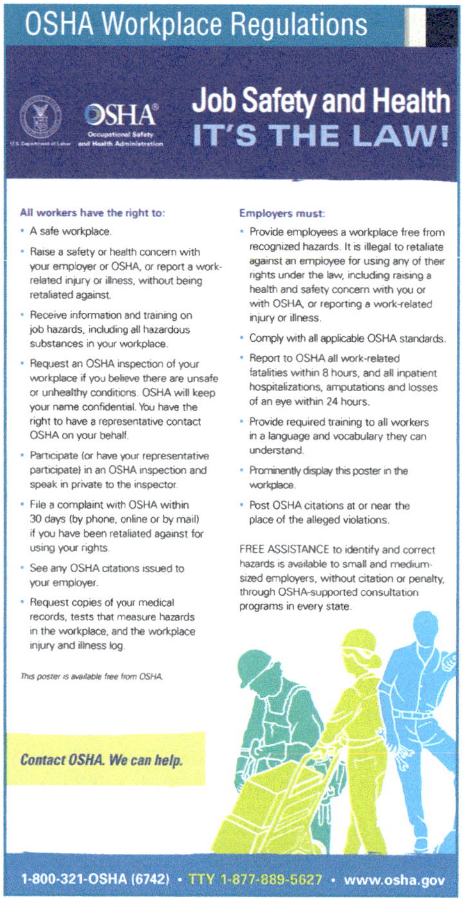

Figure 7-4. OSHA workplace regulations are required to be posted at all places of employment.

Field Superintendents

A *field superintendent* is the general contractor's site representative who is responsible for the day-to-day operations and overall supervision of the construction job site. Field superintendents are hands-on leaders with experience in working with construction personnel. They have exclusive authority over the job site during construction and work under the direction of the PM. In this way, the field superintendent is the eyes and ears for the PM. In addition to supervising job-site personnel, they must coordinate work with the GC and other subcontractors.

A field superintendent must have a clear understanding of the contract documents and scope of work. It is the responsibility of the field superintendent to ensure that the specified equipment, materials, and services for the construction project are delivered per the terms of the contract and staged in the proper location, in the correct quantity, and in good quality and condition.

A field superintendent must have strong planning and leadership skills, must continually plan in short and long term for workforce requirements, and must be able to anticipate situations that could cause possible delays or require changes to the plan. A field superintendent must interact with the foremen to determine, with their input and assistance, what their requirements are to keep the project on schedule and budget. The field superintendent also works with the PM to ensure that all work is completed as shown and specified in the contract documents.

> **TECH FACT**
>
> Injuries on job sites can have a negative effect on the workforce and cause increases to insurance requirements and costs. Research studies have proven that when people work safely, their productivity increases.

A field superintendent assigns duties and delegates tasks to the GC's workforce. In addition, the field superintendent must coordinate all information and requirements for the project, including safety requirements and compliance of regulatory codes. **See Figure 7-5.** The field superintendent also allocates equipment and tools, schedules material deliveries, makes various decisions at the job site, oversees quality control, and coordinates the sequencing of all construction activities and operations at the job site.

A field superintendent schedules workers for specific working hours or shifts. Typical working shifts start from early morning and last until late afternoon. In some instances, construction operations can occur over a 24-hour period. There is usually a considerable amount of activity at the job site during construction hours. These activities must be choreographed so as not to cause interference between workers or damage complete or newly installed work. Field superintendents are responsible for managing the contractors and their work crews. **See Figure 7-6.**

Field Superintendents

U.S. Army Corps of Engineers

FIELD SUPERINTENDENT ROLES AND RESPONSIBILITIES

- Responsible for the day-to-day activities at job site
- Assign duties and delegate tasks for workforce
- Allocate equipment and tools
- Schedule material deliveries
- Coordinate safety requirements and compliance of regulatory codes
- Maintain quality control
- Coordinate proper storage of materials and equipment
- Resolve job-site disputes

Figure 7-5. The field superintendent assigns duties and delegates tasks to the GC's workforce.

Foremen

A *foreman* is a field supervisor who is responsible for directing each labor crew and activity on a project but is not considered company management. Most foremen have completed industry-approved supervision training, have extensive industry experience, oversee five or more journeyworkers and apprentices (including themselves) on construction projects, and report directly to the subcontractor for their trade.

As hourly employees, foremen usually receive a higher wage than other tradesworkers; however, they are still considered part of the labor force. A foreman and their crew are assigned specific tasks or responsibilities on segments of work or areas of a project. Foremen directly manage the journeyworkers and apprentices that work under them and ensure that the technical work tasks are completed in a neat and professional manner within the specified schedule of the contract. Foremen are responsible for maintaining the supply of materials and equipment to their workforce by interacting with the workforce about their needs and listening to suggestions for methods of completing future tasks.

Journeyworkers

A journeyworker, journeyman, craftworker, or technician is a skilled worker who has successfully completed an official apprenticeship qualification in a building trade or craft. Journeyworkers are considered competent and authorized to work in their field as a fully qualified employee often through licensing examination. They primarily work with their hands or with tools of their trade. Most journeyworkers are compensated by an hourly wage and benefit package that can include health care, paid vacation time, and retirement.

Some examples of journeyworkers include carpenters, electricians, plumbers, welders, pipefitters, and ironworkers. Although journeyworkers have limited responsibilities from a management perspective, they are the skilled labor that completes the work. They also are tasked with training the apprentices that are learning to work in their trade. Journeyworkers that install VDV systems and equipment are typically trained as electricians.

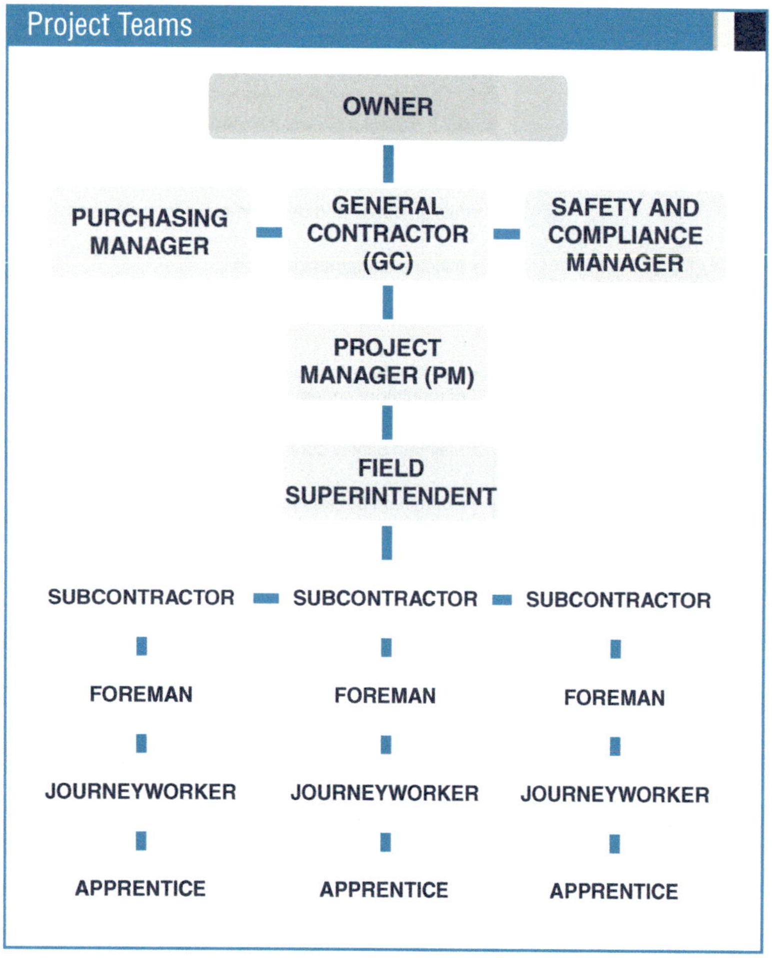

Figure 7-6. The field superintendent typically is responsible for the day-to-day activities of project subcontractors and their work crews.

Apprentices

An *apprentice* is an individual training to learn a trade through a combination of on-the-job training and related classroom instruction in an apprenticeship program. An *apprenticeship program* is a training program that provides practical hands-on training accompanied by theoretical classroom instruction. Apprenticeship programs are sponsored jointly by employer and labor groups, individual employers, and employer associations. Training is accomplished under the supervision of a skilled journey person from which the apprentice learns theoretical aspects and hands-on skills of the occupation.

Apprentices are typically paid through a step-based raise system that follows along with practical and school-related hours over the course of three to five years, depending on the trade and licensing examination requirement. Apprentice responsibilities include materials management, equipment maintenance, and repetitive jobs during the early phase of their training. As apprentices' skills evolve, they are assigned more stringent tasks and responsibilities.

PROJECT PLANS

A project plan is used to document and communicate expectations, control schedules and delivery, and manage risks. A project plan improves communication by outlining the plan for team members and allows for feedback early in the process. A project plan also helps facilitate expectations of management and project owners by updating milestones and timelines as the project progresses.

Through the implementation of a project plan, team members can quickly locate the necessary project information for every step in the process. Timelines are clearly defined to maintain the project's priorities and objectives. A project plan allows for planning the schedule of each activity and how it affects the timing of the entire project. Most project plans consist of contract documents, general conditions of the contract, specifications, addendums, POs, change orders, building permits, shop drawings, submittals, and project schedules.

Contract Documents

A *contract document* is a set of instructions created by a designer and used by a contractor to estimate project costs and build a project. Contract documents include bid information, contract agreements, general and supplemental conditions of the contract, drawings, and technical specifications. The documents are used to illustrate design intent and to provide detailed information for the construction of a project. They include the contractual procedures and obligations that become a legally binding contract.

Contract documents are often referred to as "the CDs" and establish the standard by which the performance of the contractor is measured.

If the work has been completed per the contract documents, then the project should be a success from a performance standpoint. The contract documents are the most valuable tool a PM uses to maintain the project's budget and timeline. The PM must carefully review the project documents to identify any discrepancies early in the project timeline, in order to provide the architects and engineers adequate time to assess the discrepancies and make the required corrections.

In addition to reviewing the contract documents, the PM must become familiar with the project job site. A review of the physical aspects and uniqueness of the job site prior to starting the work should be performed. Items that must be noted include utility entrances, site access for deliveries, and any existing conditions and hazards that could be present and interfere with the construction process. A report on the existing job-site conditions is helpful when preparing the project plan and must be shared with the team members.

The contract documents are the main section of the project contract. A contract is a written agreement between two or more parties. A project contract outlines performance guidelines or delivery, the party responsible for work (by type and product), a schedule of when the work is to be performed and completed, the work costs, and the consequences of work that does not meet the obligations of the contract.

The contract defines the scope of work, sets ground rules for the project, and provides guidance for dealing with unforeseen events. Contracts are never standard and most GCs and owners use several different forms of contracts, depending on the type and value of the work. The American Institute of Architects (AIA) provides a variety of standardized forms for the construction industry. AIA Document A401® is the standard agreement between contractor and subcontractor. **See Figure 7-7.** A PM must be knowledgeable and understand the contract text and use it as a tool to control the job and to recognize and resolve problems.

Contractor and Subcontractor Agreements

AIA® **Document A401® – 2017**

Standard Form of Agreement Between Contractor and Subcontractor

AGREEMENT made as of the _____ day of _____ in the year _____
(In words, indicate day, month and year.)

BETWEEN the Contractor:
(Name, legal status, address and other information)

This document has important legal consequences. Consultation with an attorney is encouraged with respect to its completion or modification.

and the Subcontractor:
(Name, legal status, address and other information)

The American Institute of Architects

Figure 7-7. The AIA provides a variety of standardized forms for the construction industry. Document A401 is the standard agreement between contractor and subcontractor.

While the construction contract is the legal document for an agreement, it does not cover enough project information on its own as a stand-alone document. Due to the complexities of the construction contract, a separate set of guidelines known as the "General Conditions of the Contract for Construction" is required. This document is designed to complement the construction contract. It recognizes the relationships between the owner, architect, and GC and the responsibilities and rights of each party. It defines key terms and procedures and mechanisms for resolving disparities and also clarifies information on the drawings or specifications. **See Figure 7-8.**

Specifications

The specifications form the legal basis for the construction contract. The specifications define the quality or grade of materials to be installed and the acceptable workmanship and standards by which the workmanship is to be evaluated. The specifications are also a guideline for resolving disputes and the basis on which the contractor can accurately estimate costs.

General Conditions of the Contract

AIA® **Document A201™–2017**

General Conditions of the Contract for Construction

for the following PROJECT:
(Name and location or address)

This document has important legal consequences. Consultation with an attorney is encouraged with respect to its completion or modification.

For guidance in modifying this document to include supplementary conditions, see AIA Document A503™, Guide for Supplementary Conditions.

THE OWNER:
(Name, legal status and address)

THE ARCHITECT:
(Name, legal status and address)

TABLE OF ARTICLES

1 GENERAL PROVISIONS

2 OWNER

3 CONTRACTOR

4 ARCHITECT

5 SUBCONTRACTORS

The American Institute of Architects

Figure 7-8. The "General Conditions of the Contract for Construction" is a set of guidelines that provides the administrative ground rules for executing the contract and performing the work.

Addenda

An addendum is the addition of omitted or updated information to a set of construction drawings, technical specifications, or a contract. Multiple addendums (addenda) issued during the bid process must be included as part of the contract. Addenda are issued when something has been omitted or modified from the original bid documents, such as any form of modification, clarification, or revision. Project addenda are issued in writing and sequentially issued with an explanation of each addendum that describes its purpose. **See Figure 7-9.**

Purchase Orders

A *purchase order (PO)* is a document with the details of a purchase that is often coded with a specific number. POs are commercial documents issued by purchasing managers who represent the owner, architect, or GC. POs indicate the product type, quantities, and agreed upon prices for materials and services. POs are used to control the purchasing of products and services from suppliers and distributors.

Change Orders

On almost every project, changes occur during construction. When changes occur, a written request, either on a standard form or in a letter from the contractor, is submitted to the designer (usually the architect) or owner. The written request gives a description of the work to be modified, the reason for the change, an itemized cost estimate, and the additional time required to complete the project if necessary. In many cases, the cost estimate is either an addition to or deduction from the original contract cost. Attachments to the proposal request can also include new or modified drawings and specifications, photographs, and/or test reports.

If the changes are accepted by the designer or owner, a change order is submitted to the contractor. A *change order* is a written form that details any variation from the original plans and specifications of a project as directed by the project owner or designer. **See Figure 7-10.** It is the only legal means available to change the contract documents after the contract has been awarded and signed. A change order is not valid until all parties have agreed and signed the form.

Building Permits

Most projects require that each trade register the project, obtain building permits, and pay fees to a local government or municipality. Permits and fees vary by county, state, or region. The fees paid to the municipality cover the cost for the review of construction documents and on-site inspections for conformance with specifications and drawings. When all the trades have completed the requirements of their specific building permits, the GC then submits a permit for occupancy. The building inspector and representatives of the fire department inspectional services must approve the permit for occupancy prior to users being allowed to occupy the building.

The GC is responsible for obtaining and paying for general building permits. Where specialty permits are required, the subcontractor is responsible for obtaining the correct and proper permits as part of the subcontractor agreement. Subcontractors are also required to notify the GC when inspections are forthcoming. In the case of hazardous material removal, such as the removal of lead paint or asbestos, a licensed and certified abatement subcontractor is required to perform the work. Subcontractors are required to file reports and documents to the appropriate agencies associated with the work.

Shop Drawings

The construction industry has a verification system for materials and equipment. A shop drawing is one part of that system. A *shop drawing* is a detailed drawing or set of drawings that shows how building elements are to be fabricated or installed and contains other pertinent information such as construction materials, finishes, and dimensions. Shop drawings are created by the contractor, subcontractor, supplier, fabricator, or manufacturer. Shop drawings are submitted to the designer for review.

Project Addenda

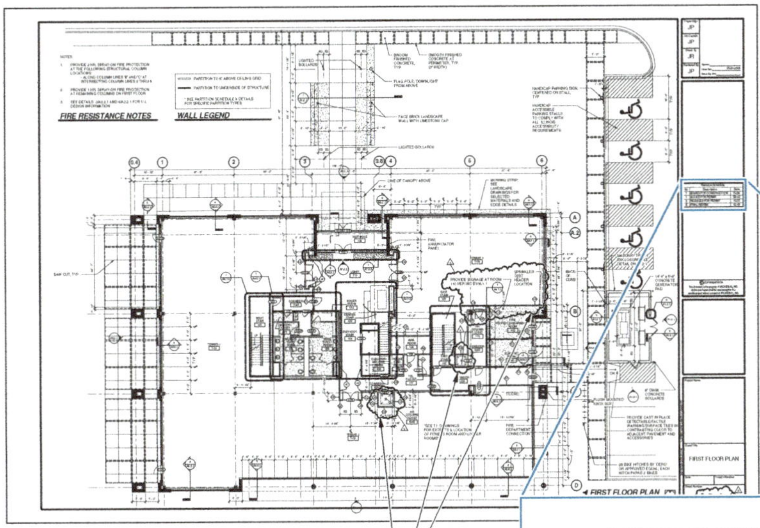

Revision Schedule		
No.	**Description**	**Date**
7	ISSUED FOR CONSTRUCTION	03.28
1	ISSUED FOR PERMIT	09.10
3	REISSUED FOR PERMIT	10.25
6	FINAL REVIEW	12.19

Figure 7-9. All project addenda are issued in writing and sequentially issued with an explanation of each addendum that describes its purpose.

Change Orders

Change Order

CO Number: C-1

Project: Address:	Cedar Falls Retail Outlet 3256 Ridge Road Creek Falls, KS	Distribution	☒ Architect ☒ Owner ☒ Cont
		Date:	Sept 22
To Contractor: Address	RA Miller & Son 451 Liberty Lane Creek Falls, KS	Project No.:	123-2017
		Contract Amount:	$285,350.00

The Contract is changed (provide and install all labor and materials) as follows:

Change Proposal No.	Description	Amount	(Add/Deduct)
CP-1	Additional Electrical Generator Stations	$ 1,350.00	Add
CP-2	Relocate door unit at Vestibule	$ 875.00	Add
CP-3	Delete wood blocking at Toilet Room 117	($ 225.00)	Deduct
**NOTE: See attached change proposals, sketches and backup information.*		$ 2000.00	Net

Not valid until signed by the Owner, Architect and Contractor.

The Original Contract Sum was:	$ 285,350.00
Net Change by previously authorized Change Orders:	$ 0.00
The Contract Sum prior to this Change Order was:	$ 285,350.00
The Contract Sum will be Increased (increased/decreased/unchanged) by This Change Order in the amount of:	$ 2000.00
The new Contract Sum including this Change Order will be:	$ 287,350.00

The Contract Time will be increased (increased/decreased/unchanged) by (-1-) days
The date of Substantial Completion as of the date of this Change Order therefore is July 14

Note: This summary does not reflect changes in the Contract Sum, Contract Time or Guaranteed Maximum price which have been authorized by Construction Change Directive.

Figure 7-10. A change order is a written form that details any variation from the original plans and specifications of a project as directed by the project owner or designer.

The work shown on shop drawings cannot begin until the drawings are returned with a review and approval stamp from the designer, indicating that the drawings fulfill the design criteria. For example, a steel fabricator prepares shop drawings that show steel beam and column sizes, connection details, plates, and bolt hole sizes. These drawings are reviewed by the structural engineer. Upon approval of the drawings, the steel fabricator can then proceed with the fabrication of the steel components.

Although the construction drawings show information regarding beams, columns, bearing walls, windows, elevators, roofing materials, and the mechanical, electrical, and plumbing (MEP) equipment, the shop drawings show the exact installation details. Shop drawings can be a drawing or set of drawings produced by a contractor for components that are usually prefabricated or they can be as simple as a sketch depicting placement and clearances around a piece of equipment. Shop drawings can also be presented in several formats, such as drawings, cut sheets, or material samples. Examples of VDV shop drawings include floor plan outlet identification sequential plans, rack and equipment elevations, and mounting heights of various AV/security and VDV equipment that are required for coordination between trades.

Submittals

As with shop drawings, a submittal is also an important part of the verification system for a construction project. Architects and engineers often specify equipment and materials by trade name or defined performance characteristics. It is the GC's responsibility to verify, prior to purchase and installation, that the materials and equipment meet the intent of the design. GCs perform this task by providing technical information and related literature for review and approval, which helps to guarantee that what is specified is what will be installed.

Information regarding the quality, quantity, and performance of materials and systems is needed to ensure that the project is built as specified. Product submittals and shop drawings illustrate functional characteristics of products and materials, any integrally related parts, and required dimensions and clearances. **See Figure 7-11.**

Project Schedules

A project schedule is a tool used by PMs to define and place in sequence a series of events to determine a timeline for the completion of each event. This is usually accomplished by using software programs that can integrate with the schedules of the GC and the skilled trades. Scheduled events for building construction projects are referred to as "activities." The amount of time needed to complete an activity is referred to as a "timeline duration." The overall timeline is the time allotted in the contract document to complete the entire project. The project schedule is part of the planning phase for a PM and requires careful thought and planning with the project team. Clear understanding of the construction processes, techniques, and sequencing of work is required when compiling a project schedule.

Submittal Reference Information

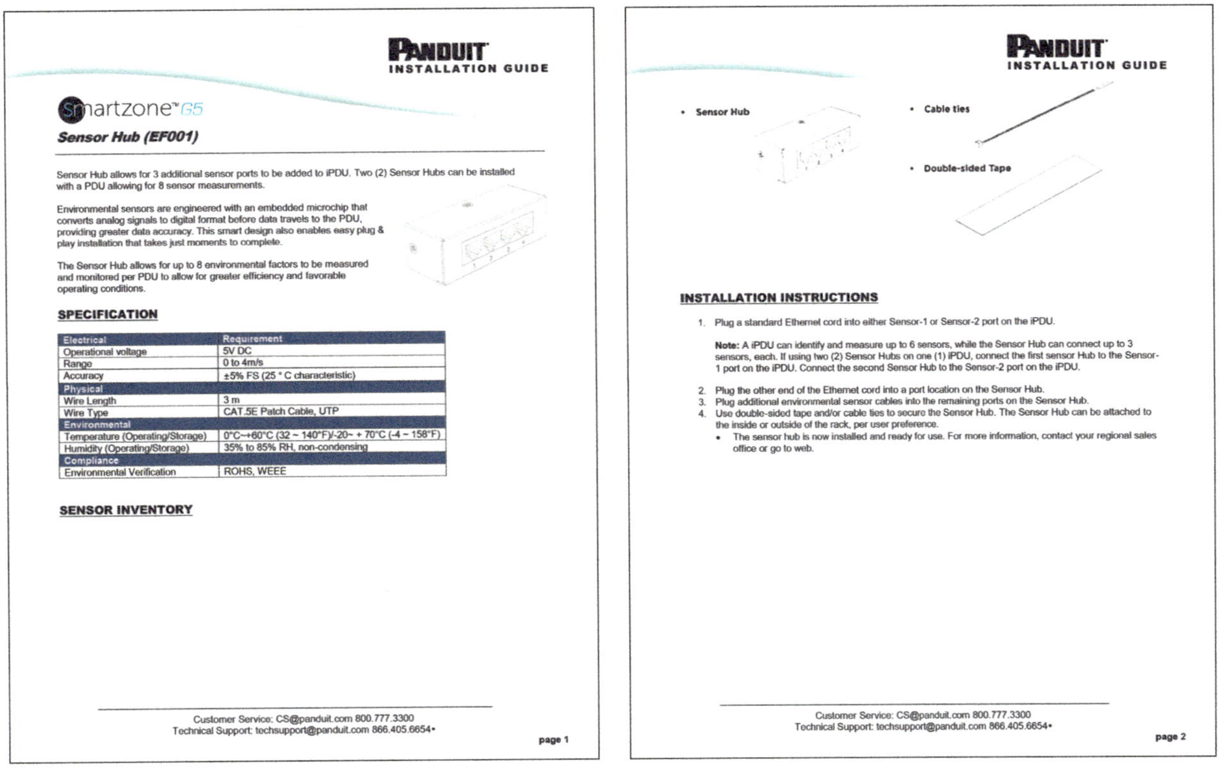

PANDUIT
INSTALLATION GUIDE

Smartzone™ G5

Sensor Hub (EF001)

Sensor Hub allows for 3 additional sensor ports to be added to iPDU. Two (2) Sensor Hubs can be installed with a PDU allowing for 8 sensor measurements.

Environmental sensors are engineered with an embedded microchip that converts analog signals to digital format before data travels to the PDU, providing greater data accuracy. This smart design also enables easy plug & play installation that takes just moments to complete.

The Sensor Hub allows for up to 8 environmental factors to be measured and monitored per PDU to allow for greater efficiency and favorable operating conditions.

SPECIFICATION

Electrical	Requirement
Operational voltage	5V DC
Range	0 to 4m/s
Accuracy	±5% FS (25 ° C characteristic)
Physical	
Wire Length	3 m
Wire Type	CAT.5E Patch Cable, UTP
Environmental	
Temperature (Operating/Storage)	0°C~+60°C (32 ~ 140°F)/-20~ + 70°C (-4 ~ 158°F)
Humidity (Operating/Storage)	35% to 85% RH, non-condensing
Compliance	
Environmental Verification	ROHS, WEEE

SENSOR INVENTORY

Customer Service: CS@panduit.com 800.777.3300
Technical Support: techsupport@panduit.com 866.405.6654•

page 1

PANDUIT
INSTALLATION GUIDE

- **Sensor Hub**
- **Cable ties**
- **Double-sided Tape**

INSTALLATION INSTRUCTIONS

1. Plug a standard Ethernet cord into either Sensor-1 or Sensor-2 port on the iPDU.

   Note: A iPDU can identify and measure up to 6 sensors, while the Sensor Hub can connect up to 3 sensors, each. If using two (2) Sensor Hubs on one (1) iPDU, connect the first sensor Hub to the Sensor-1 port on the iPDU. Connect the second Sensor Hub to the Sensor-2 port on the iPDU.

2. Plug the other end of the Ethernet cord into a port location on the Sensor Hub.
3. Plug additional environmental sensor cables into the remaining ports on the Sensor Hub.
4. Use double-sided tape and/or cable ties to secure the Sensor Hub. The Sensor Hub can be attached to the inside or outside of the rack, per user preference.
   - The sensor hub is now installed and ready for use. For more information, contact your regional sales office or go to web.

Customer Service: CS@panduit.com 800.777.3300
Technical Support: techsupport@panduit.com 866.405.6654•

page 2

Panduit Corp.

Figure 7-11. Technical information and related literature can be included with submittals for review and approval to help guarantee that the equipment specified is the equipment that will be installed.

Gantt Charts. Project schedules are typically tracked and monitored using a Gantt chart. A *Gantt chart* is a type of bar chart in which project tasks are represented by a bar or line, with one end representing the start date and the other end representing the completion date. A Gantt chart allows project participants to visualize a project timeline by transforming task names, dates, durations, start dates, and end dates into horizontal bar form. **See Figure 7-12.**

Milestones. Before construction on a project begins, the contractor fine-tunes the construction schedule by clearly identifying all significant progress points, also known as milestones. A *milestone* is a recognizable event that represents a meaningful completion or identifiable point in a project. These key points in time give the PM and the project team an idea of when vital events should occur. For example, the installation of the last piece of structural steel in a steel-framed high-rise building, also known as topping off, signifies the completion of the building's structural components and is considered a milestone. **See Figure 7-13.**

Although a milestone alone has no time duration, it is a point in the schedule that indicates significant work has been completed. A project needs to meet milestones in order to stay on track. Milestones validate how a project is progressing and provide an opportunity to evaluate the current status of the project as compared to the overall schedule.

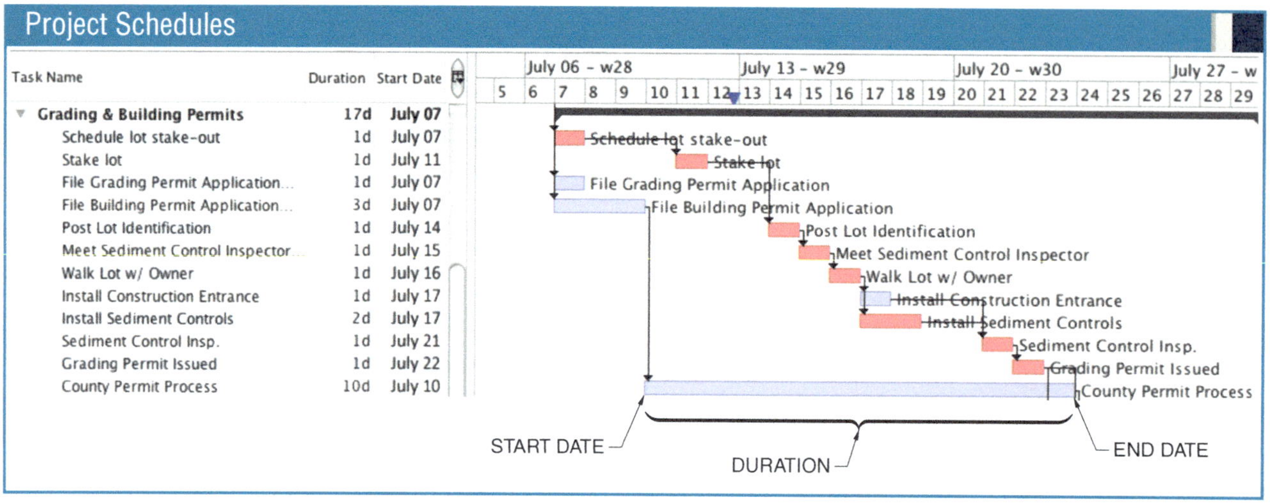

Figure 7-12. Project schedules are typically tracked and monitored through the use of a Gantt chart.

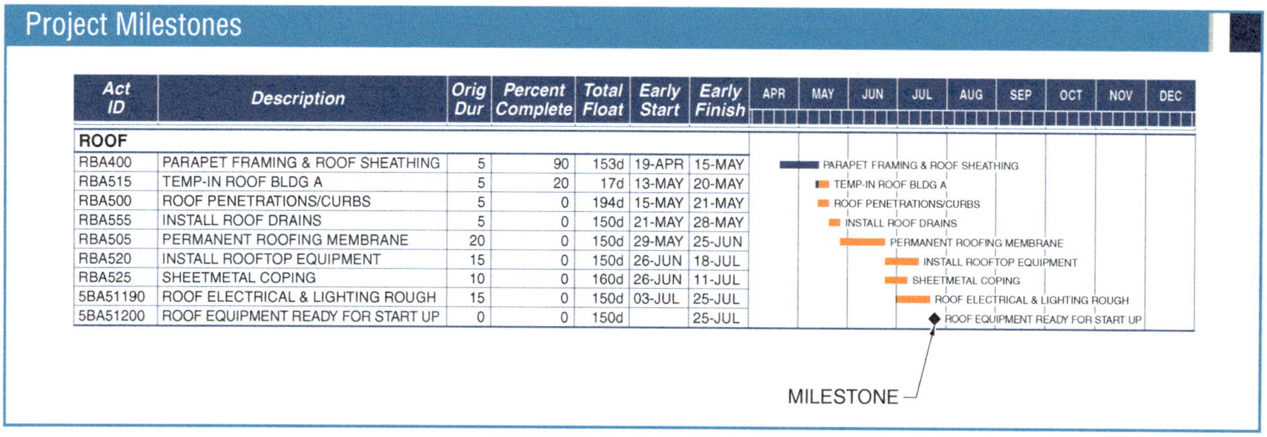

Figure 7-13. Project milestones are points in a schedule when significant accomplishments have been made.

PROJECT DOCUMENTATION AND RECORDKEEPING

Project documentation and recordkeeping is a requirement for all construction projects, no matter the size of the project. Circumstances can arise during construction activities that warrant documentation of how work was completed, how decisions were made, and who made the decisions. This documentation must be accessible in the event of legal implications with a project. In some cases, a contractor's performance, or lack of performance, can lead to legal action taken against them or their subcontractors. In the event of legal proceedings, logs and daily reports are considered a business record and can be admissible in court and included in the official record. Project documentation includes daily construction logs, requests for information (RFIs), time and material slips, communication records, cost tracking, and meeting minutes.

Daily Construction Logs

The progress of daily construction is often documented in a daily construction log. The daily construction log is a written account of the progress of a specific project or task, or for all activities accomplished and completed on a specific day. Daily construction logs are typically prepared by the field superintendent based on their supervision, along with written and oral communications with the foremen working on the project. Daily construction logs are primarily used to report to the PM, provide a daily record, and assist with forecasting the work schedule. **See Figure 7-14.**

Daily Construction Logs

DAILY LOG

DAY__________ 20___ MAY 2

CONTRACTOR____________________ JOB NAME____________________ JOB NO________________

Work Performed Today ____________________

Problems - Delays ____________________

Sub-Contractor Progress ____________________

Special Assignments ____________________

Extra Work	Authorized By	Approx. Price

Equipment Rented Today	Rented From	Rate

Supervisor's Signature ____________________

Weather __________

Temp AM ______ **PM** ______

Safety Meeting __________

Work Force	No.
Superintendent	
Clerk	
Bricklayers	
Carpenters	
Cement Masons	
Electricians	
Iron Workers	
Laborers	
Operating Eng.	
Plumbers	
Pipe Fitters	
Sheet Metal	
Truck Drivers	
Total	

Equipment	Hrs.

Material Purchased

Figure 7-14. Daily construction logs are used to report to the project manager, provide a daily record, and assist with forecasting the work schedule.

For a daily construction log to be effective, the following items must be documented daily:

- Work performed—The work performed on a specific day and any subcontractors used to perform work must be included on the daily log. The quantity of workers, a detailed summary of the work they performed, and where on the project the work took place must be documented. Any special equipment, lifts, safety concerns, and their supervision must also be noted.
- Overall progress and milestones—Schedule reports and delays caused by other subcontractors or the GC and milestones met or missed must be documented.
- Weather conditions—The weather conditions (precipitation, visibility, and temperature) must be recorded at the beginning, middle, and end of each shift. Any precipitation noted should be documented if it caused delay to the completion of any tasks that day.

Additional items that can be documented on a daily construction log include the following:

- visitors to the site, including the PM, owner, architect, inspectors, and the purpose and duration of their visit
- documentation of deliveries (or attempted deliveries) to the job site and any missing or damaged materials or equipment
- list of employees and subcontractors on site, the work they performed, and the hours they worked
- any unusual events that occurred; conversations conducted with the owner, PM, GC, architect or engineers; and reasons for any delays or work stoppages
- any industrial accidents or injuries to personnel, no matter how minor, and actions taken to prevent the accidents in the future
- forecasts and plans for future deliveries of materials and management of labor

Although handwritten daily construction logs are still used by many field supervisors, they can be complemented by software systems used to manage projects. Software systems allow the inclusion of photographs and digital videos to enhance the daily construction log. No matter which system is used, handwritten or electronic, the daily construction logs provide the project team with a guideline to document the past and assist in planning the future activities and desired milestones.

Requests for Information

A request for information (RFI) is a document used for clarifying scope and the coordination of project elements that arise during the construction process. On any construction project, most questions and queries can be solved through coordination with other trades, while others may require additional information from the architect or engineer. The documentation of RFIs is required in the construction management process and must be consistent. Each RFI must be sequentially numbered and added to the record in the form of an RFI log.

The purpose of an RFI log, or any log such as change order or submittal logs, is to document the process. The log tracks the ownership/responsibility of the request (or question) and when it has been sent, received, addressed, and returned. However, the time required for items to be answered can adversely affect the schedule. RFI logs provide a record that is accessible by all parties involved and recognizes the relationships between the owner/architect, PM, GC, and affected subcontractors.

Time and Material Slips

Time and material slips, sometimes referred to as time and materials log sheets, are documents used by GCs to help subcontractors complete work that they identify as out of the scope of the agreed upon contractual work or to authorize overtime to accelerate the schedule. Under this type of arrangement, the subcontractor is paid based on actual direct labor cost based on a predetermined rate, the cost of the materials and equipment used, and an agreed upon percentage for overhead and profit. Time and material arrangements are entered into as oral or written agreements and should be authorized on a daily/weekly basis by an authority with the ability to authorize payment to subcontractors, such as the field superintendent or PM. Time and material slips are typically converted into change orders and become part of the contract. **See Figure 7-15.**

Communication Records

Communication is the process of passing information from one party to another and is critical to the success of a project. The two forms of communication for construction projects are oral and written communications. Whether information is communicated between the PM and the project supervisors and crew or between the GC and owner, communication is a daily part of the project. For example, a PM relies on oral communication when relaying information to the project team each day.

Oral communication requires those receiving it to be careful listeners and requires repetition, feedback, and follow up. It must be positive and clearly state the objectives. Oral communication is typically in the form of face-to-face meetings or telephone conversations. Oral communication must be followed-up with written communication.

Written communication is especially important when multiple parties are involved. Email messages, text messages, hard-copy letters, memos, and other documents are considered written communication. The logs used for construction projects are also examples of written communication that provide summaries of major events, identify responsibilities, trace histories of events, and explain the reasons those events occurred.

Project Cost Tracking

Every contractor has a method for tracking and determining the costs to complete a project. Most cost-tracking methods are a component of an accounting software program that integrates with the estimating software. **See Figure 7-16.** As part of the project plan, a labor and material cost to complete the project is developed in conjunction with a schedule of values typically listed on AIA Document G702®. Percentages of completion for each line item are billed monthly for work completed within the billing time frame. Change orders and any other additional work may also be listed and billed as they are completed.

Time and Material Slips

TIME AND MATERIALS LOG SHEET

DATE:		CONTRACT:		UNIT:	USCGC
CONTRACTOR:		INSPECTOR:		SHEET:	OF
CHANGE REQUEST:	CR -	TITLE:			

EMPLOYEE NAME/TRADE	BADGE OR ID #	START TIME	STOP TIME	DESCRIPTION OF WORK

MATERIALS:	UNIT OF ISSUE	QUANTITY

Figure 7-15. Time and material slips are commonly converted into change orders and become part of the contract.

Labor and material costs are entered into the software program. Then, through analysis and by monitoring work while on site, the PM can assess the real-time cost to complete the project. Without oversight and the ability to judge the remaining work and time associated to complete the work, the cost to complete is only an estimate. To more accurately assess the real-time cost to complete a project, the PM and job-site supervisors must monitor past performance and apply it to future projections.

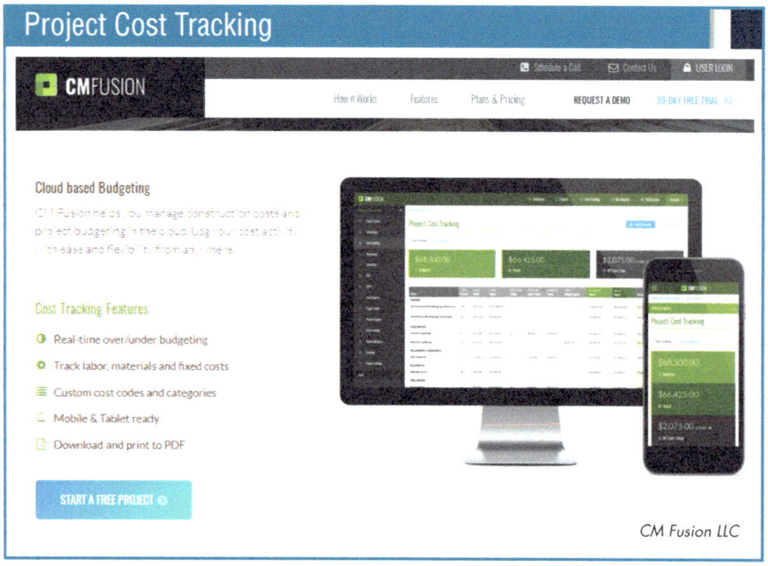

CM Fusion LLC

Figure 7-16. Most cost-tracking methods are typically a component of an accounting software program that integrates with estimating software.

Meeting Minutes

Every project typically has a weekly project status meeting on site, usually conducted by the GC and the project subcontractors. Most often, the architect, MEP engineers, and owner's representatives are present at such weekly meetings. The purpose of project status meetings is to encourage, maintain, and promote communication. Every project status meeting must be documented in the form of meeting minutes, which are a written synopsis of the meeting agenda, the names of the individuals present, the topics discussed, and any subsequent action plan or decisions related to the project. The meeting minutes are entered into the records of the project. **See Figure 7-17.**

Typical project status meetings are usually formatted and conducted as follows:

- The GC project representative proctors the meeting and controls and documents the agenda topics.
- The meeting minutes and agenda are completed immediately and distributed in advance, so corrective action can take place in a timely manner and prior to the scheduling of the next meeting.
- Meetings are scheduled for the same day and time each week and take place whether a team member is present or not.
- If a topic on the meeting agenda is not specific to most of the attendees and may require a large amount of time to resolve, such topics are typically discussed in a separate meeting between the affected parties, with the result reported back to the team.
- The minutes must be of substance and assign accountability to the parties involved.
- Unresolved items remain on the minutes until resolved.
- Minutes are formatted to identify the issue, the parties the issue affects, and the action required to resolve the issue.
- Meeting minutes are typically formatted to include the meeting number, date, time, location where they occurred, attendees, absentees, any parties affected by an issue, and any parties that should attend a future meeting.
- All meeting participants must vote to accept the previous meetings minutes, acknowledging their accuracy. Any discrepancies in the meeting minutes must be discussed and resolved, as required, at the time of the acceptance vote.

Meeting minutes of record must provide a disclaimer statement that the intentions of the meeting are intended to be true and an accurate part of the record. This encourages review and feedback, in writing, of any incorrect statements, assumptions, or omissions within a predetermined time period. Meeting minutes may also indicate that not every conversation and discussion is documented verbatim.

Meeting Minutes

Outline of Construction Meeting Notes

Item #	Date Initiated	Comments	Resolution	Responsible Party	Date Due	Date Complete	CCDs Outstanding
5		**Shop Drawings:** (Attached Shop Dwg. Log) Review of upcoming submission (by trade) Status of past submission (by trade)					
6		**Review of RFIs:** (Attach updated RFI log). Information required (by trade)					
7		**Quality of Work/Inspections:** Field Inspection Report Mock-ups Review of workmanship Other Inspections Walk-throughs QA Report Card findings					
8		**Document Maintenance/Updating:** Updated drawings Audit documentation Controlled inspection documentation					
9		**Change Order Status:** New CO conditions, completion of Discovery Memos Status of existing COs					
10		**Payment Status:** Delays in payments					
11		**Other Issues or Items:**					

Figure 7-17. Meeting minutes set the tone for accountability on a project and are an official record of the project, from start to finish.

SUMMARY

The project manager (PM) is the individual with the responsibility of managing a construction project. The PM is ultimately responsible to the project owner and maintains the schedule and budget of the project. The PM relies on the field superintendent and other members of the project team to fulfill the goals of the contract. The PM follows the project plan, which is written documentation of items that are required to successfully complete the project. Each aspect of the project must be documented as work progresses and entered into the project records. The project records are maintained daily and are the official documentation of the project from start to finish.

Project Management

CHAPTER 7

Review

Name ______________________________ Date ____________________

True-False

T F **1.** In many cases, project managers (PMs) are the only link between labor and management.

T F **2.** For construction projects, the purchasing manager is often the architect's representative.

T F **3.** A safety program is a requirement for commercial and industrial construction job sites only.

T F **4.** In some instances, construction operations can occur over a 24-hour period.

T F **5.** Apprentice programs typically occur over the course of three to five years, depending on the trade and licensing examination requirement.

T F **6.** The construction contract is the legal document for an agreement and covers all project information.

T F **7.** It is the field superintendent's responsibility to verify, prior to purchase and installation, that the materials and equipment meet the intent of the design.

T F **8.** Shop drawings can be a drawing or set of drawings produced by a contractor for components that are usually prefabricated or they can be as simple as a sketch depicting placement and clearances around a piece of equipment.

T F **9.** Construction status meetings are usually scheduled for the same day and time each week and take place whether a team member is present or not.

T F **10.** The purchasing manager rarely specifies a window of time or a specific date for when products can be delivered to a job site.

Multiple Choice

________________ **11.** Purchasing managers base their purchases on the ___, as well as any items noted during careful review of prints and drawings.
- A. best vendor quotes
- B. bill of materials (BOM)
- C. GC's preference
- D. PM's preference

________________ **12.** A safety ___ is the process of critically examining and evaluating the safety and occupational health hazards of a job site.
- A. audit
- B. inspection
- C. program
- D. summary

________________ **13.** A ___ is the general contractor's site representative who is responsible for the day-to-day operations and overall supervision of the construction job site.
- A. field superintendent
- B. foreman
- C. project manager
- D. safety manager

________________ **14.** The ___ form(s) the legal basis for the construction contract.
- A. bill of materials
- B. owner's equity
- C. print schedule
- D. specifications

________________ **15.** The daily construction log is typically prepared by the ___ based on their supervision along with written and oral communications with the foremen working on the project.
- A. engineer
- B. field superintendent
- C. GC
- D. project manager

Completion

________________ **16.** A(n) ___ is the contractor's representative who oversees the overall planning and execution of a construction project.

________________ **17.** The ___ for a project is the staff member(s) responsible for coordinating the purchases of various products and services in the form of purchase orders (POs).

________________ **18.** An experienced ___ is an asset to the project team because they understand how to negotiate a lower cost and reduce overall risk by ensuring timely deliveries and management of relationships with suppliers.

_______________ **19.** The ___ is a federal regulatory agency that develops and publishes safety and occupational health regulations and has legal authority to inspect workplaces, enforce regulations, and propose fines and penalties for violations.

_______________ **20.** A(n) ___ assigns duties and delegates tasks to the GC's workforce.

_______________ **21.** A(n) ___ is a training program that provides practical hands-on training accompanied by theoretical classroom instruction.

_______________ **22.** A(n) ___ is a written agreement between two or more parties.

_______________ **23.** A(n) ___ is a written form that details any variation from the original plans and specifications of a project as directed by the project owner or designer.

_______________ **24.** A(n) ___ drawing is a detailed drawing or set of drawings that shows how building elements are to be fabricated or installed and contains other pertinent information such as construction materials, finishes, and dimensions.

_______________ **25.** A(n) ___ is a recognizable event that represents a meaningful completion or identifiable point in a project.

Short Answer

26. List the skills and traits that a successful project manager (PM) must possess.

27. List the staff members on a typical construction project team.

28. List three key items concerning a job site that a PM must be familiar with and note prior to starting a construction project.

29. List six items that must be documented daily in an effective daily construction log.

30. Explain how time and material slips are used on a construction project.

CHAPTER

8

Using Drawings and Specifications to Develop Project Estimates

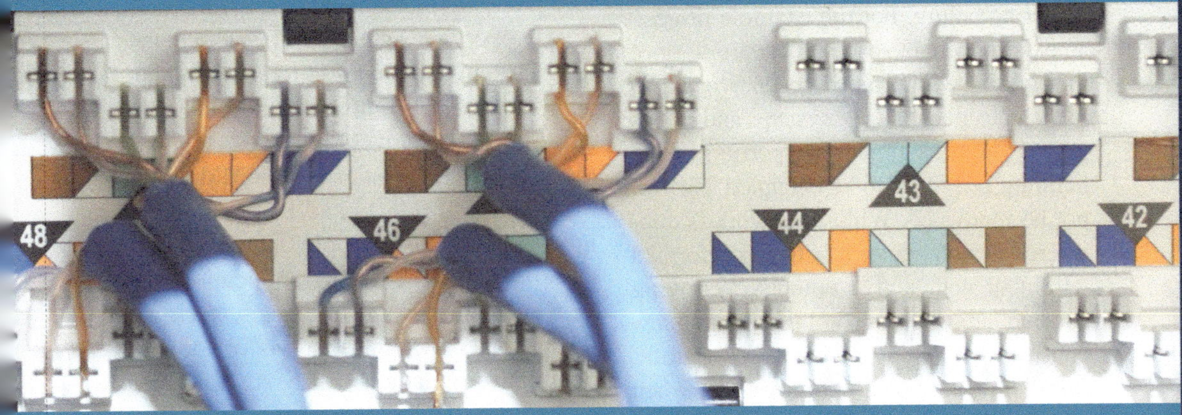

The ability to accurately read and interpret drawings, specifications, and other project instructions is required to construct commercial projects. By cross-referencing drawings specific to various trades, an estimator can visualize the intents of the architect, owner, engineers, and other members of the project team. By reviewing the areas within the building to be constructed and being knowledgeable with all elements provided for a project, an estimator can obtain the necessary information to formulate the cost to complete the work.

In addition to the estimator, the project leaders and journeyworkers for the VDV portion of a project must have the ability to interpret the scope of work in order to properly plan and construct the space in coordination with other trades and within the cost parameters established by the estimator. This is accomplished through analyzing drawings and specifications, applying known labor units, and factoring in past labor performance and known material costs.

Objectives

- Define "invitation for bid (IFB)" and describe the bid documents.
- Explain the requirements for the example commercial building renovation project, its IFB and project schedule, and the scope of work review.
- List the prints to be used for the VDV pathway plans of the example commercial building renovation project.
- Describe the site survey and observations for the example commercial building renovation project.
- List the project specifications for the example commercial building renovation project.
- Describe how the estimate is developed for the example commercial building renovation project.

Learner Resources

atplearningresources.com
access code 783821

PROJECT: COMMERCIAL BUILDING RENOVATION

A commercial building renovation project is a typical example of a VDV installation project. This type of project can be used to emphasize the possible applications involving printreading. For the example project, an existing tenant space on the 3rd floor of the building will be renovated in two phases to include updated VDV systems and equipment, while the remaining portion of the building continues to be occupied and will not be renovated. While this work is considered new, much of it is to be performed within or near active tenants occupying adjacent areas. This is not considered new construction, but it is considered retrofit. Installers must use good judgement when performing obstructive work such as drilling and other loud activities adjacent to occupied areas.

Commercial building construction is described as either core and shell or tenant fit-out. Core and shell construction includes the exterior structure and interior construction of the building. Interior construction includes the stairwells, elevators, and utility services, such as MEP/fire alarm and communication systems, which are distributed vertically to the tenants that occupy the building. The completed core and shell allows occupants to move between the floors of a building and distributes services efficiently to the tenant floors.

Tenant fit-out construction is the process of constructing interior tenant space suitable for occupancy. The tenants who lease commercial office space often have the spaces constructed to their specifications and desired finishes. Some tenants may occupy a previous tenant's already-constructed space and reuse the space "as is" to minimize relocation costs. The work for new tenant fit-out spaces includes constructing walls, partitions, doorways, and ceilings; installing systems furniture; painting; and installing MEP, VDV, security, and access control systems.

In a commercial building, the premises distribution system is a preplanned network of cabling within the building designed for the transmission of VDV network services. The system connects either to external local exchange carrier services for voice and data bandwidth connections or to a customer/owner centrally located main distribution location.

A properly designed VDV system must meet current network requirements and be able to easily accommodate both expansion and future moves, adds, and changes (MACs). A VDV system must also support multivendor products and be application independent. For example, when a VDV system is designed to be application independent, hardware interfaces, such as RJ-45 for copper installations or LC for fiber-optic installations, can be standardized and are able to support multiple platforms and OEMs.

For the example commercial building renovation project, a bid must be submitted to be considered for a contract to perform the work. An *invitation for bid (IFB)* is a request for contractors, subcontractors, and suppliers to submit pricing for services and/or materials for a project. An IFB includes a description of the work, bid opening information, legal requirements, and any specific bidding instructions. IFBs are used for various private and public projects with clearly defined specifications and scopes of work.

A *bid* is an offer from a contractor, subcontractor, or supplier to perform construction work, provide a construction service, or provide materials or supplies at a stated price. The first step in developing a bid or cost estimate for any project is to thoroughly review the contract bid documents. The bid documents include instructions to bidders and bid forms, which provide necessary information for the bidding contractor to estimate and bid a project. **See Figure 8-1.** Bid drawings, which are construction drawings used to bid a project, are detailed graphic representations of the work.

Construction drawings include basic information such as dimensions, notes, and schedules. The project manual includes written information such as bidding information, technical specifications for the construction products and systems, and terms and conditions of the contract. The project manual may also include a section that outlines the procedures for bidding the project. This section is often titled "Instructions to Bidders."

The bid documents include a bid form. A *bid form* is a document that contractors are required to use when bidding a project. It includes space to enter project identification, prices, quantities, substitutions, and bidder contact information. The bidder completes the bid form, signs and seals it, and then submits it to the owner or construction manager at the designated time and location as described in the instructions to bidders. This provides the owner with a simple and uniform method for comparing and evaluating competing bids. The bid can be used to develop the project schedule and review the project estimate and the scope of work.

Commercial Building Renovation Project Requirements

For the example commercial building renovation project, the space to be renovated is the third floor of a three-story existing building with steel framing and 4″ thick concrete floors. The floor-to-deck height is 16′-0″, with steel beams located at a height no lower than 14′-0″. The finished acoustical ceiling height is 10′-0″ above finished floor (AFF). The plans provide an overview of the existing conditions and depict the final layout of the space. On-site parking is available, and shared access to a freight elevator during normal business hours is provided. Occasionally, tenants are required to perform some work after normal business hours.

Bid Documents

INSTRUCTIONS TO BIDDERS

PLEASE READ ALL INSTRUCTIONS CAREFULLY BEFORE PREPARING AND SUBMITTING YOUR BID.

All bids shall be prepared and submitted in accordance with the following requirements. Failure to comply with any requirements shall cause the bid to be considered irregular and shall be grounds for rejection of the bid.

1. The bid form furnished by MUNICIPALITY NAME with the proposal shall be used and shall not be altered in any manner. DO NOT SEPARATE THE BID FORM FROM THE PROPOSAL.
2. All entries on the bid form, including signatures, shall be written in ink.
3. The Bidder shall submit a unit price for every item on the bid form. The unit prices for the various contract items shall be written in figures. UNIT PRICES MUST BE LIMITED TO TWO DECIMAL PLACES.
4. An amount bid shall be entered on the bid form for every item. The amount bid for each item shall be determined by multiplying each unit bid by the quantity for that item, and shall be written in the "Amount Bid" column of the form.
5. The total amount bid shall be written in figures in the proper place on the bid form. The total amount shall be determined by adding the amounts bid for each item.
6. Changes in any entry shall be made by marking through the entry in ink and making the correct entry adjacent thereto in ink. A representative of the Bidder shall initial the change in ink. Do not use "White Out" or similar product to make corrections.
7. The bid shall be properly executed. All bids shall show the following information:
   a. Name of individual, firm, corporation, partnership, or joint venture submitting bid.
   b. Name of individual or representative submitting bid and position or title.
   c. Name, signature, and position or title of witness.
   d. Federal Identification Number
   e. Contractor's License Number (if available)
8. Bids submitted by corporations shall bear the seal of the corporation.
9. The bid shall not contain any unauthorized additions, deletions, or conditional bids.

INSTRUCTIONS TO BIDDERS

Division 00 - Procurement and Contracting Requirements
Section 00 41 43 - BID FORM

PROJECT TITLE: ______________________________

BIDDER'S NAME: ______________________________

BIDDER'S CONTACT: ________________ PHONE #: ______________

The undersigned bidder declares that it has read the specifications, understands the conditions, has examined the site, and has determined for itself all situations affecting the work herein bid upon. Bidder proposes and agrees, if this proposal is accepted, to provide at bidder's own expense, all labor, machinery, tools, materials, etc., including all work incidental to, or described or implied as incidental to such items, according to the Contract Documents and that the bidder will complete the work within the time stated, and that bidder will accept in full payment therefore the unit price(s) set forth below:

PROPOSED PRICE.

ITEM NO.	QTY	DESCRIPTION OF ITEMS	UOM	UNIT PRICE	AMOUNT

BID ALTERNATES

ITEM NO.	QTY	DESCRIPTION OF ITEMS	UOM	UNIT PRICE	AMOUNT

Evaluation of Bids. The evaluation of bids and determination of the low responsive bid will be based on prices bid for items ______ in the above table.

Addenda. Bidder acknowledges review of all Addenda through No. ______ .

Bid Security. A certified check, cashiers check, or other obligation of a bank, or a bid security bond in substantially the form set forth in section 00 43 13, for at least 5% of the total bid without sales tax, accompanies this bid.

Contract Time. The Contract, in its entirety, shall be Substantially Complete within ______ calendar days, dating from date of award of the Contract, and shall be Finally Complete within 30 calendar days of Substantial Completion.

Project No.
Contract No.

BID FORMS

Figure 8-1. Bid documents include instructions to bidders and bid forms.

The example commercial building renovation project has specific VDV system requirements. These include horizontal wiring VDV outlets in drywall partitions modular furniture VDV outlets, floor-mounted VDV outlets, and ceiling- and wall-mounted WAPs. Equipment rooms must be equipped to house cable runways, network and equipment racks, patch panels, cable management hardware, and other connecting hardware.

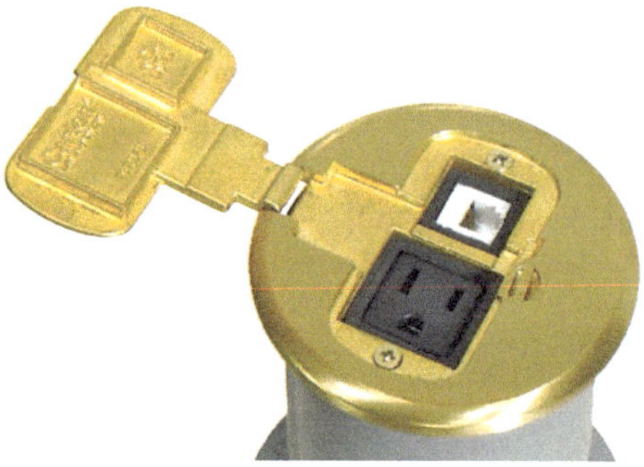

Carlon®-Thomas&Betts

A building renovation project may include the installation of items such as floor outlets.

The IFB and Project Schedule

The first step for the example commercial building renovation project is to review the IFB, scope of work, and project schedule. A copy of the scope of work and project schedule is included in the IFB as a reference for the bidders to analyze the time frame to complete various tasks compared to the estimate and in coordination with other trades. It is assumed that all related work required by other trades to complete the installation is on schedule, with no delays. The project schedule for completion of the example commercial building renovation project is 24 weeks.

Scope of Work Review

Prior to submitting a bid for any project, an estimator must perform a thorough review of the scope of work. For the success of the overall project, the estimator must be familiar with the scope of work and identify any inconsistencies or mistakes between the scope of work and the project schedule. The review begins by verifying that the dates marked on the prints and specifications match. For example, if it is determined that other trade prints, such as those used for carpentry and drywall installations, are dated differently than the prints for the electrical contractors, an inquiry for the reason of the discrepancy must be conducted prior to commencing work on the fabrication and installation of components.

VDV SYSTEM PRINTS

For the commercial building renovation project, the detail drawings within the VDV system prints contain information such as OEM part numbers, descriptions, and quantities for the wiring, hardware, and finish materials to be installed. For example, there may be a print for specific details, such as outlet and faceplate details. **See Figure 8-2.** As with any trade prints, the specific dimensions for wiring, hardware, and finish materials are not always provided on the detail drawings. Rather, dimensions are obtained from floor plans, riser diagrams, elevation drawings, and information details only available from other trade prints.

The example commercial building renovation project is a two-phase project to renovate the third floor of an existing three-story commercial office building. Phase one (P1) requires the installation of all new wiring back to the existing equipment room located in the Phase 2 (P2) space to be renovated. The existing (P2) equipment room is to stay active during P1. The following criteria applies to P2:

- P2 is the construction of a new second equipment room and starts after the completion and decommissioning of the original equipment room in P1.
- The construction of P2 requires that all new wiring for P1 be relocated from the existing equipment room during P2 to the

newly constructed equipment room used during P1.

- The building owner occupies the existing space on the first floor, while the second floor is occupied by a different tenant.
- All work performed in the ceiling of the second floor must be completed during off hours as not to interrupt or interfere with the existing tenant.

The P1 space is approximately 6000 sq ft and the P2 space is 9500 sq ft, for a total project square footage (area) of 15,500 sq ft. The area of any space can be calculated by referencing the distances on the floor plan scale. An alternative method to calculate lengths and other measurements is to measure the distance between columns.

The structural columns for the example commercial building renovation project are identified as A through D (listed downward from the top left of the drawing) and 1 through 10 (listed from left to right at the top of the drawing). The area between columns 1A and 2B can be calculated by first measuring the dimensions on the print listed for these columns: 21′-9″ and 27′-6″. These dimensions are then multiplied together for a square footage of roughly 598 sq ft (21.75 × 27.5 = 598.125). **See Figure 8-3.**

The VDV system prints include floor plans, furniture plans, cable pathway layouts, and notes that require referencing during the project. Other written items that must be referenced throughout the project include VDV pathway plans, a site survey, and project specifications.

VDV Drawing Sheets, Legends, and Notes

For the example commercial building renovation project, the first drawing sheet, *Drawing T-001 Tel Data Legend and Notes*, provides information regarding the owner's name, architect's name, drawing list, and drawing revision column. Two important features of the drawing are the "General Pathway Requirements" and the "Communications Legend." It can be determined by outlining the pathway requirements on the first sheet that the pathways for the wiring are part of the scope of review. **See Appendix.**

It can also be determined by reviewing Notes 1 through 20 that a variety of different types and sizes of conduit are required for this project. The communication legend provides information regarding the outlet and equipment room (ER) symbols used on the floor plans and a description of what each symbol represents. The outlet and ER requirements for the project are listed on the drawing and must be verified against the requirements outlined in the project specification. **See Figure 8-4.**

VDV System Prints

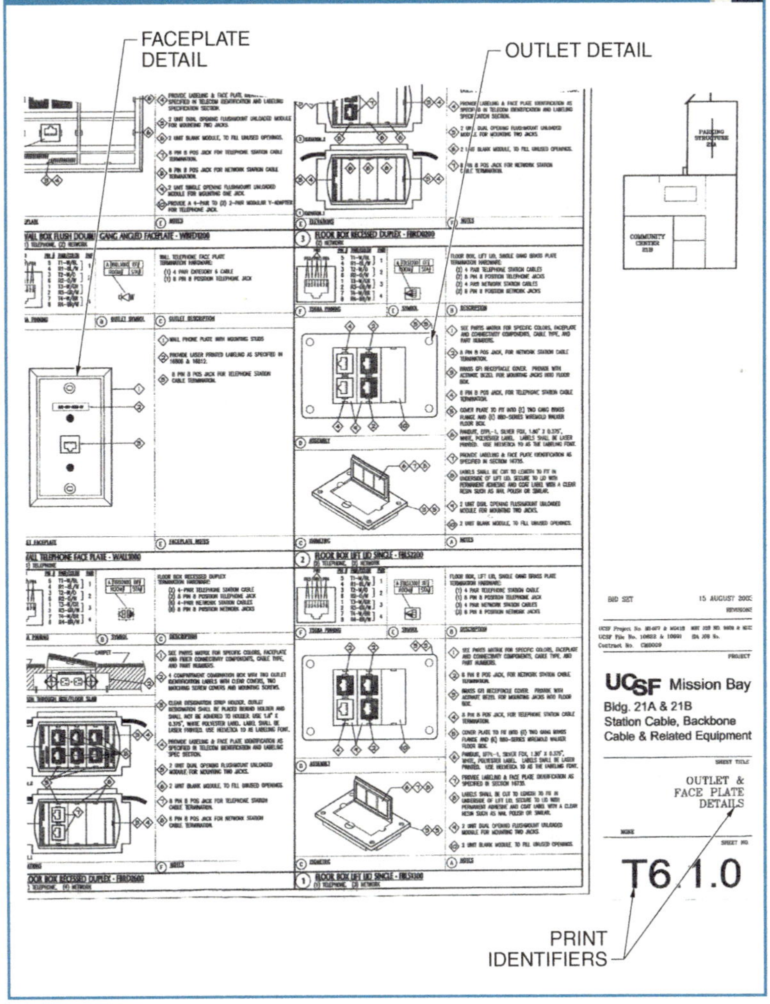

Figure 8-2. The detail drawings within the VDV system prints for the example commercial building renovation project may contain specific information such as outlet and faceplate details.

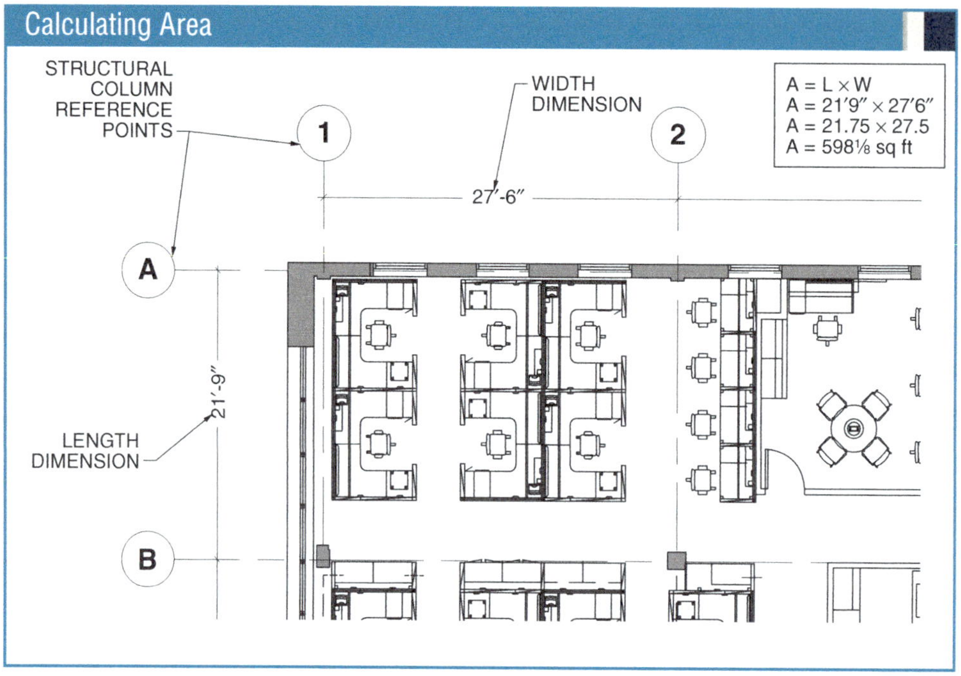

Figure 8-3. The area of any space can be calculated by referencing the prints and using the distances between structural columns to multiply the length of a space by its width.

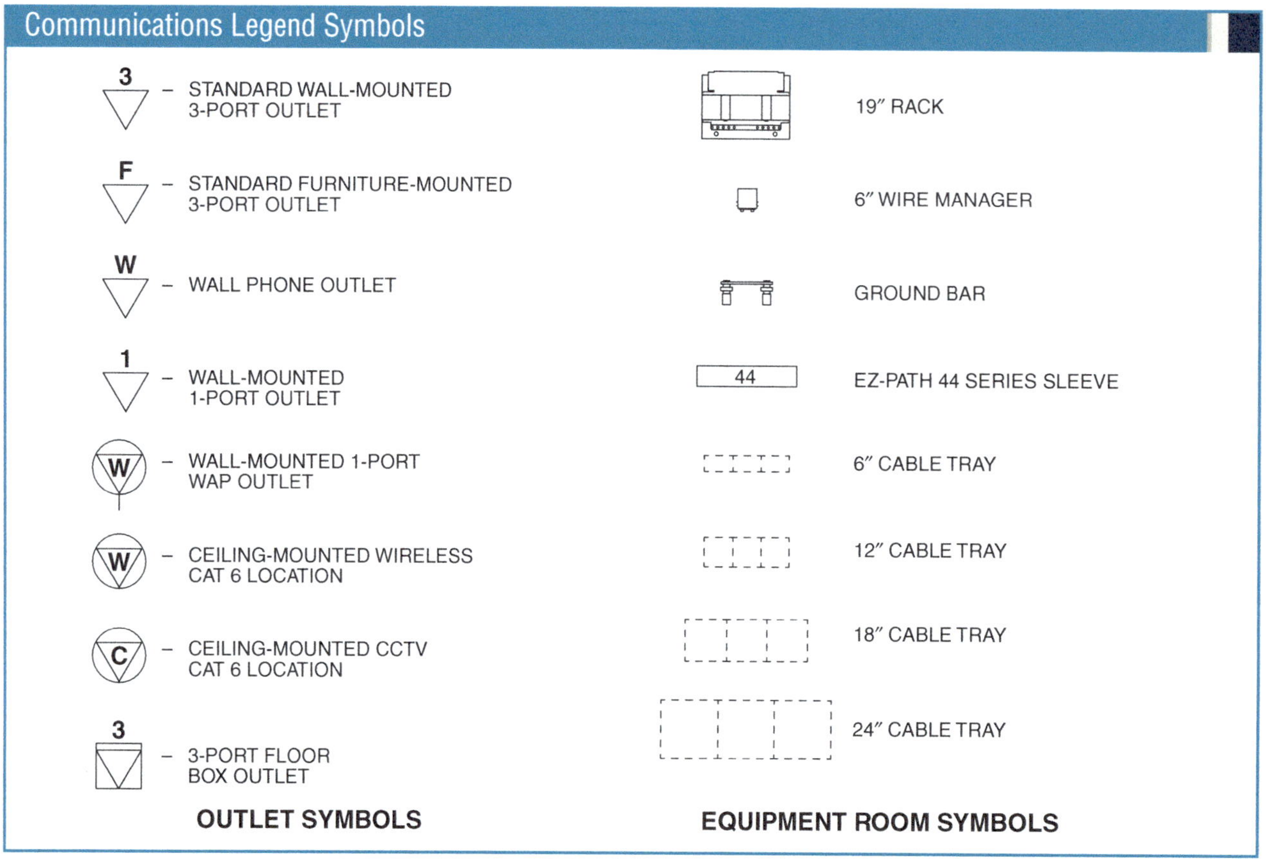

Figure 8-4. The communication legend provides information regarding the symbols used on the floor plans and a description of what each symbol represents.

VDV Pathway Plans

For the example commercial building renovation project, there are five prints concerning VDV pathways to be reviewed by the estimator. These five prints are *Drawing T-103 Tel Data 3rd Floor Pathway Plan, Drawing T-203 Tel Data 3rd Floor Coordination Plan, T-303 Tel Data 3rd Floor Existing Plan, T-403 Tel Data 3rd Floor Plan*, and *T-503 Tel Data 3rd Floor Details*. These prints must be reviewed in detail and referenced throughout both phases of the project for the proper installation of pathway systems.

> **TECH FACT**
>
> When communications cables are to be installed in or near plenum spaces, NEC® Section 300.22 and Subsection 725.3(N) should be consulted for further clarification on compliance.

T-103 Tel Data 3rd Floor Pathway Plan. The first VDV pathway print to be reviewed, *T-103 Tel Data 3rd Floor Pathway Plan*, provides an overview of the finished space dimensions and has a vertical demarcation line between column lines 5 and 6 that segregates the P1 and P2 sequences of work. This plan also provides a view of the horizontal cable tray to be installed above the finished ceiling (which provides a pathway for P1 and P2 wiring). *Note:* In addition to providing the location for the tray, *T-103 Tel Data 3rd Floor Pathway Plan* also notes that the tray changes in depth size from one end to the other. Additionally, a symbol for fire-rated sleeves is depicted for areas where the cable tray penetrates through rated walls.

Other items to note on this drawing include the following:

- Multiple conduit stub-ups varying in size provide pathways from the drop-in ceiling, through walls, and down to the floor below to serve floor boxes and systems furniture access, where there are no walls to provide access. This also indicates that to perform this work, access is required in the tenant space on the floor below.
- A 2″ conduit from a junction box originating in Mechanical Room 3007 (between column lines 4 and 5) and routed to the cable tray is to be installed. Coordination between the installer and building owner is required before installing the 2″ conduit. **See Figure 8-5.**

Because of these notations, additional study of the documents is required. This must be noted as a scope of work review item to verify when reviewing the project schedule.

T-203 Tel Data 3rd Floor Coordination Plan. The second VDV pathway print to be reviewed, *T-203 Tel Data 3rd Floor Coordination Plan*, includes details for multiple devices to be installed on the third floor. The details are as follows:

- Detail 01 – Typical Cable Tray Support and Clearance Detail depicts the installation of the cable tray intended for use above the acoustical ceiling. It provides notes on installation, as well as side and isometric views of the raceway. Mounting hardware, threaded rod dimensions, and miscellaneous hardware are also detailed.
- Detail 02 – Premise Pathway Diagram illustrates the vertical distribution pathway required for the backbone cabling between the renovated tenant space and the existing main distribution frame (MDF) located on the first floor (level 1). Symbols are used to represent grounding busbars, conduit sections, fire-rated sleeves, and cable trays.
- Detail 03 – Series 44 3 Gang Wall Penetration and Detail 04 – Series 44 Single Wall Penetration depict isometric views and details of the quantity and type of wall-penetrating fire-rated sleeves.
- Detail 05 – Poke Thru Detail – 8AT or Equal and Detail 06 – Poke Thru Detail – RC9 or Equal illustrate the specified poke thru details for floor-mounted outlets and floor-mounted access for systems furniture outlet locations.
- Detail 07 – Wall Box Details illustrates the rough-in wall boxes for telecommunications outlets and systems furniture outlets wall access.
- Detail 08 – Typical Telecommunications Ground Bus Detail (TGB) & (SGB) provides an isometric view of the grounding busbars, busbar mounting guidelines, and additional notes to the installer.

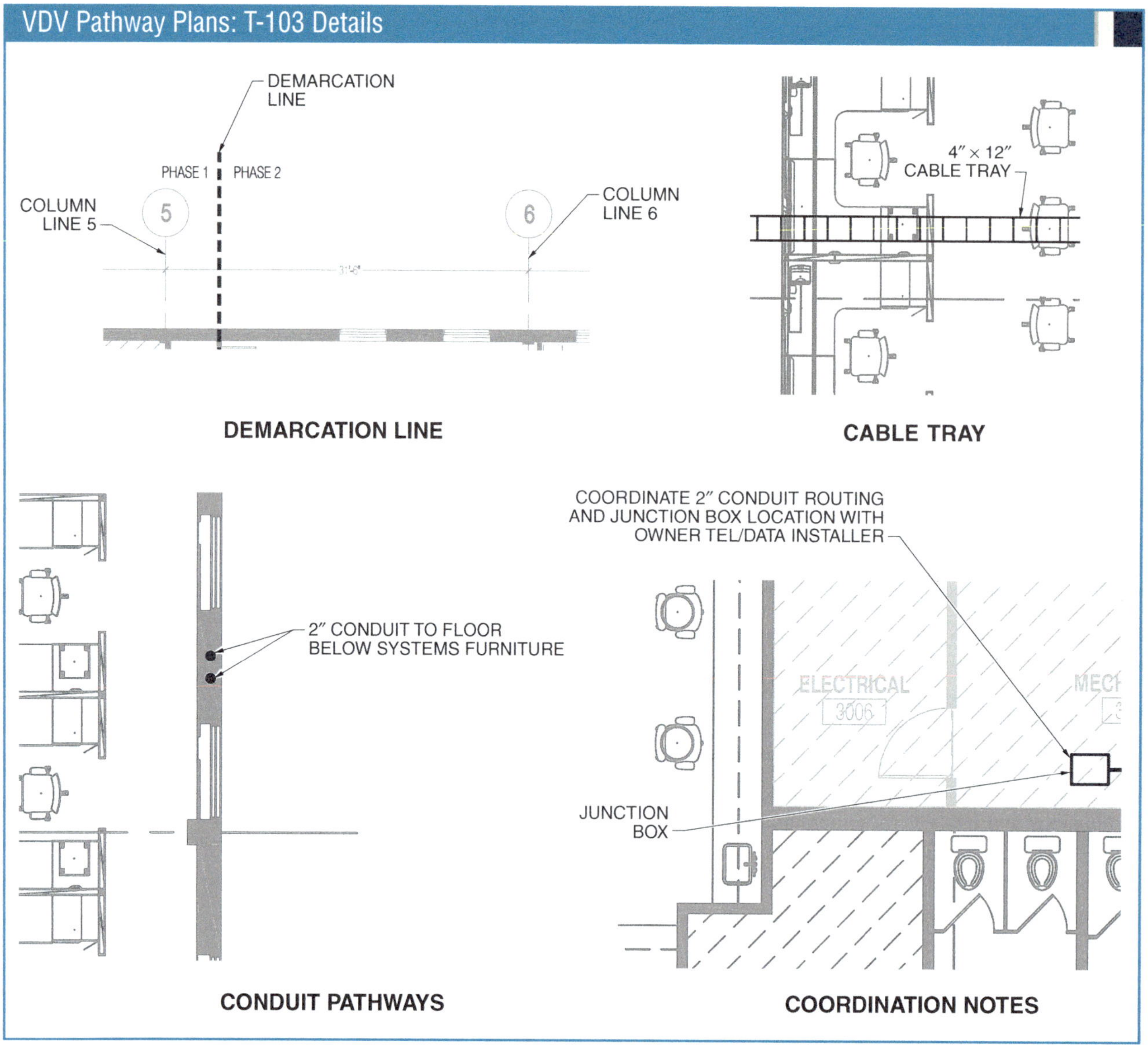

Figure 8-5. Drawing T-103 in the VDV pathway plans for the example commercial building renovation project includes details on items such as the demarcation line, cable tray locations, conduit pathways, and coordination notes.

The poke thru details and wall box details shown on the prints indicate that these items penetrate floors and provide a fire rating to maintain the overall fire rating of the floor. These devices are available in a variety of sizes and finishes and are designed to accommodate multiple applications. The conduit, boxes, and rings provide support for outlets and wall feeds, are covered by gypsum wallboard, and are not commonly noticed or visible to the owner or end user. **See Figure 8-6.**

T-303 Tel Data 3rd Floor Existing Plan. The third VDV pathway print to be reviewed, *T-303 Tel Data 3rd Floor Existing Plan*, represents the scope of work for the demolition to be performed for the project. This plan also provides the location of the existing equipment room, identified as IDF (intermediate distribution frame) Room 3033, to be utilized for the P1 wiring. Additionally, demolition and construction instructions are included on the drawing as "drawing notes."

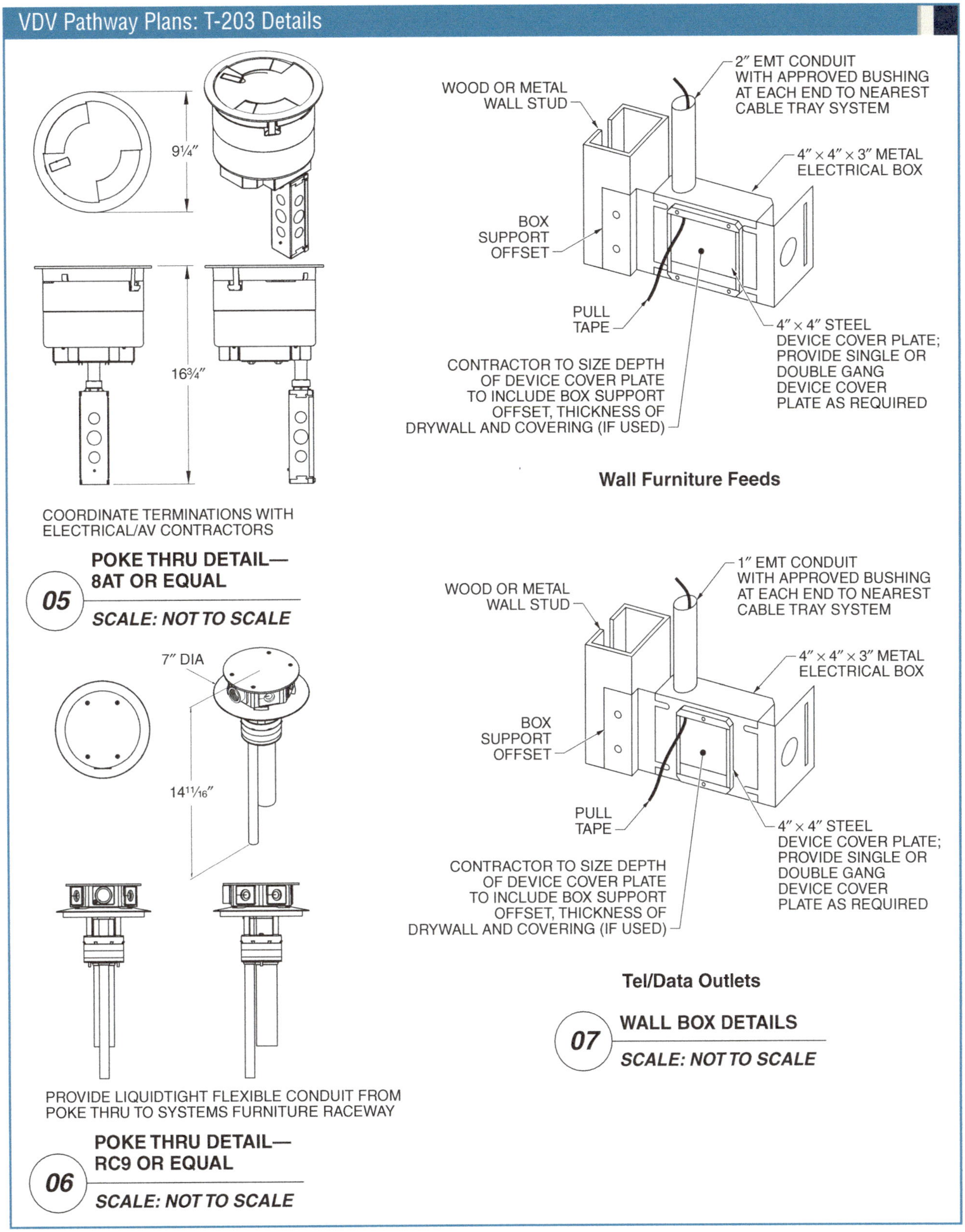

Figure 8-6. Drawing T-203 in the VDV pathway plans for the example commercial building renovation project illustrates the specified poke thru details for floor-mounted outlets, the floor-mounted access for systems furniture outlets, and the rough-in wall boxes for telecommunications outlets and systems furniture outlets wall access.

T-403 Tel Data 3rd Floor Plan. The fourth VDV pathway print to be reviewed, *T-403 Tel Data 3rd Floor Plan*, provides installation details indicating the locations of wall-mounted, floor-mounted, ceiling-mounted, and systems furniture communications outlets. This print also depicts the P1 and P2 sequencing, shows the location of the new IDF room serving the renovated space, and illustrates the quantity of outlets required for reference purposes during the development of the estimate and the installation. **See Figure 8-7.**

VDV Pathway Plans: T-303 and T-403 Details

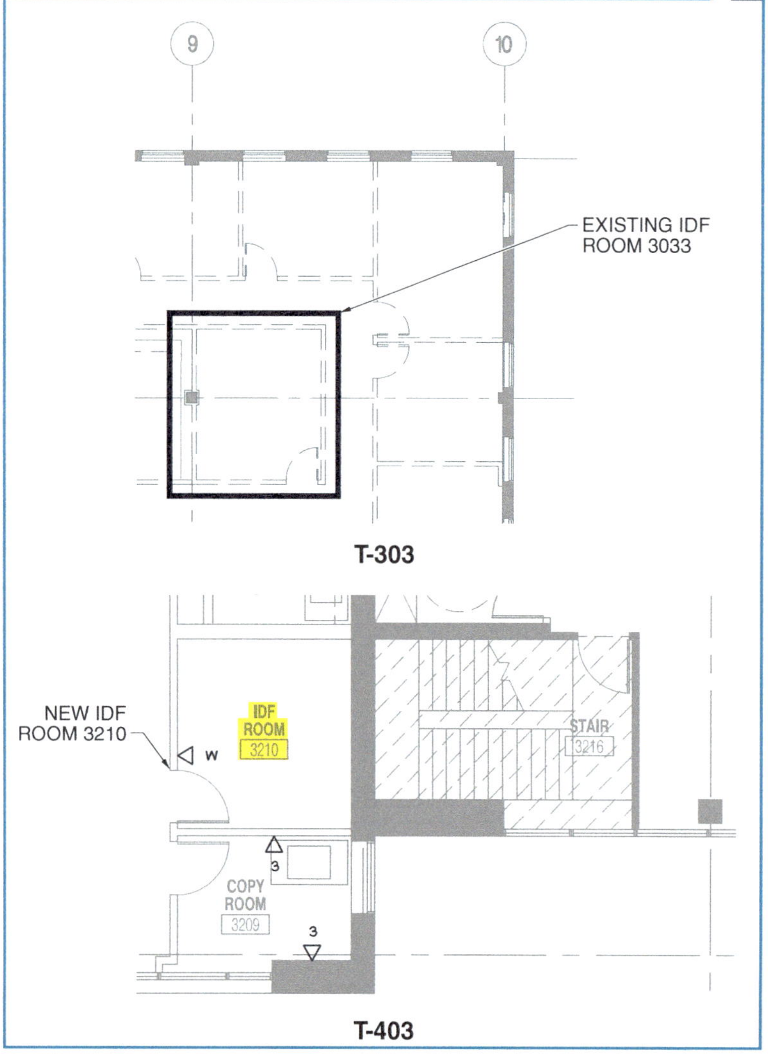

Figure 8-7. Drawing T-303 in the VDV pathway plans for the example commercial building renovation project includes the scope of work for demolition, while Drawing T-403 includes details for the new installation work to be performed.

T-503 Tel Data 3rd Floor Details. The fifth VDV pathway print to be reviewed, *T-503 Tel Data 3rd Floor Details*, provides a visual reference for the finished equipment with additional details of the outlets and an overview of new IDF Room 3210. The details are as follows:

- Detail 01 – TracJack 3 Port Plate shows a standard three-port wall plate, categorized by performance (CAT) type, color, use, and specific part number. (TracJack is a product manufactured and distributed by Legrand North America, LLC.)
- Detail 02 – TracJack provides, in detail, the outlet part number, description, and icon color and type.
- Detail 03 – Furniture Plate illustrates the furniture outlet assemblies to be installed in the systems furniture raceways and finish systems by type, color, use, and specific part numbers. **See Figure 8-8.**
- Detail 04 – Wall Phone Plate shows the type of outlet and standard identification number for wall phones.
- Detail 05 – Wireless Access Point Plate and Detail 07 – Surface Mount Wireless/CCTV Outlet show the wall- and surface-mounted wireless access point (WAP) and CCTV outlets.
- Detail 06 – Outlet Identification depicts the outlet identification sequence used within the premises. Detail 06 is also used to provide information and instructions for the preinstallation and postinstallation of outlets for the installer and end user.
- Detail 08 – J-Hook Cable Support Detail shows an isometric view of the J-Hook cable support for all wiring installed outside of the cable tray raceway.
- Detail 09 – IDF Room 3210 Equipment Layout is a plan view of new IDF Room 3210. Detail 09 provides locations of equipment, cable runways, and fire-stop sleeves. Equipment clearances are noted so that proper space can be provided after the installation of the owner's network equipment. **See Figure 8-9.**
- Detail 10 – IDF Room 3210 Racks Layout provides an equipment rack elevation for new IDF Room 3210 that indicates the placement of all copper and fiber-optic

connecting hardware, as well as space allocated for the owner's equipment.

- Detail 11 – IDF Room 3210 Fiber Panel Layout shows the fiber-optic termination unit that is to be installed in rack #3 in IDF Room 3210 at rack units U41 and U40. The owner's fiber-optic backbone enclosure provides space for existing 12-strand primary distribution fiber and future secondary distribution fiber for miscellaneous future applications.
- Detail 12 – IDF Room 3033 Racks Layout shows the location of the P1 wiring that is to be temporarily serviced out of existing IDF Room 3033. Upon completion of P1 IDF Room 3033 and the completion of P2 IDF Room 3210, the outlets are to be relocated to the new room as marked in the detail drawing.
- Detail 13 – IDF Room 3033 Fiber Panel Layout provides useful information about existing and future fiber-optic backbone (primary and secondary) cabling required during each phase of the project.

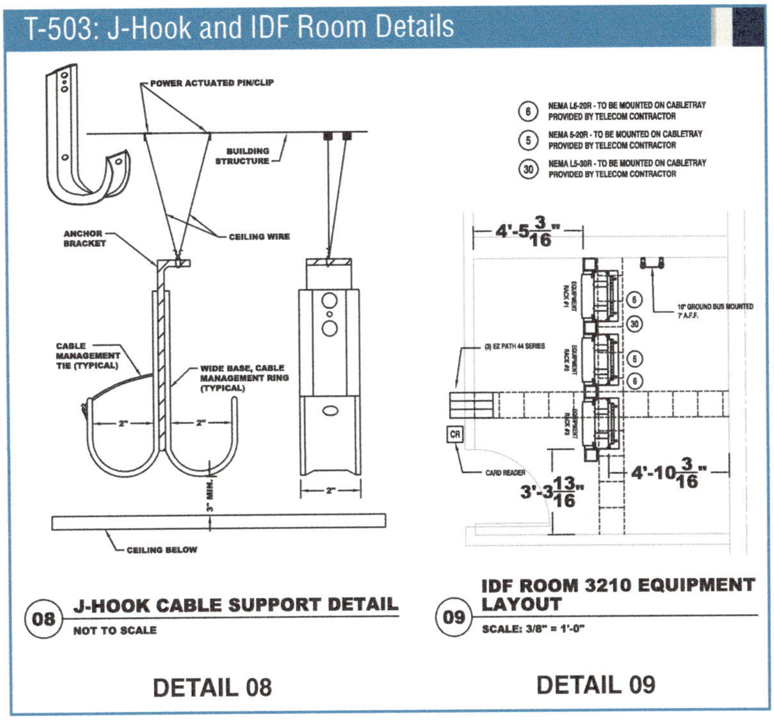

Figure 8-9. Details 08 and 09 in Drawing T-503 for the example commercial building renovation project illustrate the details for J-Hook cable supports and new IDF Room 3210.

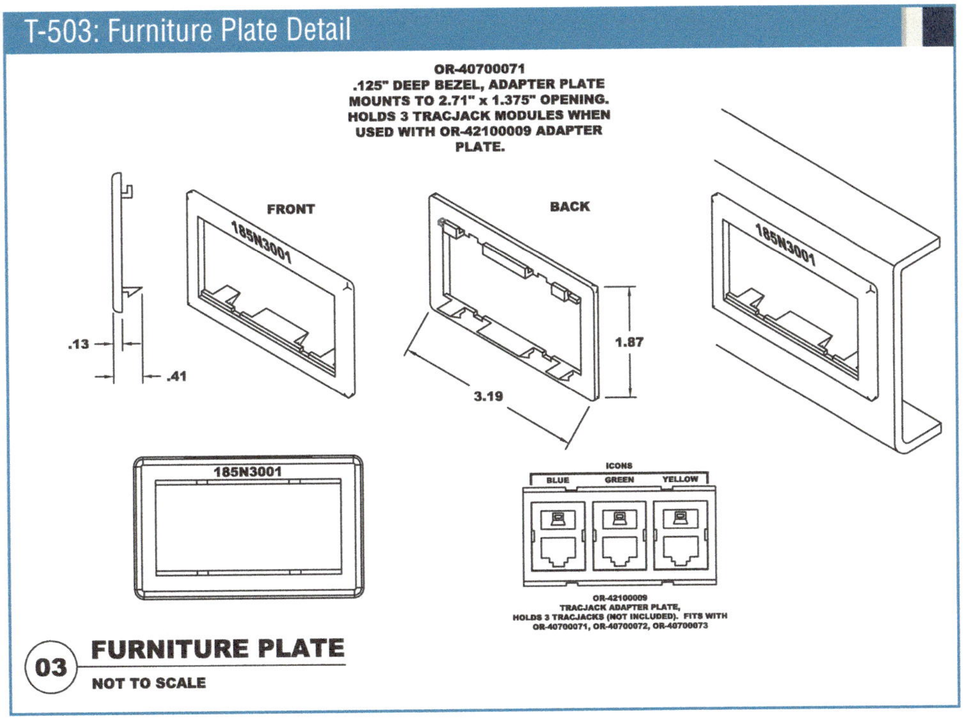

Figure 8-8. Detail 03 in Drawing T-503 for the example commercial building renovation project illustrates the furniture outlet assembly that will fit in the systems furniture raceway and finish systems.

Once a thorough review of the drawings has been completed by the estimator, the next step is a review of the design narrative that the engineer (based on the owner's requirements) submitted to the architect and MEP design team for the coordination between the different trades on the project. A *design narrative* is a brief description of a project and its coordination process. The design narrative is not considered part of the drawings or specifications and is provided only as an example of the project's coordination process. Once the review of the design narrative is completed, then the specifications can be reviewed. **See Figure 8-10.**

Site Surveys and Observations

Site surveys illustrate the existing building site conditions and the layout of the planned final construction. Detailed measurements of a building site and the planned building are necessary to ensure the proper placement of the building on the site. For example, a site survey for the example commercial building renovation project would result in the following observations:

- The backbone cabling extends connectivity from MDF Room 1010 on the first floor to IDF Room 3210 on the third floor. The cabling is to be installed in 2″ conduit that runs 20′ horizontally on the first floor to a mechanical room vertically aligned riser, at which point it is installed vertically from the first floor to the third floor. The conduit traverses vertically from the first floor through Mechanical Rooms 2013 and 3013 and requires core drilling on each floor that it traverses through. The conduit then terminates in a junction box in Mechanical Room 3013 and continues horizontally to the cable tray as part of the P1 scope. There are no obstructions vertically in the riser and enough space for the conduit to fit in each mechanical room.
- The existing horizontal wiring in the P1 and P2 areas on the third floor must be removed. Approximately 150 cables require removal in P1, and approximately 200 cables require removal in P2. The acoustical ceiling tiles in P1 have previously been removed, but they must be removed in P2 by other tradesworkers.
- The existing equipment room on the third floor has enough space to house the P1 temporary patch panels.
- A walk-through of the existing tenant space on the floor below indicates that the perimeter of the space contains hard-wall offices and an open area for furniture similar to the new layout of the third floor. Work in the existing space requires the removal of acoustical ceiling tiles and cleanup of debris from the removal process, and must be performed after normal business hours in order to not disrupt the tenant's normal operation.
- Measurements of the existing walls to remain for the third floor are taken and the accuracy of the dimensions on the drawing scale is confirmed. The distance from IDF Room 3033 and the furthest outlet in P1 does not exceed distance limitations per industry standards.

Project Specifications

A *project specification* is a written supplement to construction prints that provides additional construction information. Project specifications contain information related to building materials, construction procedures, quality control, and the legal issues of building construction. Bidding information, contract requirements, and information on construction phases are also provided in project specifications in an organized, standardized manner. The project specification must be completely and carefully reviewed by the estimator during the bidding and construction process to completely understand and accurately complete a project. The project specification is formatted into three distinct parts: Part 1 – General, Part 2 – Products, and Part 3 – Execution.

Project Specifications, Part 1 – General.

In Part 1 – General, a brief description of the work is outlined. This part describes the intent, suitable OEMs for materials and equipment, and required warranties. Also included are the product submittal process, obtaining information, codes and standards that must be referenced, coordination with trades, record drawings, identification, and documentation.

Design Narratives. . .

DESIGN NARRATIVE
To: Architect
From: Technology Engineer
CC: Owner
Re: **3rd Floor Renovation Architect/MEP Requirements for Telecommunications System**

The following represents the project requirements for the equipment room and tenant space:

Existing IDF Room 3033:
Houses the new P1 cabling. Power and cooling will need to remain active during the P2 demolition and during the construction of the P2 new IDF 3210. Upon owner IT commissioning of Room 3210, the construction manager will be required to coordinate with the systems contractor and owner the removal and relocation of new P1 cabling from Room 3033 to Room 3210. Upon completion, Room 3033 will bev decommissioned.

Room 3210 Equipment Room Intermediate Distribution Frame (IDF):
The IDF room is to provide LAN connectivity to support the renovated tenant space. The IDF room will connect to the existing Main Distribution Frame (MDF) located on level one.

Electrical Requirements:

1. The IDF will require the following dedicated circuits:
   a. Two (2) NEMA L6-20R
   b. One (1) NEMA 5-20R
   c. One (1) NEMA L5-30R
   d. All conductors shall be sized #10 AWG.
2. Electrical circuits will be mounted on overhead cable tray provided by systems contractor.
3. The IDF will require convenience/utility outlets in the room per the NEC.
4. Provided a minimum of 50 fc brightness measured 3′ AFF for lighting.
5. One (1) #4 AWG ground wire with a 10″ ground bus from an approved building ground. (Electrical panel ground is preferred).
6. Label all services with circuit number and panel location.
7. The IDF will require three (3) EZ PATH Series 44+ (PN#-EZDP44) fire-rated sleeves.

Electrical Requirements:

1. The IDF will require the following dedicated circuits:
   a. Two (2) NEMA L6-20R
   b. One (1) NEMA 5-20R
   c. One (1) NEMA L5-30R
   d. All conductors shall be sized #10 AWG.
2. Electrical circuits will be mounted on overhead cable tray provided by systems contractor.
3. The IDF will require convenience/utility outlets in the room per the NEC.
4. Provided a minimum of 50 fc brightness measured 3′ AFF for lighting.
5. One (1) #4 AWG ground wire with a 10″ ground bus from an approved building ground. (Electrical panel ground is preferred).
6. Label all services with circuit number and panel location.
7. The IDF will require three (3) EZ PATH Series 44+ (PN#-EZDP44) fire-rated sleeves.

Mechanical Requirements:

1. The IDF should be designed to provide an ambient temperature range of 68°F minimum to 72°F maximum.
2. The relative humidity of the IDF should be kept between 30%-50% RH.
3. The IDF will house one (1) Cisco 4507 or equal LAN Switch at 3580 BTU/hr.
4. The IDF will house one (1) Cisco 3845 or equal Router at 1485 BTU/hr.
5. The IDF will house one (1) Cisco VG224 at 485 BTU/hr.
6. The IDF will house one (1) APC Smart-UPS 3000VA RM at 348 BTU/hr.
7. Drip pans will be required to protect LAN equipment from water if applicable.

Figure 8-10. (*continued on next page*).

. . .Design Narratives

Architectural Requirements:

1. The IDF must be a minimum of 8′ × 10′.
2. Doors should be lockable–owner requires card access.
3. All walls should be painted white to reflect light inside the room.
4. Floors should be anti-static tile.
5. Ceiling finish should be two-foot by two-foot (2′ × 2′) acoustical ceiling tile mounted to no lower than nine-feet (9′) above finished floor.

Tenant Fit-out Requirements:

1. The Telecom Premise Backbone Distribution will require a dedicated two-inch (2″) conduit from the ceiling of MDF 1010, thru the vertical Elect./Mech. Building Riser Rooms. Provided a 12″ × 12″ × 4″ deep junction box with cover in the 3rd Floor Elec/Mech. Room. Continue two-inch (2″) conduit to cable tray within tenant space.
2. Each Tel/Data location will require a four-inch (4″) square box with a single-gang mud ring and require one (1) dedicated one-inch (1″) EMT conduit stubbed to an accessible ceiling. Each conduit will have a pull string provided.
3. Each modular systems furniture wall feed will require a four-inch (4″) square box with a double gang mud ring and require one (1) dedicated two-inch (2″) EMT conduit stubbed to an accessible ceiling. Each conduit will have a pull string provided. No more than four workstations per wall Tel/Data conduit.
4. Each modular systems furniture floor feed will require one (1) dedicated two-inch (2″) EMT conduit stubbed to accessible ceiling on floor below. Each conduit will have a pull string provided. No more than four workstations per floor conduit.
5. Each modular systems furniture floor location will require one (1) dedicated RC-900FFTCGY poke through (or equivalent).
6. Each Tel/Data flush poke through/floor box will require one (1) Combination Tel/Power RC4ATCBK poke through or floor box device (or equivalent).
7. All Tel/Data locations shall be located per the architect's direction and at heights in accordance with ADA.
8. Horizontal cable tray will be required to support Tel/Data Cabling above finished ceiling.
9. Horizontal Sleeves EZ PATH Series 44+ (PN#-EZDP44) will be required for all rated wall penetrations to support cable tray pathways above ceiling. Locate additional sleeves as required.

END OF NARRATIVE

Figure 8-10. A design narrative is provided as an example of a project's coordination process.

Project Specifications, Part 2 – Products. In Part 2 – Products, most of the products required for the installation are specified by OEM name and part number. In most cases, this is to establish a minimum product performance requirement in the event that an OEM's product is substituted for a product from a different OEM.

For the example commercial building renovation project, the products specified in Part 2 – Products include various cable types, connecting hardware, pathway types, fire protection systems, and equipment room hardware. **See Figure 8-11.** The information in this part is the basis for identifying the materials list and creating a bill of materials (BOM). By accounting for the materials and equipment specified in Part 2 – Products, the estimator can quantify the materials and labor units applied for each specific task.

Project Specifications, Part 3 – Execution. The details outlined in Part 3 – Execution can vary by engineer or designer. In the example commercial building renovation project, the installation is summarized by task to provide a complete written scope of work. The task summary is outlined from sections 3.1.A through 3.1.BB and includes all of the work shown on the drawings and specified in the document. Part 3 – Execution concludes with sections for adjusting and cleaning, testing, unit pricing, and the project financial summary.

DEVELOPING THE ESTIMATE

Once the prints and project specifications have been reviewed by the estimator, the amount of materials required for construction must be determined. A *takeoff* is the practice of reviewing contract documents, including the prints and specifications, to determine the quantities of material to be included in a bid. A takeoff is prepared to begin the process of developing an overall cost for the construction. Printreading and calculation skills are required to develop an accurate and comprehensive list of materials necessary for the project. Once the takeoff is complete, the total amount is calculated and priced to prepare the estimate. Takeoffs can be prepared manually or with the assistance of software programs. *Note:* For the example commercial building renovation project, the manual method is to be used.

Takeoff sheets are formatted in a specific manner for each individual project and trade. For example, one of the takeoff sheets for the example commercial building renovation project quantifies the different outlets or devices by phase (P1 or P2). Upon determining the quantity and types of devices on *Drawing T-403 Tel Data 3rd Floor Plan* by phase, the estimator can determine the quantity of cables to be installed and termination hardware required for the entire project. **See Figure 8-12.** By determining the quantities of cables and types of devices to be installed, the estimator can then estimate the required labor and factor it into the project cost.

Project Specifiations: Part 2–Products

PART 2–PRODUCTS

2.1 COMMUNICATION HORIZONTAL CABLING

A. Unshielded Twisted Pair Copper Category 6 cable shall be CMP type manufactured by Belden and provided in the following type and colors:

1. Cable #1 Blue PN#3613D15U1000
2. Cable #2 Green PN#3613005A1000
3. Cable #3 Yellow PN#3613004A1000

B. All ground wire distribution within equipment rooms shall be insulated #6 AWG stranded wire.

C. All ground wire distribution Telecommunications Bonding Backbone between equipment room shall be insulated #4 AWG stranded wire. (By Electrical Contractor)

2.2 COMMUNICATION COPPER BACKBONE CABLING

A. Voice multi-pair premise backbone cable shall be manufactured Essex PN#18-499-36 25 pair Cat. 3 CMP.

2.3 PREMISE COMMUNICATION BACKBONE FIBER OPTIC CABLING

A. Twelve (12) strands Optical Fiber Nonconductive Plenum (OFNP) (armored) Single Mode Fiber Optic Cable manufactured by Belden PN B9W241.

Figure 8-11. Most of the products required for installation in the example commercial building renovation project are specified by an OEM name and part number in Part 2 – Products of the project specifications.

Takeoff Sheet (T-403 Tel Data 3rd Floor Plan)

INDICATES THREE CABLES PER OUTLET

Device >	Hardwell 3	Furniture 3	Flr. Box 3	W-Phone 1	W-less 1	CATV/SEC-1
Drawing # TD Room						
T-403 IDF RM.3303 (Phase 1)	24	24	2	3	5	4
T-403 IDF RM.3210 (Phase 2)	24	60	7	2	8	8
TOTAL CAT.6 UTP OUTLETS	48	84	9	5	13	12
TOTAL CAT.6 UTP CABLES	144	252	27	5	13	12

TOTAL AMOUNT OF OUTLETS

TOTAL CABLE REQUIREMENTS (3 × 48 = 144)

Figure 8-12. Upon determining the quantity and types of devices in Drawing T-403 by phase, the estimator can determine the quantity of cables to be installed and termination hardware required for the entire project.

The scale of the drawing must be confirmed and the amount, pathway types, and the amounts and types of cable required must be calculated to create the BOM. The BOM provides descriptions of required materials and equipment and their OEM part numbers, quantities of each, unit costs, and labor units. Takeoff quantities are recorded when generating the BOM in order to establish takeoff quantity totals and by taking a physical inventory of the units in question.

A standard practice when verifying the scale on a print is to measure the dimension of a standard door opening. A standard door opening, per ADA requirements, is 36″ wide. The distances between the column lines shown on a print also have dimensions and can be measured to verify the accuracy of the scale. *Note:* Reference *Drawings T-103 Tel Data 3rd Floor Pathway Plan* and *T-403 Tel Data 3rd Floor Plan* takeoff sheets to determine types, quantities, and lengths of pathways. The various pathways must be calculated, and the required parts determined and then added to the BOM.

Once the pathways have been identified, the amount of cable that is required can be determined. The P1 cabling originates in existing IDF Room 3033. However, IDF Room 3033 is to be decommissioned during P2 and moved to new IDF Room 3210. The cabling must be routed in the cable tray to the P1 space. The estimated distance from IDF Room 3033 to the end of the cable tray in the P1 space is approximately 150′. Additional cabling length must be added within the P1 space to allow for 20′ of slack in IDF Room 3033. The furniture outlets located along column lines 1A appear to be the furthest distance away from IDF Room 3033. **See Figure 8-13.**

The items needed to create the BOM for P1 include the following:

- The estimated length of CAT 6 cable required to be run to seven single-cable outlets for furniture locations is roughly 265′ from IDF Room 3303, or a total cable quantity of about 2000′ (7 × 265 = 1855). The additional 145′ of cable is to be used for incidental single runs for certain wall-mounted telephones, WAPs, and CCTV/security outlets.
- The estimated length of CAT 6 cable required to be run to 50 three-cable outlets for wall, furniture, and floor locations is roughly 220′ from IDF Room 3303, or a total cable quantity of 11,000′ (50 × 220 = 11,000).

Some estimators record the length of each cable run and add them together, while other estimators use an average distance to determine cable quantities. An architect's scale used on prints or distance-measuring wheels used on the actual job site can be used to calculate lengths of cable runs. Software programs that read prints and automatically determine distances can also be used. Estimates must be as accurate as possible, as inaccurate estimates can cause the bid to be lost or result in a cost overrun due to underestimating material requirements on active projects.

Once the BOM for P1 is calculated, the next step is to continue to build the BOM for P2. The amount of backbone cabling that is required for the P2 BOM can be calculated by referencing the total length of 2″ conduit on the print, including the distance adjacent to the cable trays on the first and third floors, the slack required within new IDF Room 3210 and the existing MDF Room 1010, and any damaged cable from pulling and termination operations. **See Figure 8-14.**

Pathway and Outlet Locations

Pathway >	1" Stub ups	2" Stub ups	12" x 2" Tray	12" x 4" Tray	44-Sleeves	2" conduit	1" conduit	4" F-Box	3" core Furn	2" conduit	J-Box
Drawing #											
T-103 Pathway Plan (Phase 1)	0	0	0	50	3	4	2	0	0	110′	1
T-403 IDF RM.3210 (Phase 2)	0	0	50′	120′	5	11	7	0	0	0	0
TOTAL CAT.6 UTP OUTLETS	65	7	0	0	0	0	0	9	2	0	0
TOTAL CAT.6 UTP CABLES	161	4	0	0	0	0	0	27	12	0	0

T-103 AND T-403 TAKEOFFS

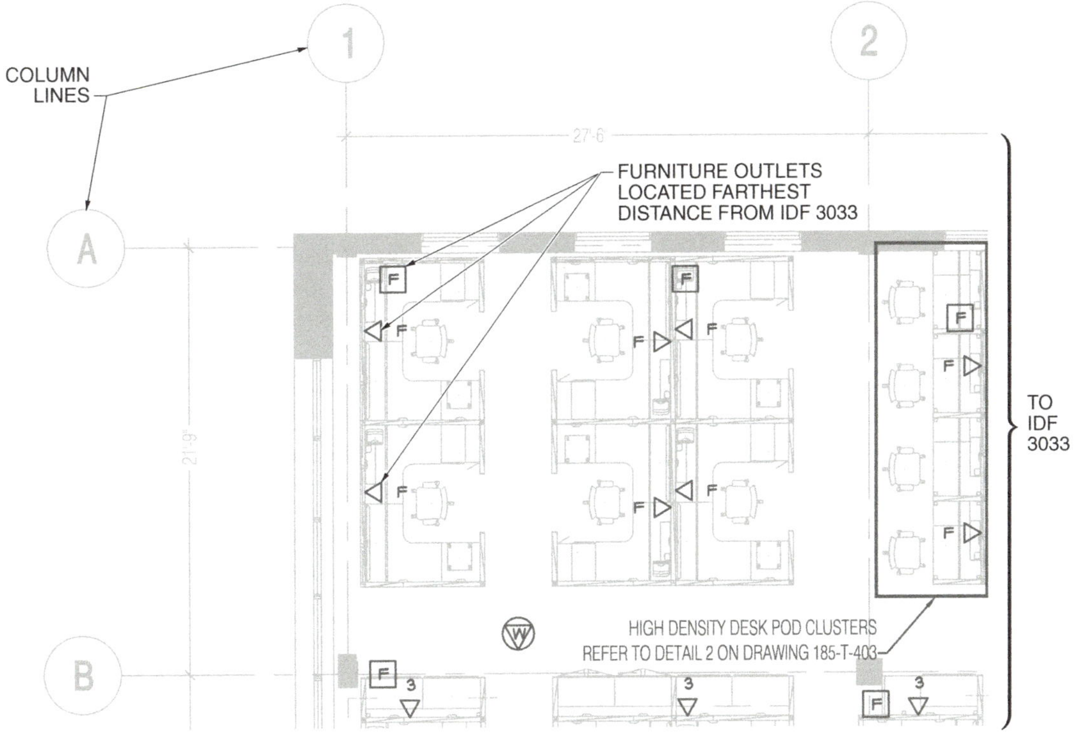

FURNITURE OUTLET LOCATIONS

Figure 8-13. Once pathways are identified and outlet locations determined, the amount of cable needed for the project can be calculated.

The cable section of the P2 BOM concludes with determining the amount of No. 6 AWG ground wire required and referencing Part 2 – Products from the project specifications for the required quantities of copper and fiber-optic patch cords. The termination hardware for the various outlet types and equipment rooms can now be identified on the prints and added to the BOM. This includes horizontal cabling, connectors, components, and other hardware such as power strips, cable trays, racks, and rack mount equipment.

Upon completion of the BOM, pricing and unit price summaries must be applied to determine the total cost of materials. Contractors typically solicit pricing from multiple distributors to develop the most competitive material price. The remainder of the pricing for labor units must be added to the BOM and totaled.

Backbone Cabling Required for Phase 2 BOM	
Item	**Amount***
Conduit from takeoff calculation, floor 1	110
Additional conduit, floor 1	20
Conduit, floor 3 cable tray	110
Equipment room slack consideration	40
Damaged cable consideration	20
Total Backbone Cabling Requirement	**300**

* in ft

Figure 8-14. The amount of backbone cabling required for P2 can be calculated by referencing the total length of the pathways in which the cabling is to be installed.

For project estimates, estimators must carefully review the specified products from the project specifications for counterfeit items when purchasing products through the internet. Even if a product is marked with a logo from an industry or standard organization, it does not mean that it is genuine. Counterfeit items sold through the internet can be marked with logos that resemble authentic logos, even though they are not actually certified by an industry organization. Counterfeit items may resemble legitimate products but often have a much lower cost than a legitimate product. The lower cost of such a product may help win a bid, but a counterfeit product damages the integrity of the final installation and does not meet the performance requirements of adopted standards and the specifications.

BOM Labor Units

Once material pricing has been established, labor units are added to the BOM to determine the final cost estimate of the project. The BOM can be expanded to include a labor unit category and a total labor category. The labor to install multiple cables concurrently is more efficient than installing cables individually. When the labor to install multiple cables is estimated, much of the labor time can be applied to cable number 1 per foot (104 hours, for a labor unit of 0.008); while the labor to install cables number 2 and number 3 per foot is significantly less (11 hours each, for a labor unit of 0.001), since they are being installed at the same time with cable number 1.

For the BOM of the example commercial building renovation project, labor units for cable installation are priced per foot. For example, since cable number 1 is 13,000′, that length is multiplied by a labor unit of \$0.82 per foot for a total cost estimate of \$10,660.00 (13,000 × 0.82 = 10,660). Cable numbers 2 and 3 have a total cost estimate of \$9020.00 each (11,000 × 0.82 = 9020.00). Labor units must be applied to each line item on the BOM. **See Figure 8-15.**

Labor Costs

Once the labor units have been established, the actual labor unit rate must be determined. While the costs of most materials on a project are fixed, labor unit rates vary. Items that can adversely affect the labor unit rate include poor labor performance, lack of labor oversight, and the sequencing of work by the construction manager. The labor breakdown for the workforce is separated by worker classification (foreman, journeyworker, and helper) and includes hourly wages, benefits, and insurance. **See Figure 8-16.**

An estimator must determine the ratio of journeyworker-to-helper for the project workforce to perform the installation. In some cases, licensing laws or collective bargaining agreements determine the journeyworker-to-helper ratio for a particular project. *Note:* For the example commercial building renovation project, a required journeyworker-to-helper ratio of 1-to-1 is used.

BOM Labor Units							
	Cable	**Part Number**	**Qty**	**Unit**	**Total**	**Labor Unit**	**Total Hours**
1	Cat. 6 CMP Belden Blue (Phase 1)	3613006A1000	13,000	$0.82	$10,660.00	0.008	104
2	Cat. 6 CMP Belden Green (Phase 1)	3613005A1000	11,000	$0.82	$9,020.00	0.001	11
3	Cat. 6 CMP Belden Yellow (Phase 1)	3613004A1000	11,000	$0.82	$9,020.00	0.001	11

Figure 8-15. Once material pricing has been established, labor units are added to the BOM to determine the final cost estimate of the project.

Labor Costs				
Classification	**Hourly Rate**	**Benefits**	**Insurance**	**Total Package**
Foreman	$42.00	$24.00	$5.00	$71.00
Journeyworker	$38.00	$22.00	$5.00	$65.00
Helper Step 1	$18.00	$14.00	$5.00	$37.00
Helper Step 2	$22.00	$16.00	$5.00	$43.00
Helper Step 3	$26.00	$18.00	$5.00	$49.00
Helper Step 4	$28.00	$20.00	$5.00	$53.00

Figure 8-16. The labor breakdown for the workforce is separated by worker classification and includes hourly wages, benefits, and insurance.

An estimator can calculate a blended labor rate based on the required journeyworker-to-helper ratio. A blended labor rate can be competitive based on anticipated workforce available at the time of the project. For example, a blended labor rate can be calculated by averaging a single helper's labor rate against a single journeyworker's labor rate with any other helpers' labor rates averaged against the foreman's labor rate. This assumes that the crew makeup is exactly as listed on the BOM. Further analysis of the labor specified for the project is required to determine the cost of overtime to perform work in other tenant spaces after hours, including establishing the pathways and installation of cables.

Once the material quantities are determined with the labor rates applied, the estimator can analyze the data, determine the final cost of the project, and prepare a proposal. The estimator also adds the markup (profit margin) to the final cost. The takeoff sheet and BOM are referenced to separate the various sections of the work requirements and prepare the unit prices.

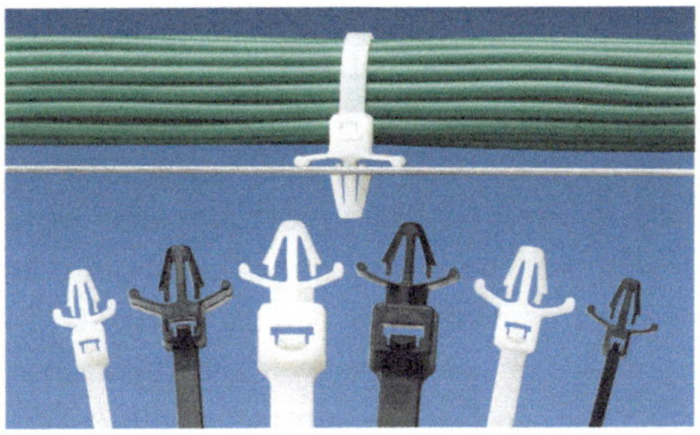

Panduit Corp.

BOMs for VDV systems may include small commodity items such as cable ties.

Unit Price Summaries

Most bid proposals require a unit price summary so when changes are required by the GC, the owner has an initial cost to reference for comparison purposes. A unit price summary is based on the unit price agreement in the contract and includes the costs of the materials needed for the project, the labor at the journeyworker rate, and a 20% markup. **See Figure 8-17.** The process for determining unit prices varies by contractor. For the example commercial building renovation project, the cost is to be for the materials and markup only (no labor rate).

Unit Price Summaries									
	Description	**Part Number**	**Qty**	**Unit Cost**	**Total Unit Cost ($)**	**Labor Units**	**Total Labor Units (Hrs)**	**Labor Unit Cost**	**Total Item Cost**
1	Cat. 6 CMP Belden Blue	3613006A1000	150	$0.82	$123.00	0.008	1.2	$78.00	$201.00
2	Cat. 6 CMP Belden Green	3613005A1000	150	$0.82	$123.00	0.001	0.15	$9.75	$132.75
3	Cat. 6 CMP Belden Yellow	3613004A1000	150	$0.82	$123.00	0.001	0.15	$9.75	$132.75
4	Trac-Jack 3 port faceplate	OR-403STJ13	1	$5.96	$5.96	0.1	0.1	$6.50	$12.46
5	Trac-Jack Cat. 6 module	OR-TJ600-08	3	$9.91	$29.73	0.088	0.264	$17.16	$46.89
6	Cable Lace & Dress	Generic-Labor	3	$1.25	$3.75	0.117	0.351	$22.82	$26.57
7	Test & Identification	Generic-Labor	3	$2.00	$6.00	0.1	0.3	$19.50	$25.50
8	Misc. Coordination	Generic-Labor	1	$2.50	$2.50	0.25	0.25	$16.25	$18.75
9	Cat. 6 Copper 9′ Blue	OR-MC609	1	$12.40	$12.40	0.033	0.033	$2.15	$14.55
10	Cat. 6 Copper 7′ Blue	OR-MC607	1	$11.01	$11.01	0.033	0.033	$2.15	$13.16
11	Total Material & Labor Cost								$624.38
12	Overhead & Profit @ 20%								$124.87
13	**Grand Total:**								**$749.25**

Figure 8-17. A summary of materials and labor required includes the costs of the materials, the labor at the journeyworker rate at $65.00/hr, and a 20% markup.

SUMMARY

The preparation of a labor and material cost estimate requires an experienced estimator who can apply analytical thinking to the process. The estimator must understand the methods and processes that the labor force utilizes during the installation, factor in intangibles that can affect the final cost, and account for the project's level of difficulty. Estimators must determine the amounts and types of materials that are required, regardless of whether specified they are, so that a complete installation is provided per the prints, plans, site surveys, specifications, and applicable codes and standards. Estimators must review the project contract for the scope of work and project scheduling to develop an accurate labor estimate as well as solicit suppliers for competitive bids. A unit price summary is included on bids so that the owner has an initial cost to reference for comparison purposes in the event of changes by the GC.

Using Drawings and Specifications to Develop Project Estimates

CHAPTER 8

Review

Name____ Date ____________________

True-False

T F **1.** For the example commercial building renovation project, a bid must be submitted to receive the contract to perform the work.

T F **2.** After submitting a bid for any project, the estimator must perform a thorough review of the scope of work.

T F **3.** As with any trade prints, the specific dimensions for wiring, hardware, and finish materials are not always provided on the detail drawings.

T F **4.** The design narrative is always considered to be part of the drawings or specifications.

T F **5.** Takeoff sheets are formatted in a standard manner for each individual project and trade.

Multiple Choice

__________________ **6.** The floor-to-deck height for the example commercial building renovation project is ___′.

A. 10
B. 12
C. 16
D. 20

__________________ **7.** Detail 13 – IDF Room 3033 Fiber Panel Layout provides useful information about existing and future ___ backbone cabling required during each phase of the example commercial building renovation project.

A. copper
B. fiber-optic
C. hybrid fiber-coax
D. OSP

__________________ **8.** Approximately ___ cables require removal in P2 of the example commercial building renovation project.

A. 100
B. 200
C. 300
D. 500

_______________ **9.** A(n) ___ is prepared to begin the process of developing an overall cost for the construction of a project.

A. bid
B. IFB
C. RFP
D. takeoff

_______________ **10.** The amount of backbone cabling that is required for the P2 BOM can be calculated by referencing the total length of ___ on the print.

A. 2″ conduit
B. copper cabling
C. fiber-optic cabling
D. horizontal cabling to work areas

Completion

_______________ **11.** A(n) ___ is a request for contractors, subcontractors, and suppliers to submit pricing for services and/or materials for a project.

_______________ **12.** A(n) ___ is an offer from a contractor, subcontractor, or supplier to perform construction work, provide a construction service, or provide materials or supplies at a stated price.

_______________ **13.** A(n) ___ is a document that contractors are required to use when bidding a project.

_______________ **14.** The ___ details and wall box details shown on the prints indicate that these items penetrate floors and provide a fire rating to maintain the overall fire rating of the floor.

_______________ **15.** A(n) ___ is a brief description of a project and its coordination process.

_______________ **16.** A(n) ___ is a written supplement to construction prints that provides additional construction information.

_______________ **17.** A(n) ___ is the practice of reviewing contract documents, including the prints and specifications, to determine the quantities of material to be included in a bid.

_______________ **18.** A standard practice when verifying the scale on a print is to measure the dimension of a standard ___.

_______________ **19.** For the example commercial building renovation project, a required journeyworker-to-helper ratio of ___ is used.

_______________ **20.** Drawing ___ represents the scope of work for the demolition to be performed for the example commercial building renovation project.

Short Answer

21. List the five prints concerning pathways for the example commercial building renovation project that must be reviewed.

22. Briefly explain the purpose of a site survey.

23. Briefly explain why at least an extra 20′ of cable must be added within the P1 space.

24. Explain why the labor needed to install one cable (cable number 1) is the same as installing two cables (cable numbers 2 and 3) in the P1 space.

25. What is the basis of unit price summaries?

CHAPTER

9

Security Drawings and Specifications

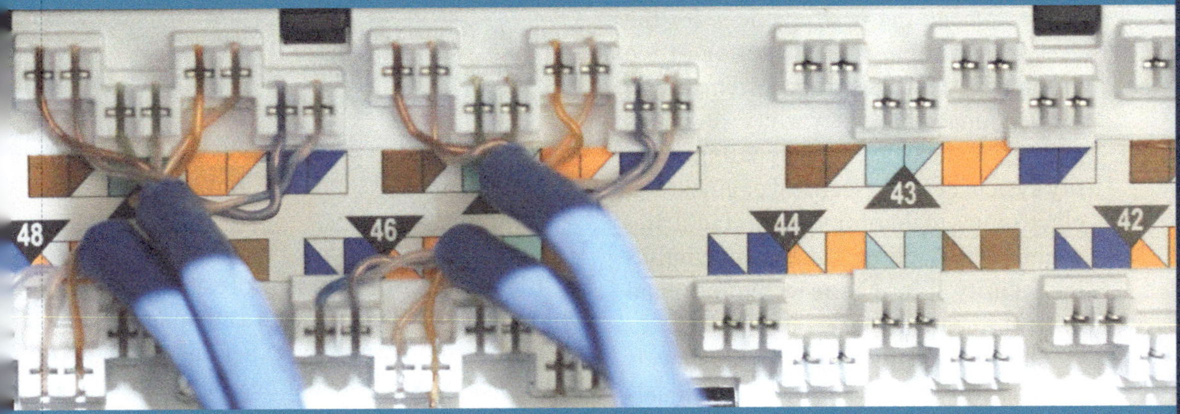

Property owners rely on security systems to protect personnel and property by locking access points such as doors and windows, providing video surveillance of the property, and sounding alarms in the event of a breach of the property, among other measures. Security-related systems are an integral part of modern construction project designs. Historically, systems that provided security to occupants and facilities were separate, stand-alone systems that did not have the capability to integrate with other types of security and alarm systems.

Security systems are designed by industry professionals who are certified to design such systems. The equipment, software, and materials needed to install security systems are procured through specifications developed during the design phase of the project. There are several types of security systems that can be specified and each type must operate within the recommended practices of NFPA 101, *Life Safety Code*.

As with other VDV systems, security system applications are often overlooked during the design and planning phase, which can cause significant cost increases to owners during the design and construction process.

Objectives

- Describe the different types of certification for security system design professionals.
- Describe the process for procuring security system proposals.
- List the different types of security systems.
- Describe NFPA 101, *Life Safety Code*.
- Describe the different types of specialty security system applications.
- Describe the pathways and spaces used for security systems.
- Explain how to interpret the abbreviations and symbols on drawings for proper security system requirements.

Learner Resources

atplearningresources.com
access code 783821

SECURITY SYSTEM DESIGN PROFESSIONALS

When developing drawings and specifications for security systems, owners and architects work with security system design professionals who are certified to recommend the best practices for the design, operation, and integration of the different systems. Security system design professionals have a variety of backgrounds and experience, including security system design, electrical engineering, electronics technology, manufacturing engineering, facility management, building security, information technology, and law enforcement.

The organization that provides levels of certification for security professionals based on education and industry experience is ASIS International. *ASIS International (ASIS)* is a professional organization for security professionals that issues certifications, standards, and guidelines for the security profession. While ASIS provides certifications for several professional designations, the most common certifications for security system designers include the Physical Security Professional (PSP) and the Certified Protection Professional (CPP) certifications.

Physical Security Professional (PSP) Certification

The *Physical Security Professional® (PSP®) certification* is a professional certification for individuals with a minimum of four years of experience in the physical security field who have successfully passed a board certification licensing exam. An individual with the PSP certification demonstrates knowledge in physical security assessments; in the application, design, and integration of physical security systems; and in the implementation of security measures.

Eligibility requirements include either a high school diploma, GED equivalent, or associate degree plus six years of progressive experience in the physical security field or a bachelor's degree or higher from an accredited post-secondary institution plus four years of progressive experience in the physical security field. Applicants for the PSP certification must be employed full-time in a security-related role. Applicants must not have been convicted of any criminal offense that would reflect negatively on the security profession, ASIS, or the certification program.

Certified Protection Professional (CPP) Certification

The *Certified Protection Professional® (CPP®) certification* is a professional certification for individuals with a minimum of seven years of security experience in the physical security field, three of which must be in responsible charge of a security function, who have successfully passed a board certification licensing exam. The CPP certification validates the complete knowledge of security management.

Eligibility requirements include either nine years of security experience (with a minimum of three years in responsible security function position) or a bachelor's degree or higher from an accredited post-secondary institution and a minimum of seven years of security experience (with a minimum of three years in a responsible security function position). Applicants for the CPP certification must be employed full-time in a security-related role. Applicants must not have been convicted of any criminal offense that would reflect negatively on the security profession, ASIS, or the certification program. **See Figure 9-1.**

TECH FACT

Acquiring an ASIS professional certification requires not only years of relevant experience and education but also strong study habits and comprehension skills, as each certification is attained by passing a timed exam. Study materials are available from ASIS for each type of certification.

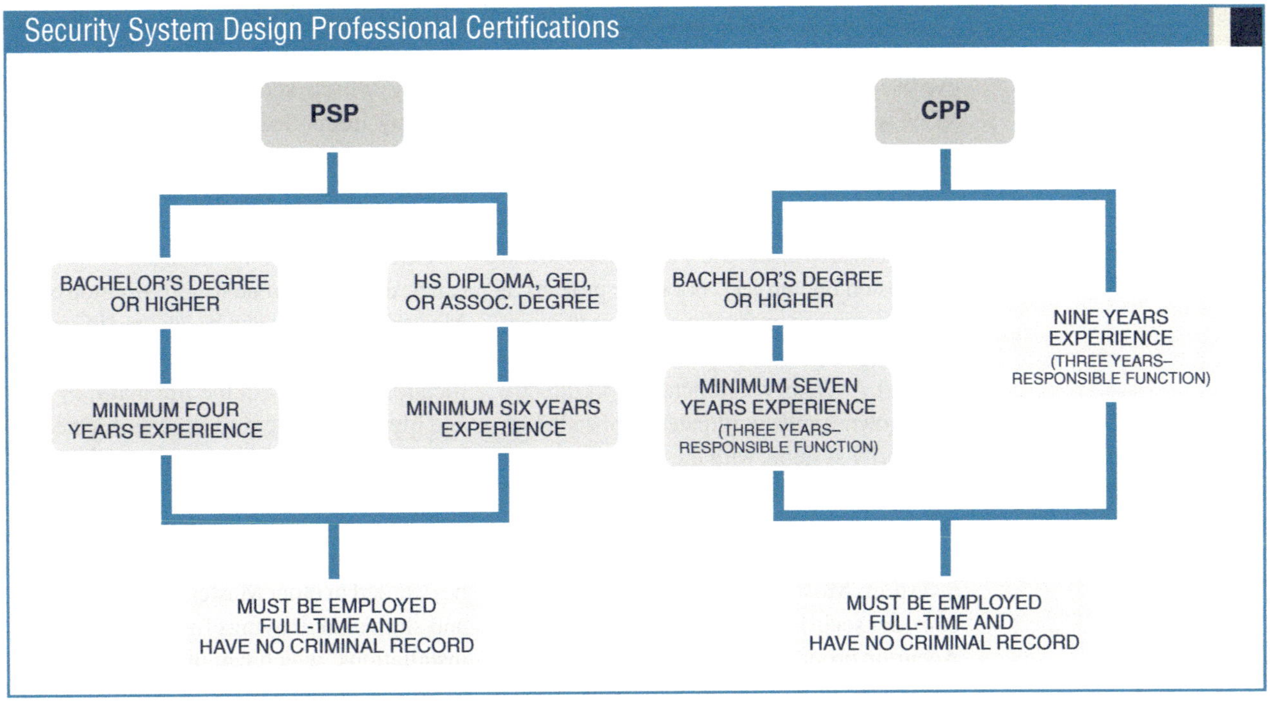

Figure 9-1. The most common certifications for security system designers include the Physical Security Professional (PSP) and the Certified Protection Professional (CPP) certifications.

SECURITY SYSTEMS PROCUREMENT

The process for determining the requirements for the security systems is similar to the process for determining the requirements for other systems (electrical, mechanical, etc.). The requirements are determined through an analysis and risk assessment performed by a PSP- or CPP-certified security professional. Input from the owners and end users for areas such as facilities management, information technology, and physical security is compiled and incorporated into the system requirements. The designer includes these requirements in the design process so that the architectural, mechanical, electrical, and communications systems required to support the security system are included in the final design.

Security system requirements may include coordination for interior/exterior cameras and camera types, room and rack sizes for housing equipment, power equipment, cooling equipment, door and window types, raceways for wiring, and network architecture. A properly coordinated design eliminates confusion during building construction and provides the owner with clear and concise drawings and specifications to obtain competitive pricing for the work.

The practice of procuring security system proposals can vary by geographic region and owner requirements. Factors that can affect security system proposals include licensing and background verification of the installers contracted to install the system. OEMs typically influence the owner and determine which contractors can install certain types of systems. The security system industry is proprietary, and OEMs typically partner with only selective contractors and integration firms that can warranty their workmanship and service their equipment.

Some owners have preapproved contractors that provide end-to-end solutions and they source the entire design and installation of the security system(s) to one vendor. Most owners hire design professionals that oversee the design, bid, and installation process of the building's security systems.

The bidding process can vary, with some projects procured through the GC's competitive bid process and some packaged with the VDV work or electrical work. The security systems contractor can also be a subcontractor to the VDV or electrical contractors. For some projects, the electrical contractor may have the expertise to install the VDV and electrical systems on one project. For other projects, a separate contractor for the installation of security systems (a security contractor) needs to be hired. A security contractor works directly for the owner and coordinates the installation of security systems with the GC.

The owner and the GC must perform a detailed scope review of the security system proposal when specifying and purchasing a system. Most construction professionals consider security systems a specialty trade. Accurate proposals qualify unknown items such as the limit of scope as it relates to the integration of the system to the owner's network and to other systems. Proposals should be broken down by subsystem to clarify which scope has been included and any items that are excluded. Modern security systems must communicate and integrate with other previously installed security systems.

Often, the scope of the security systems can be dedicated to one specific division of the MasterFormat or it can be covered in several different divisions. The security system specifications, along with the fire alarm system and life safety system specifications, are typically outlined in MasterFormat Division 28. **See Figure 9-2.** Due to the convergence and separation of construction disciplines, security requirements and scope can also be detailed in other MasterFormat Divisions and shown on drawings for other trades and installations, which can include items such as wiring, hardware, pathways, equipment, software, and network hardware.

CSI MasterFormat Division 28

DIVISION 28– ELECTRONIC SAFETY AND SECURITY

28 00 00 ELECTRONIC SAFETY AND SECURITY

- **28 01 00 Operation and Maintenance of Electronic Safety and Security**
  - 28 01 10 Operation and Maintenance of Electronic Access Control and Intrusion Detection
    - 28 01 10.51 Maintenance and Administration of Electronic Access Control and Intrusion Detection
    - 28 01 10.71 Revisions and Upgrades of Electronic Access Control and Intrusion Detection
  - 28 01 20 Operation and Maintenance of Electronic Surveillance
  - 28 01 30 Operation and Maintenance of Electronic Detection and Alarm
    - 28 01 30.51 Maintenance and Administration of Electronic Detection and Alarm
    - 28 01 30.71 Revisions and Upgrades of Electronic Detection and Alarm
  - 28 01 40 Operation and Maintenance of Electronic Monitoring and Control
    - 28 01 40.51 Maintenance and Administration of Electronic Monitoring and Control
    - 28 01 40.71 Revisions and Upgrades of Electronic Monitoring and Control
- **28 05 00 Common Work Results for Electronic Safety and Security**
  - 28 05 05 Selective Demolition for Electronic Safety and Security
  - 28 05 13 Conductors and Cables for Electronic Safety and Security
    - 28 05 13.13 CCTV Communications Conductors and Cables
    - 28 05 13.16 Access Control Communications Conductors and Cables
    - 28 05 13.19 Intrusion Detection Communications Conductors and Cables
    - 28 05 13.23 Fire Alarm Communications Conductors and Cables
  - 28 05 26 Grounding and Bonding for Electronic Safety and Security
  - 28 05 28 Pathways for Electronic Safety and Security
    - 28 05 28.29 Hangers and Supports for Electronic Safety and Security
    - 28 05 28.33 Conduits and Backboxes for Electronic Safety and Security
    - 28 05 28.36 Cable Trays for Electronic Safety and Security
    - 28 05 28.39 Surface Raceways for Electronic Safety and Security
  - 28 05 48 Vibration and Seismic Controls for Electronic Safety and Security
  - 28 05 53 Identification for Electronic Safety and Security
- **28 06 00 Schedules for Electronic Safety and Security**
  - 28 06 10 Schedules for Electronic Access Control and Intrusion Detection
  - 28 06 20 Schedules for Electronic Surveillance
  - 28 06 30 Schedules for Electronic Detection and Alarm
  - 28 06 40 Schedules for Electronic Monitoring and Control
- **28 08 00 Commissioning of Electronic Safety and Security**

28 10 00 ELECTRONIC ACCESS CONTROL AND INTRUSION DETECTION

- **28 13 00 Access Control**
  - 28 13 13 Access Control Global Applications
  - 28 13 16 Access Control Systems and Database Management
  - 28 13 19 Access Control Systems Infrastructure
  - 28 13 26 Access Control Remote Devices
    - 28 13 26.11 Wireless Access Control Devices
  - 28 13 33 Access Control Interfaces
    - 28 13 33.16 Access Control Interfaces to Access Control Hardware
    - 28 13 33.26 Access Control Interfaces to Intrusion Detection
    - 28 13 33.33 Access Control Interfaces to Video Surveillance
    - 28 13 33.36 Access Control Interfaces to Fire Alarm
  - 28 13 43 Access Control Identification Management Systems
  - 28 13 53 Security Access Detection

Figure 9-2. Division 28 of the CSI MasterFormat covers fire alarm, life safety, and security systems.

SECURITY SYSTEM TYPES

There are various types of separate, stand-alone systems that are considered part of a building's entire security system. The separate types of systems that are commonly part of physical building security include access control, video surveillance, and intercom systems. **See Figure 9-3.**

The separate systems are part of an intrusion detection and notification system. In the event of a breach of any one of these systems, a notification is sent to security and building management personnel and to command centers through a notification on a smartphone, tablet, or PC. Security systems are typically used in commercial businesses and institutions. However, modern residential installations also include security systems that are often integrated into smart home technology.

The functions of a surveillance system and an access control system are unique. Since end users often require that these systems be easy to use, software OEMs typically offer integrated systems that can share data and even control both systems from a common platform. For example, an integrated system can detect when a door is opened and then use a camera to record an image of the person who opened it.

Access Control Systems

An *access control system* is a security system that controls door locks. Access control systems are used to regulate access to exterior and interior areas of a premises using credentials. Access control systems include card readers, biometric readers that allow access through fingerprint or facial recognition, push-to-exit panels, and communication keypads.

The most common type of access control system employs a device adjacent to a door for reading a badge, proximity card, smartphone app, or fob. The device either allows or refuses passage based on preprogrammed access levels. **See Figure 9-4.** These systems are independently cabled and separate from other structured cabling systems in a facility. Although most access control systems are controlled from a dedicated computer and require the use of an Ethernet (LAN) connection via a data network, the individual devices have dedicated and specific cables. *Note:* Modern access control technology uses smart cards and mobile credentials that utilize wireless Bluetooth technology.

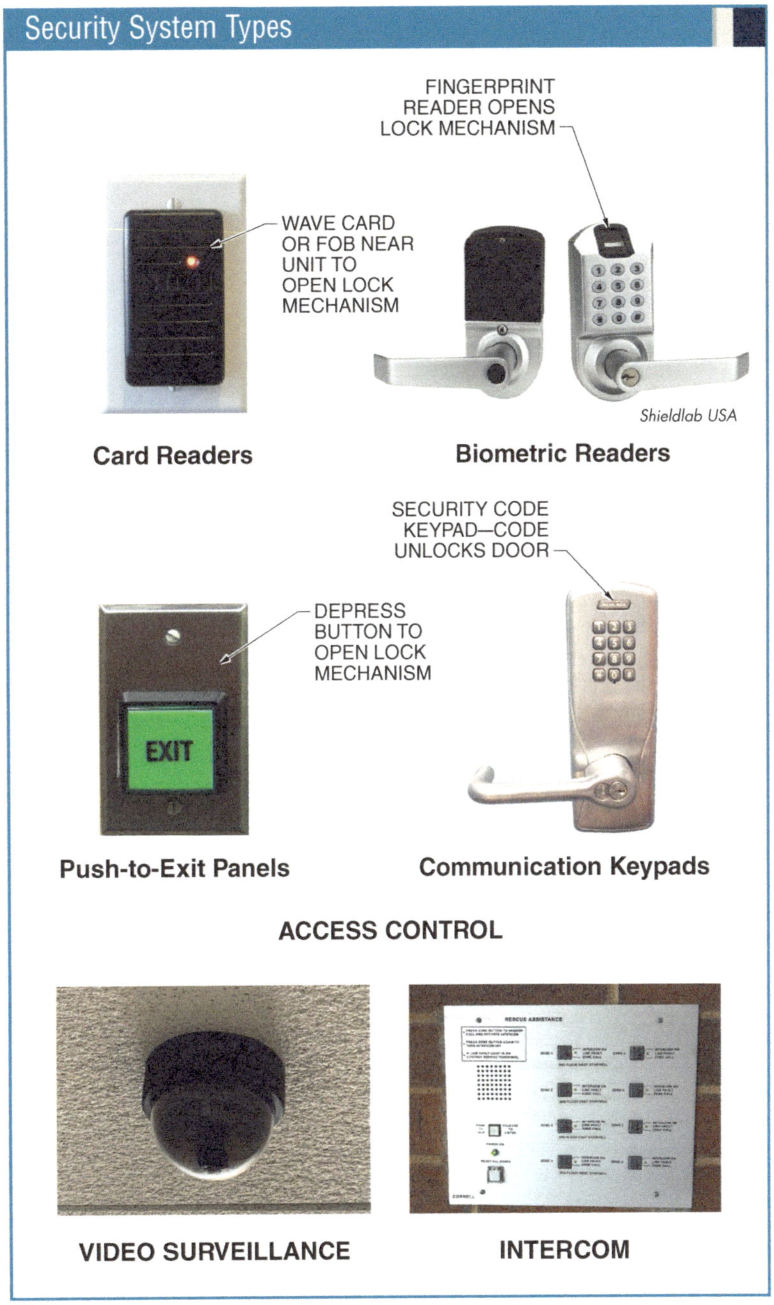

Figure 9-3. The separate types of systems that are commonly part of physical building security include access control, video surveillance, intrusion detection and notification, and intercom systems.

Access Control Systems

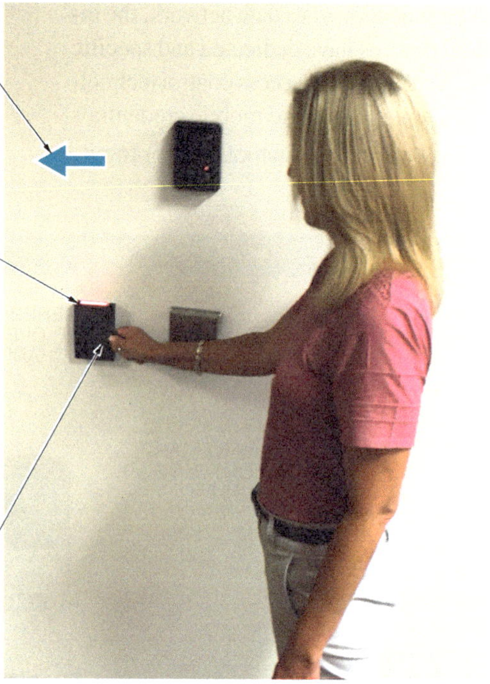

Figure 9-4. The most common type of access control system employs a device adjacent to a door for reading a badge, proximity card, smartphone app, or fob. The door either unlocks or refuses passage based on preprogrammed access levels.

In addition to building or premises access, access control technology is also used for other applications, such as time and attendance reporting, toll road technology, and asset tracking of individuals and physical objects. For example, infant abduction prevention systems in maternity wards, tracking devices for patients in assisted living facilities, inventory trackers in retail stores, and criminal evidence storage in law enforcement facilities may use access technology.

Traditional hardware and connections for card access control systems require multiconductor wiring to devices such as motion sensors, door release buttons, card readers, electromechanical door strikes, and door contacts. A card access control system uses a head-end access control module with multiple circuit boards to accommodate various amounts of doors and devices. These systems are expandable by floor, building, and other sites. Software provides the ability to add and delete users and assign rights and permissions to certain areas, based on the allowed levels of access.

Modern access control systems can be wireless. The amount and types of wire in the hardware for wireless access control systems is substantially reduced, is category type (CAT 5, CAT 5e, CAT 6, etc.), connects to the owner's network through Wi-Fi, and is identified by an IP address. When power is required to lock or unlock a door, rather than requiring power locally at the point of entry, it is delivered via power over Ethernet (PoE) or digital electricity. *Digital electricity* is a method of transferring high amounts of power over nonpower cabling. Digital electricity is accomplished by sending extremely short, high-voltage pulses for long distances over standard low-power cabling (such as limited-power or Ethernet cable) to a unit that accumulates and then transforms the energy into a substantial power source.

A *limited-power (LP) cable* is a cable that can carry both power and data in a single cable to supply power to equipment such as IP video surveillance cameras, intelligent LED lighting systems, and devices that use PoE or digital electricity. The pulses cease within 3 ms if there is a breach or unwanted contact in the cable. The horizontal category cabling is connected via the direct-attached method, where the cable is terminated with an RJ-45 modular head and connected directly to devices.

Video Surveillance Systems

A *video surveillance system* is a system that controls security cameras. Most legacy surveillance camera systems are connected with standard coaxial cables. Modern video surveillance cameras are network-ready cameras or IP cameras. IP cameras use the same type of four-pair category cables as a data network. The IP signaling used is the same as that used with a computer, which allows IP cameras to use the same types of switches, cabling, patching, and other components as a data network. However, because the volume of data involved with a video signal is much greater than the amount of data involved with other systems, the network used for video systems is usually separate from the data network.

Video surveillance systems record activity from motion, including unwanted intrusion, at the locations where they are installed to provide detection, deterrence, and observation. When video surveillance systems are integrated with other security systems, they enhance the overall protection of people and property. **See Figure 9-5.** However, the use of cameras varies by state and city privacy ordinances. Video surveillance systems must be designed by professionals with knowledge of the various installation requirements.

Video surveillance systems are used in virtually every location, including retail, residential, commercial, industrial, and institutional locations, as well as for health care facilities, transportation centers, public roads and walkways, and places of assembly. Modern camera technologies utilize artificial intelligence (AI) high-definition video analytics, which allows security professionals to detect, verify, and act in real time. AI software is designed to accurately recognize the movement of people and vehicles, differentiating them from motion that is not relevant.

Modern video surveillance systems are designed to reduce false alarms and send alerts via mobile applications directly to smartphones, tablets, and PCs. AI software for video surveillance equipment quickly sorts through hours of video footage and locates specified targets, such as individuals or motor vehicles, across an entire building, transportation system, institutional campus, or municipality. AI software continuously programs the system to identify typical activity, detecting and automatically sending alerts for unusual motion. AI software can also recognize, capture, and record motor vehicle license plates, as well as provide facial recognition of individuals within the range of the camera lens. AI software can also integrate with access control systems, allowing it to automatically act on door events or grant access and unlock doors directly from a camera view.

Video Surveillance Systems

Figure 9-5. Video surveillance systems can include interior and exterior cameras that can be observed from a separate video monitor.

TECH FACT

All targets of thermal image cameras radiate energy that is measurable on the infrared spectrum. As the target's heat level increases, it radiates more energy.

Cameras for video surveillance systems are high definition and provide enhanced image quality in a variety of lighting environments, such as those in parking lots, garages, building entrances, and hallways. The cameras can be surface mounted, recessed in a ceiling, mounted to an arm, or mounted in box-type housing, bullet-type housing, or mini-dome housing. Viewing options for cameras include pan-tilt-zoom (PTZ) or 360° fish-eye lenses that provide a panoramic view without blind spots. These types of cameras provide more coverage than standard cameras. In areas where lighting conditions are poor or dark, thermal image cameras, which provide thermal signatures of subjects based on their body heat, can be installed. **See Figure 9-6.**

Thermal Image Cameras

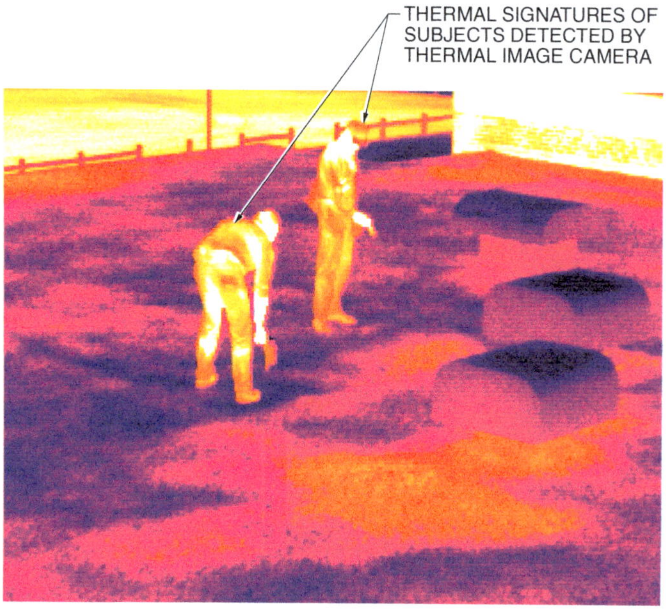

Figure 9-6. In areas where lighting conditions are poor or dark, thermal image cameras, which provide thermal signatures of subjects based on their body heat, can be installed.

When a security camera digitally captures an image, it is recorded by individual points known as pixels. A *pixel (picture element)* is the smallest unit area of a video screen that can be turned on or off or varied in intensity. Cameras are often classified by the resolution of the captured video, measured in megapixels (MP). *Resolution* is the raw number of pixels contained within an image. Security cameras are available with resolutions ranging from low to high. The higher a camera's MP rating, the higher the quality of the image captured by the camera. Camera resolutions typically range from a low resolution of 2 MP, or 2,000,000 pixels, to a high resolution of 40 MP, or 40,000,000 pixels. High-resolution cameras can also compress video data and transmit it to a network. The more advanced the quality and capabilities of a product, the higher its cost.

A *network video recorder (NVR)* is a video recorder that uses an IP camera and server to view, record, and play back an IP video stream over a network. The performance of a video surveillance system depends on the NVR and the software that controls it. An NVR functions as the management and storage system and allows for nonstop recording. NVRs reside on their own network or they can be part of the owner's IT network. NVRs allow different locations throughout a property to be recorded, locally or remotely, and viewed remotely on monitors or mobile applications. **See Figure 9-7.** Modern video surveillance systems are no longer hardwired with coaxial cable because they utilize category cable (CAT 5, CAT 5e, CAT 6, etc.) along with PoE, digital electricity, or fiber-optic cable.

Intrusion Detection and Notification Systems

An *intrusion detection and notification system* is a security system that monitors, detects, and sends notifications in the event of a breach or another security event in commercial, institutional, and industrial facilities in order to protect life and property.

Intrusion detection and notification systems are integrated with the video and access control systems. The hardware installed with intrusion detection and notification systems includes the following:

- control panel
- keypads
- door and window contacts
- motion sensors
- glass-break detectors
- interior/exterior sirens
- hold-up panic buttons
- photoelectric beam detectors
- water/temperature/humidity sensors
- gas and carbon monoxide (CO) detectors
- silent alarms
- voice (audio) alarm verification
- duress code systems
- driveway/exit alarm systems

The hardwiring for intrusion detection and notification systems includes various multiconductor cables constructed with a fire-rated sheath, sometimes a metallic shield or braid, and wire sizes ranging from 16 AWG to 22 AWG. The current trend for VDV cabling installation has it combined with the VDV scope. Typically, the cabling is composed of category (CAT) cable, connected with RJ-45 connectors to the network, and integrated with other security systems.

The control panel for an intrusion detection and notification system controls both notification and detection and, depending on the size of the system, can support a combination of hardwired and wireless devices for installation flexibility in multiple areas with perimeter and interior control. The control panel has an Ethernet port for connection to universal plug-and-play modules for IP alarm communication and remote programming, which is compatible with modern IP networks.

Most intrusion detection and notification system control panels provide remote access, which provides end users with the ability to control their security systems and view system cameras remotely from mobile devices such as smartphones, computers, and tablets. Intrusion detection and notification system control panels require a power supply with auxiliary or back-up power supplies and modules for different applications and future expansion. The modules are circuit-board type with wire terminal strips or interconnect wiring connectors for connections to the various input and output ports.

Network Video Recorders (NVRs)

Figure 9-7. NVRs allow local or remote recording and remote viewing on monitors or mobile applications.

Keypads used for intrusion detection and notification systems are typically more advanced than keypads used for singular systems and provide information beyond just an intrusion alarm. **See Figure 9-8.** Intrusion detection and notification system keypads are designed to be connected to cellular, Ethernet, and/or basic telephone networks. With modern keypads, addressable and wireless expansion is possible to meet application requirements.

Intrusion Detection and Notification System Keypads

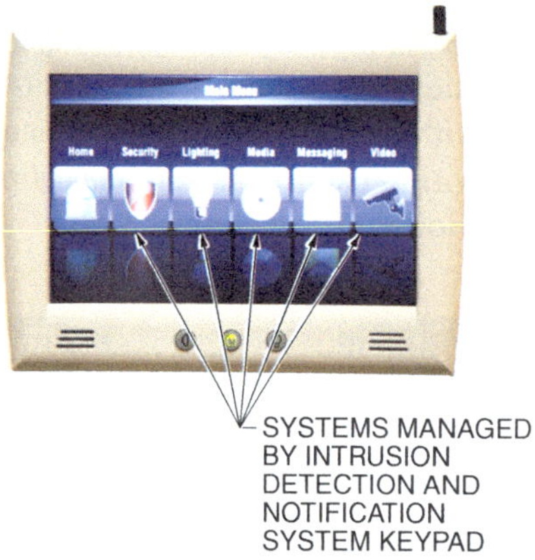

Figure 9-8. Keypads used for intrusion detection and notification systems are typically more advanced than keypads used for singular systems and provide information beyond just an intrusion alarm.

Because keypads are designed with extensive memory and hardware, an intrusion detection and notification system keypad is classified as a smart device that, when combined with software, can store relevant data and use field events to make intelligent decisions. For example, intrusion detection and notification system keypads can integrate and operate similar to a smartphone to provide the ability to arm, disarm, or verify the status of a security system from virtually any location through mobile apps. Keypads can also control smart devices such as HVAC thermostats, lighting, door locks, and garage doors. Smart keypad technology provides electronic alerts through emails or cellular text messages when a door is opened or when the security system is disarmed or tampered with.

Intercom Systems

An *intercom* is an internal communication system that allows direct intercommunication with another connected device within a building, building complex, campus, factory, or residential dwelling. Intercoms are typically accessed through an intercom panel that can be push-tone (user depresses telephone pushbuttons to connect with an extension device), automatic (once the telephone handset is removed from the cradle, another extension is automatically connected), or manual (the user connects with another device by depressing a single button, such as with a door buzzer entry system). Intercom panels can have speakers for communication purposes and a status panel that indicates the status of the connection to the user.

An intercom system can be integrated into a security system to regulate access to building entrances by delivering efficient communication. If access is allowed, the intercom system can also provide confirmation through two-way audio and close-up video to further secure the premises. Modern intercom system cameras often have a fish-eye lens. **See Figure 9-9.**

Intercom Systems

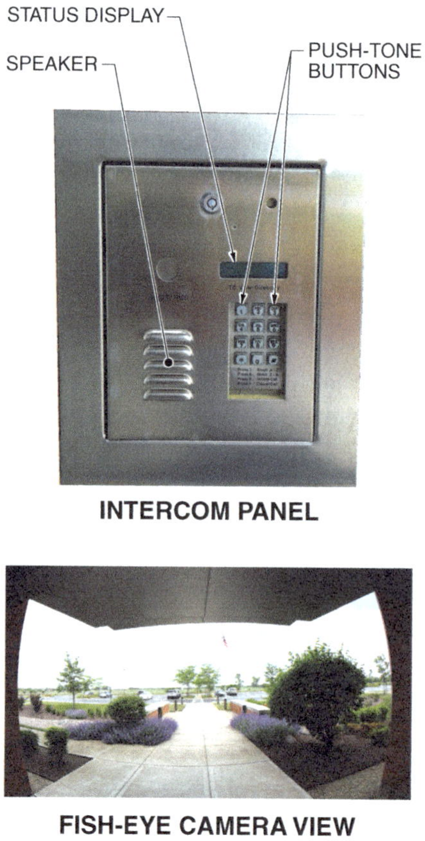

Figure 9-9. Modern intercom systems often have two-way audio and video capability with a fish-eye camera view.

Visitors seeking access to a building initiate the intercom call button, which is connected to a staffed master station. Individuals at the master station can then conduct two-way conversations with the visitors to determine whether building access should be allowed prior to remotely unlocking the entryway door. The video component protects against additional visitors from gaining entrance without permission. Intercom systems can also be integrated with mobile devices to provide remote monitoring without the need to staff a master station.

An intercom can also function as an emergency communication device after normal business hours. Emergency intercom devices can also be installed in public areas, such as institutional campuses and parking facilities. Emergency intercom devices can be stand-alone types or mounted to a facility wall or structural column. They can provide emergency strobes and beacons for visual communication and may also be connected with a video camera to provide additional information to first responders. Often, emergency intercom devices are connected directly to local law enforcement agencies for after-hours monitoring and to allow faster response times.

FIRE ALARM SYSTEMS

A *fire alarm system* is a building system used to notify building occupants of a possible fire upon the detection of smoke, heat, or both. Fire alarm systems are code driven, which means that the installer must follow strict guidelines when installing, programming, and testing the system. Installed fire alarm systems are approved and inspected by the AHJ.

Fire alarms have dedicated power-limited fire protection (PLFA) cables, devices, control panels, and communications systems for the announcement and reporting of a detected smoke or fire event. **See Figure 9-10.** A *power-limited fire protection (PLFA) cable* is a nonplenum-rated cable used for fire alarm installations. PLFA cables are not suitable for installation in risers, ducts, plenums, or other environmental air spaces.

Figure 9-10. Fire alarms have dedicated PLFA cables, devices, control panels, and communications systems for the announcement and reporting of a detected smoke or fire event.

THE LIFE SAFETY CODE

NFPA 101®, Life Safety Code®, is a standard used for strategies to protect building occupants based on building construction, protection, and occupancy features that minimize the effects of fire and related hazards. The *Life Safety Code* is the only document that covers life safety in both new and existing structures. It specifically covers protection from hazards to human life in buildings as well as in other public and private human occupancies (such as vehicles and vessels), but only when those occupancies are permanently fixed to a foundation, attached to a building, or permanently moored for human habitation. Insurance companies also use the *Life Safety Code* to evaluate risks and set rates as well as assess compliance after an incident. **See Figure 9-11.**

Security systems must be designed to meet the *Life Safety Code* and are required to integrate with additional systems, such as mass notification systems, fire alarm systems, elevators, and information technology networks connected through internet protocol (IP). When a revision to the *Life Safety Code* is adopted into local law, existing structures may have a grace period but must eventually comply with the *Life Safety Code*. In some cases, the AHJ can permit previously approved features to be used under specified conditions.

NFPA 101®, Life Safety Code®

Figure 9-11. The *Life Safety Code* is a standard used for strategies to protect building occupants based on building construction, protection, and occupancy features that minimize the effects of fire and related hazards.

The *Life Safety Code* is coordinated with hundreds of other building codes and standards, such as electrical, fuel-gas, mechanical, plumbing (for sprinklers and standpipes), energy, and fire codes. The *Life Safety Code* is also integrated with local codes and often requires that the individuals installing life safety systems comply with local public safety laws and be licensed or registered in order to protect the consumer.

Public safety licenses are often required by states and counties for security system installers and are required in addition to electrical and systems technician licenses. Public safety licenses often require criminal background checks and must be renewed at intervals of one to three years, depending on the issuing municipality. Also, the journeyworkers involved with security installations require extensive knowledge of building and electrical codes as well as strong network technologies skills. Installations may also require coordination with local inspectors, law enforcement, fire marshals, AHJs, and information technology personnel.

SPECIALTY SECURITY SYSTEM APPLICATIONS

Most security system applications can be applied to construction projects that require public safety, such as governmental buildings, homeland security checkpoints, health care facilities, commercial buildings, places of assembly, entertainment venues, retail facilities, residential dwellings, critical infrastructure, and educational facilities. However, certain applications may require specialty security systems, such as active shooter response and lockdown systems and cyber security systems.

Active Shooter Response and Lockdown Systems

Due to the increase of active shooter incidents, a relatively new security application known as an "active shooter response and lockdown system" has been developed. While these systems can be installed in any facility, they are typically installed in school buildings and campuses, from the elementary through college levels. Through the integration of security systems and real-time active shooter hardware and software, law enforcement can receive real-time alerts, rather than waiting for multiple 911 emergency calls. This allows first responders to arrive simultaneously at a specific location. In addition to notifying first responders, these systems can detect gunshots, both inside and outside of a building, and automatically lock down the affected buildings.

Active shooter response and lockdown systems work with video surveillance systems, which can also be used to provide law enforcement with the ability to gather evidence and prevent further shootings and violence. Active shooter response and lockdown systems can also provide mass notification and mobile alerts to the general public. The features of active shooter response and lockdown systems include the following:

- capability for full-facility automated lockdown when gunfire is detected
- integration with most intercom systems and mass notification systems for alerting building occupants of potential threats
- easy to install, with the capability to put the entire building security under one system
- provides critical information to law enforcement and building officials to enable immediate and appropriate actions
- ability to control multiple buildings with multiple scenarios
- enables facility access control to deny access to unwanted individuals

Cyber Security Systems

Cyber security, also referred to as information technology security, is the convergence of the technologies, processes, and practices designed to protect networks, devices, programs, and data from attack, damage, or unauthorized access. With security systems converging to the network platform and the physical components utilizing network IP access, securing them is vital to performance and stability. Technology and cyber security experts estimate that by the year 2022 there will be over 200 billion devices connected to the internet and billions more connecting to customer-owned networks.

Cyber security project specifications must have detailed management software to complement the integrated security software solutions. Cyber security systems are typically designed by software engineers and programmers tasked with protecting critical information systems. The three main entities that must be protected are endpoint devices (such as computers, smart devices, and routers), network systems, and the Cloud. The technology and systems used to protect these entities include firewalls, domain naming system (DNS) filtering, malware protection, antivirus software, and email security solutions. While most VDV installers are not directly involved in cyber security, they should be familiar with the systems and methods used to prevent cyber attacks.

SECURITY SYSTEM PATHWAYS AND HARDWARE

The pathways and spaces used for security systems are similar to the pathways installed for VDV systems, often sharing space with other VDV systems in cable trays, conduits, and other pathways. To prevent installation errors, careful examination of drawings and specifications for pathways, spaces, and connecting hardware must be performed. *Note:* Many requirements for security system pathways and hardware are often specified and shown in other specification sections and other nonelectrical trade drawings.

Security System Pathways

A *pathway* is a set of devices or spaces used to route telecommunications cables from a telecommunications room or equipment room to a work area. Pathways include conduit (metal and plastic), cable trays, ducting, plenums, J-hooks, metal and plastic surface raceways, ladder trays, wire mesh cable trays, and cellular metal floor raceways. The type of pathway used varies with the type of construction and the customer's requirements. Often, more than one type is used. For example, wire-mesh cable trays and J-hooks serve the same purpose.

A cable support system must be in place before a cable can be installed. **See Figure 9-12.** Normally, VDV technicians install cable support systems. However, in some projects, electricians may install the system on jobs that require extensive use of electrical metallic tubing (EMT). *Note:* The NEC® does not allow ceiling support wires or any part of a ceiling support system to be used to support VDV cables. Cables cannot be laid directly on ceiling tiles.

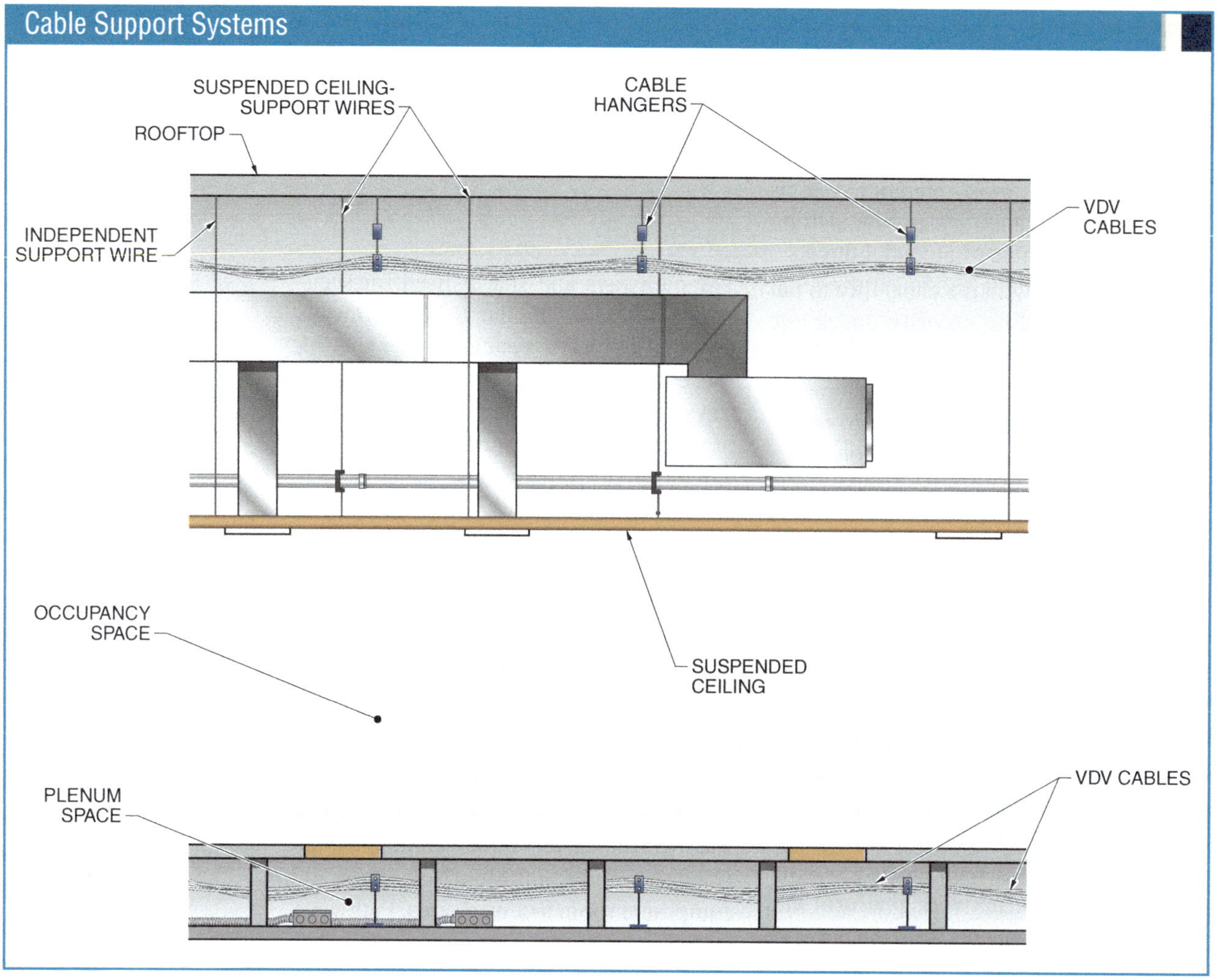

Figure 9-12. A cable support system must be in place before a cable can be installed.

Security System Hardware

A major component of any security specification covers the doors and hardware to be installed in the building under construction. The coordination of the installation of doors and hardware may affect several trades and be specified in several MasterFormat divisions. For example, doors and hardware are specified in Division 08, conduits and raceways in Division 26, wire and cable in Division 27, and network equipment in Division 40; or the complete list of hardware could be specified in Division 28 only. The methods used for installation can vary by region, OEM recommendations, codes, and the level of experience and expertise of the VDV system installer.

Common designs used with electromechanical door strikes and door contacts are the failsafe and fail-secure designs. With a failsafe design, the door's locking mechanism defaults to an unlocked state when the power is lost or a fire alarm is triggered, requiring power to lock the door. With a fail-secure design, the door's locking mechanism defaults to a locked state when power is lost or a fire alarm is triggered, requiring power to unlock the door. **See Figure 9-13.** Whether the door strike should be the failsafe or fail-secure design depends on the function of the door lock, its application, the locations of the building's egress exits, and whether it is in a stairwell exit.

When an electric strike is used on an egress stairwell fire-rated door exit, it must be a fail-secure design in the event of a fire alarm, according to NFPA 80, *Standard for Fire Doors and Other Opening Protectives*. Therefore, fire-rated doors on stairwells must be latched closed in the event of a fire and the electric strikes (but not the locksets themselves) on those doors must default to a locked state (fail-secure design). The fail-secure design allows the door's lock to latch closed but also allows an occupant to use the door's lockset, which is failsafe, to freely egress.

As a standard practice in the event of power loss or a triggered fire alarm, most doors with electromechanical locks are typically of the failsafe design. For doors where an electromechanical double-cylinder is used, the failsafe design should not be used with any door that is required for egress. *Note:* Double-cylinder classroom (institutional) security locksets designed to allow free egress are exceptions to this practice. Institutional-function locksets are sometimes double-cylindered and specialized to be fail-secure; however, their proper applications can vary. The local AHJ and/or building codes must be consulted for proper specifications.

Failsafe vs. Fail-Secure Electromechanical Door Strikes

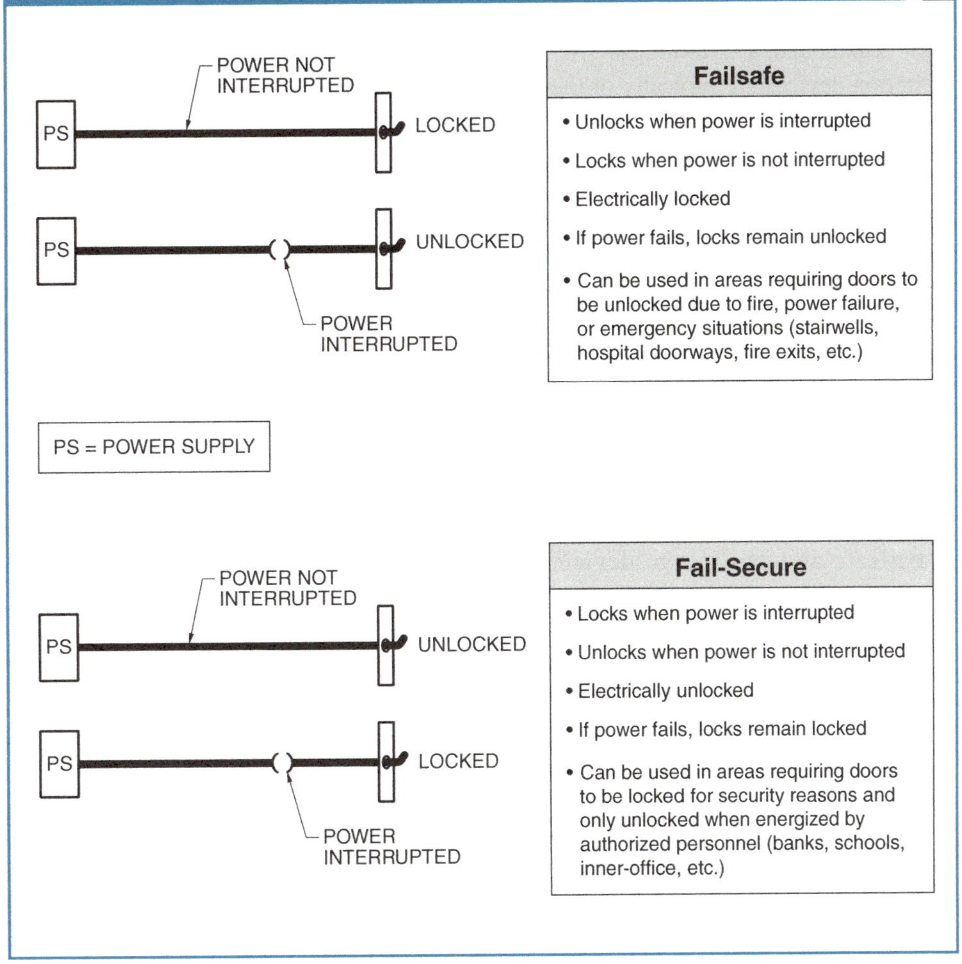

Figure 9-13. With a failsafe design, the door's locking mechanism defaults to an unlocked state when the power is lost or a fire alarm is triggered, requiring power to lock the door. With the fail-secure design, the door's locking mechanism defaults to a locked state when power is lost or a fire alarm is triggered, requiring power to unlock the door.

Electric panic hardware trim usually works on the opposite side of panic or fire exit hardware, as lever-operated trim. However, since it is on the opposite side of the door, its requirement to be failsafe or fail-secure is different from application to application. For example, on a re-entry door for a stairwell, failsafe electrified panic hardware trim is used because of the requirement to move freely from either side of the door in the event of a power failure or triggered response. Alternately, any other use of electric panic hardware is typically fail-secure.

An electrified latch retraction (crash bar) with electromagnetic locks is a panic hardware device or fire exit hardware device that allows for latch retraction when power is applied. When power is not applied, the latch is extended. Therefore, electric latch retraction devices are typically of the fail-secure design.

Electromagnetic locks are typically a failsafe design only. When power is removed from an electromagnetic lock, it changes to an unlocked state. Because an unlocked state only occurs when power is removed from the electromagnetic lock, it requires a release mechanism for egress. Examples of release mechanisms include request-to-exit switches in panic hardware, pushbutton actuator and sensor combinations, and fire alarm and sprinkler system triggers.

Failsafe and fail-secure devices are subject to approval from the AHJ. Building codes must be researched prior to the installation of failsafe or fail-secure devices. In addition to electric door strikes, other components that are part of failsafe or fail-secure devices are electromagnetic locks, electric mortise locks, electric cylindrical locks, and IP point-of-entry integrated locks. The wires for electromechanical door strikes and door contacts are typically connected using wire nuts, Type B connectors (this connector is often referred to as a "beanie" or by the tradename "Dolphin"), or modular RJ-45 connectors. **See Figure 9-14.**

SECURITY SYSTEM ABBREVIATIONS AND SYMBOLS

A separate set of abbreviations and symbols is used to depict security system devices and components. Typically, the security system abbreviations and symbols used on a print are part of ANSI standards. Security system abbreviations and symbols represent cameras, card readers, motion detectors, security control panels, and other security-related equipment. **See Figure 9-15.**

Although standards exist for security system symbols, some security system designers and manufacturers prefer to use their own sets of symbols. While some OEM symbols resemble standard symbols, most possess minor differences. **See Figure 9-16.** Security system abbreviations and symbols with definitions are located on the legend sheet of a security system project.

Security System Floor Plans

The size and complexity of a security system determines whether a separate floor plan must be provided. While a large, complex security system requires a separate security system floor plan, a less-complex, basic security system may be depicted on a fire alarm and life safety floor plan or on a power floor plan. A security system is more likely to be shown on a fire alarm and life safety floor plan than a power floor plan because the fire alarm and life safety floor plans are similar to security system designs.

A security system floor plan uses the plan view of the building floor. In addition to showing elevator shafts, doors, plumbing fixtures, rooms, walls, and windows, the security system floor plan also shows the location of door contacts, motion detectors, cameras, security control panels, and items that interface with the security system. **See Figure 9-17.** Other drawings and information related to the security system, such as detail drawings, elevation drawings, and/or schedules, may also be shown on the security system floor plan.

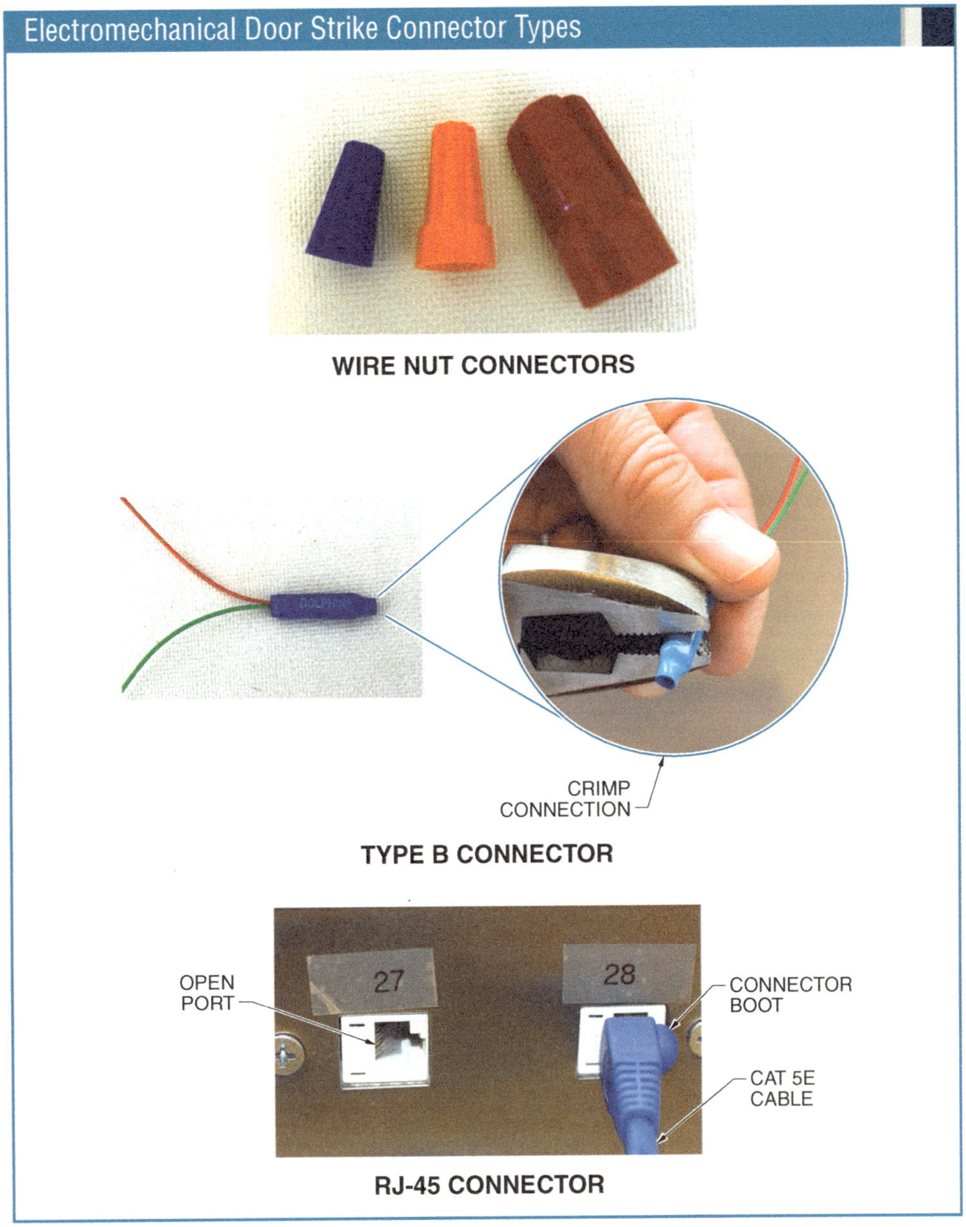

Figure 9-14. The connection of wires for electromechanical door strikes and door contacts typically utilize wire nuts, Type B wire connectors, or modular RJ-45 connectors.

A security system floor plan provides information about the location and mounting of devices and components that are connected to the security system. Often, the security system floor plans do not have lines drawn from the devices back to the security system control panel. Special care must be taken to verify that devices and components are properly mounted according to the print specifications. The effectiveness of the security system depends upon the proper placement of the devices. For example, a motion detector must be properly positioned in order to detect the presence of an intruder. Additional information that may be on a security system floor plan includes security camera viewing angles, security system zones, the manufacturer's name and part numbers for security-related items, security hardware for doors, windows, roof-top hatches, and outdoor floodlight locations.

Security System Abbreviations	
CPS	Camera power supply
CR	Card reader
MTV	CCTV monitor outlet
DB	Doorbell
DBZ	Door buzzer
DRCN	Door contact
DVR	Digital video recorder
ELDO	Electric door opener
ELDS	Electric door strike
ELMG	Electromagnetic lock
FPMPV	Flat panel monitor public view
GB	Glass break
HU	Hold-up button
IC	Intercom unit
ICM	Intercom master station
ICS	Intercom substation
KS	Keyswitch
MI	Master intercom and directory unit
ML	Magnetic lock-door alarm
MON	Monitor
MD	Motion detector
RB	Release button
DC	Security door contacts
SP	Surge protector
TA	Trailer alarm
TS	Touch sense bar

Figure 9-15. Security system abbreviations depict card readers, motion detectors, security door contacts, and other security-related equipment.

Security System Symbols		
Device	**Abbrev.**	**Symbol**
CCTV CAMERA 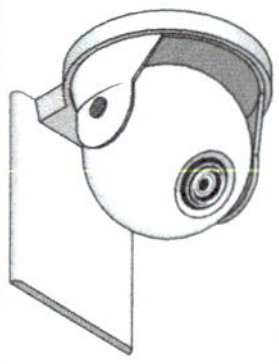	CCTV	
	TVWP	
CCTV COAXIAL CABLE OUTLET 	CCCO	
DOOR CHIME 	DC	
SIREN 	SRN	
FLAT PANEL MONITOR 	FPM	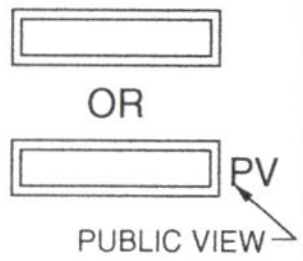
KEYPAD 	KPD	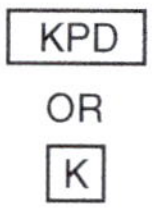
CONTROL PANEL 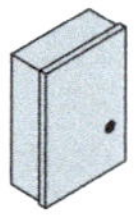	SCP	
CARD READER 	SCR	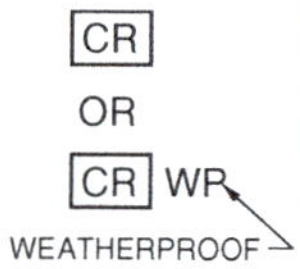

Figure 9-16. Symbols for access control systems and other types of security systems that come from the manufacturer can have minor differences.

Security System Detail Drawings

Security system detail drawings provide additional information that cannot be shown on a security system floor plan. Security system detail drawings include elevations, installation instructions, and schedules. **See Figure 9-18.** Elevation drawings provide a view of how security devices, such as door contacts and electric door strikes, interface with building components. *Note:* Elevation drawings may or may not be drawn to scale.

Security System Floor Plans

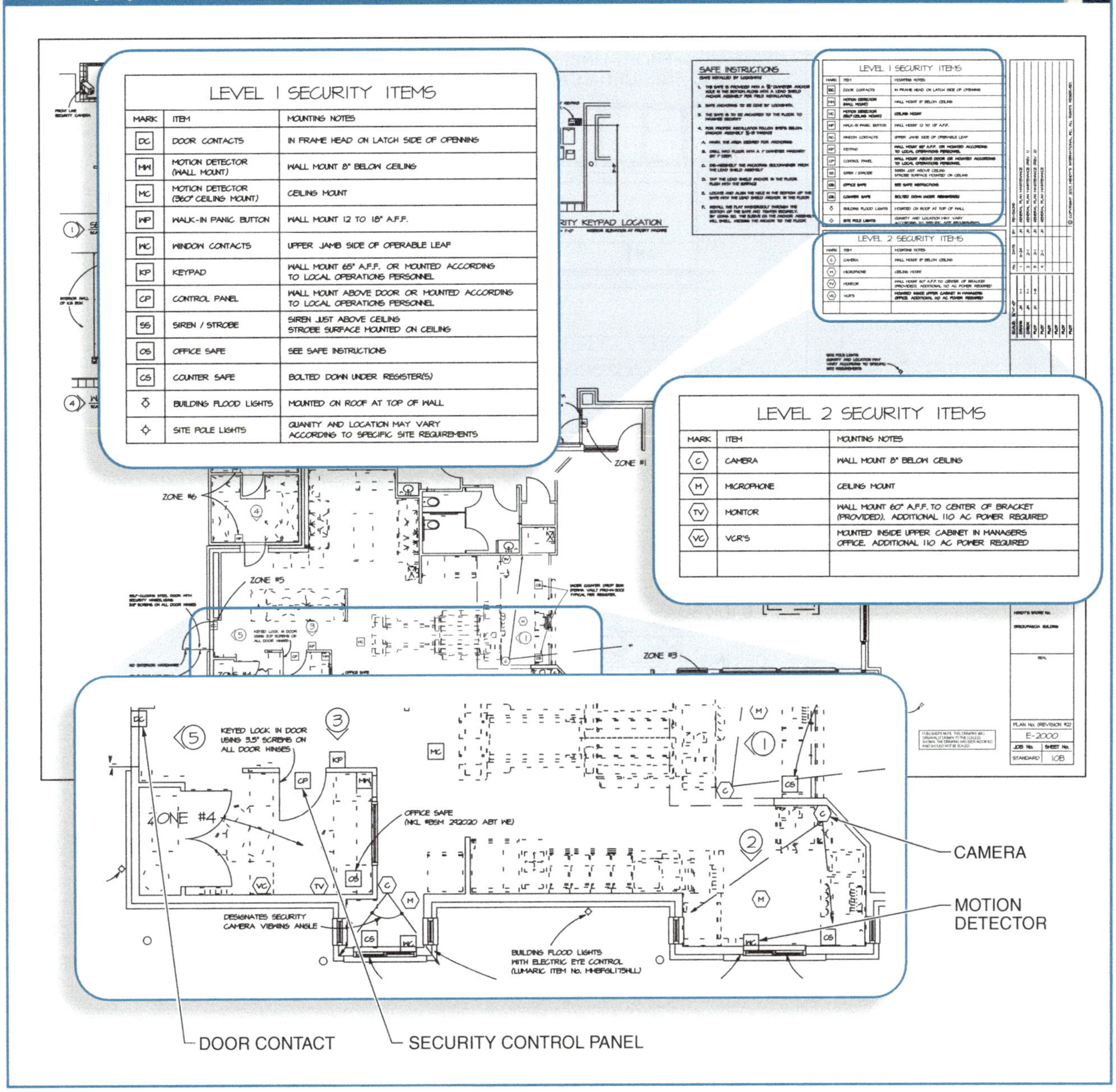

Figure 9-17. Security system devices and components are shown on security system floor plans.

Installation instructions for security systems provide a step-by-step guide for installing a specific security-related item. A tradesperson must not deviate from the installation instructions unless receiving authorization to make changes from the security system designer or architect.

Schedules provide a summary of the installation information for security devices and components. A schedule includes the symbols for devices and components, symbol definitions, and relevant installation and mounting information per the manufacturer's instructions.

Security System Detail Drawings

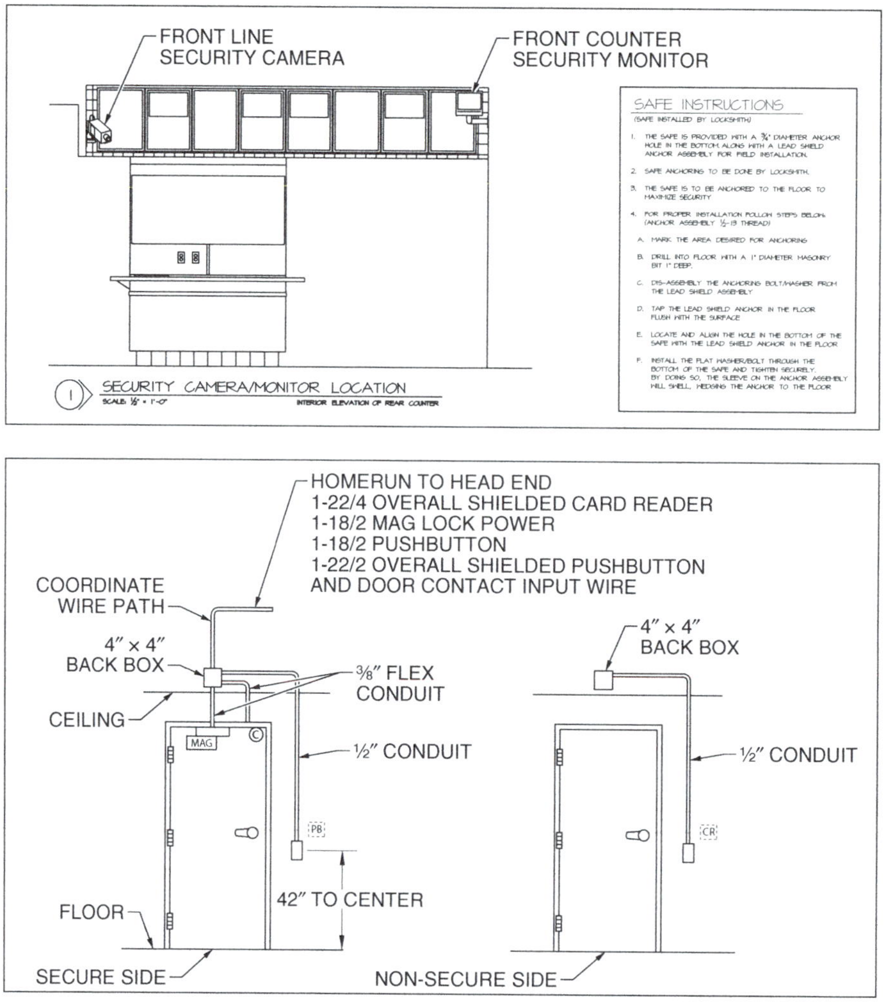

Figure 9-18. Security system detail drawings provide additional information to assist when installing or troubleshooting security systems.

SUMMARY

Security systems are used in every type of application, from industrial facilities to residential buildings. Security systems are designed by professional designers with certifications from ASIS, such as the PSP or CPP certification. Professional designers properly procure the equipment needed for the different types of security systems specified. Security systems are designed to meet the *Life Safety Code* and can include specialty systems such as active shooter response and lockdown and cyber security systems. Security systems require dedicated pathways and electromechanical hardware for doorways and other entry points.

Security Drawings and Specifications

CHAPTER 9

Review

Name____ Date ____________________

True-False

T F **1.** The Physical Security Professional (PSP) certification is a professional certification for individuals with a minimum of two years of experience in the physical security field who have successfully passed a board certification licensing exam.

T F **2.** The Certified Protection Professional (CPP) certification is a professional certification for individuals with a minimum of five years of security experience in the physical security field, three of which must be in a responsible charge of a security function, who have successfully passed a board certification licensing exam.

T F **3.** The practice of procuring security system proposals can vary by geographic region.

T F **4.** The most common type of access control system employs a device adjacent to a door that reads a badge, proximity card, smartphone app, or fob; while the door either allows or refuses passage based on preprogrammed access levels.

T F **5.** The higher a camera's MP rating, the higher the quality of the image captured by the camera.

T F **6.** Modern video surveillance systems are only hardwired with coaxial cable.

T F **7.** An intrusion detection and notification system keypad is classified as an analog device.

T F **8.** The Life Safety Code is the only document that covers life safety in both new and existing structures.

T F **9.** The coordination of the installation of doors and hardware may affect several trades and be specified in several MasterFormat divisions.

T F **10.** Electric latch retraction devices are typically fail-secure design.

Multiple Choice

______ **11.** ___ is a professional organization for security professionals that issues certifications, standards, and guidelines for the security profession.

A. ASIS International (ASIS)
B. The Global Security Exchange (GSX)
C. The International Code Council (ICC)
D. The Securities and Exchange Commission (SEC)

______ **12.** ___ is a method of transferring high amounts of power over nonpower cabling.

A. Bluetooth
B. Digital electricity
C. Fiber optics
D. Wireless fidelity

______ **13.** ___ is a standard used for strategies to protect building occupants based on building construction, protection, and occupancy features that minimize the effects of fire and related hazards.

A. The International Building Code (IBC)
B. The International Fire Code (IFC)
C. NFPA 101, *Life Safety Code*,
D. NFPA 70, *National Electric Code*,

______ **14.** When an electric strike is used on an egress stairwell fire-rated door exit, it must be a fail-secure design in the event of a fire alarm, according to NFPA ___ requirements.

A. 70E
B. 80
C. 95
D. 101

______ **15.** A(n) ___ is a set of devices or spaces used to route telecommunications cables from a telecommunications room or equipment room to a work area.

A. cable cluster
B. equipment rack
C. pathway
D. splitter

Completion

______ **16.** A(n) ___ system is a security system that controls door locks.

______ **17.** Most legacy surveillance camera systems are connected with standard ___ cables.

______ **18.** A(n) ___ cable can carry both power and data in a single cable to supply power to equipment such as IP video surveillance cameras, intelligent LED lighting systems, and devices that use PoE or digital electricity.

______ **19.** A(n) ___ is the smallest unit area of a video screen that can be turned on or off or varied in intensity.

_______________ **20.** A(n) ___ cable is a nonplenum-rated cable used for fire alarm installations.

_______________ **21.** ___ is convergence of the technologies, processes, and practices designed to protect networks, devices, programs, and data from attack, damage, or unauthorized access.

_______________ **22.** With a(n) ___ design, the door's locking mechanism defaults to an unlocked state when the power is lost or a fire alarm is triggered and requires power to lock the door.

_______________ **23.** Electromagnetic locks are typically only of the ___ design.

_______________ **24.** A PSP certification can be attained with a bachelor's degree or higher from an accredited post-secondary institution plus ___ years of progressive experience in the physical security field.

_______________ **25.** A CPP certification can be attained with a bachelor's degree or higher from an accredited post-secondary institution, a minimum of ___ years of security experience (with a minimum of three years in a responsible security function position), and having been employed full-time in a security-related role.

Short Answer

26. List ten hardware components installed as part of intrusion detection and notification systems.

27. List five active shooter response and lockdown system features.

28. Explain where access technology is used in addition to buildings or premises.

29. List five types of hardware components that are part of failsafe or fail-secure devices.

30. List five types of information that may be located on security system floor plans.

Security Drawings and Specifications

CHAPTER 9

Activity

Name ______________________________ Date ______________

SECURITY SYSTEM

Instructions

Create a list of materials that would be included in a proposal by using the drawings, legends, and specifications to determine material and connecting hardware types and quantities for the security system. List any qualifications or exclusions required that may not be on any of the drawings. Categorize the work by system type.

Project Overview

The security system is to be installed in a workplace day-care center in a commercial office facility. The security system has the following requirements:

- Building size: 25,000 sq ft
- System types: access control, video surveillance, intrusion detection, and intercom
- Final system to be integrated with all system types to provide a single security management solution
- System to reside on company network to integrate with other non-security building systems
- System to be monitored locally, remotely, and by municipal authorities
- Project is design-assist, with drawings and specifications to be used as a guideline for proposal development
- Review of drawings and specifications required to determine scope of work
- Proposal to include description of each system with quantities
- Proposal to comply with local building codes
- Marked-up drawings showing locations and elevations for equipment room (ER) to be included
- All devices shown shall be considered new

Drawings:

- TY001 – Security Legend and Abbreviations
- TY200 – Security Details
- TY101 – Security Floor Plan

Specifications:

- Security – 28000, Part 1 – General Conditions
- Security – 28000, Part 2 – Products
- Security – 28000, Part 3 – Execution

Note: Drawings and specifications can be accessed in the online learner resources.

Access Control System

1. Devices/Quantity:

2. Conductors/Quantity:

3. Connecting hardware/Quantity:

4. System components/Quantity:

5. Integration required:

Video Surveillance System

6. Cameras/Quantity:

7. VDV cable/Length:

8. Connecting hardware/Quantity:

9. System OEM components:

10. Integration required:

Intrusion Detection System

11. Devices/Quantity:

12. Wire/Length:

13. Connecting hardware/Quantity:

14. Software:

15. Integration required:

Intercom System

16. Devices/Quantity:

17. Wire/Length:

18. Connecting hardware/Quantity:

19. System OEM components:

20. Integration requirements:

Main Distribution Frame (MDF) Room

21. Sleeves/Quantity:

22. Cable runways:

23. Equipment racks:

24. Plywood size:

25. Ducts and covers:

26. Grounding system:

27. Firestopping system:

28. Electrical/Quantity:

29. MDF room layout, wall, and rack elevations:

30. Additional questions/qualifications/exclusions:

CHAPTER

10

Audiovisual System Drawings and Specifications

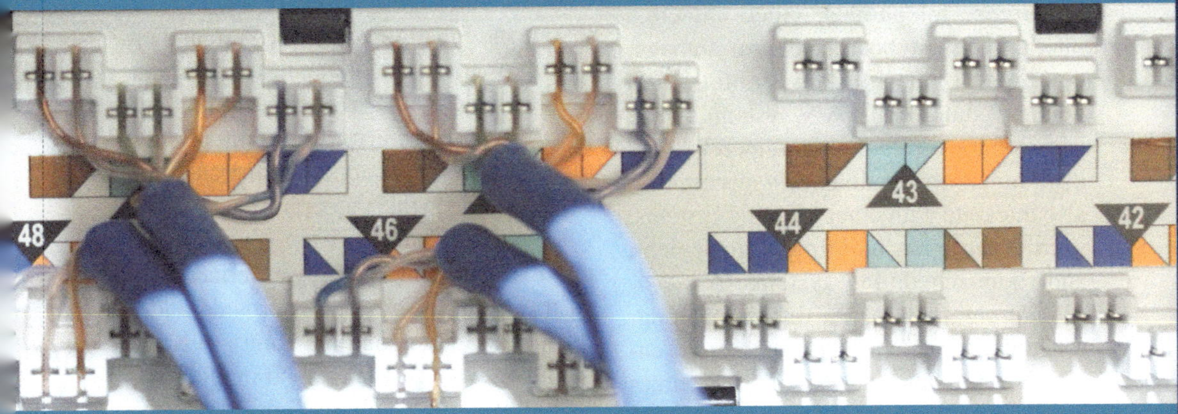

Audiovisual (A/V) systems have evolved from use in educational applications only to use in common residential and commercial applications. A/V systems are designed by professionals with specialized credentials and training. A/V system designs are created to adhere to specific system standards and network integration through pathways and spaces. A/V systems are installed in accordance with specific drawings and specifications. Upon completion of the A/V system installation, the system is verified and an inventory is taken to verify that the system has been installed properly and within specifications.

Objectives

- Describe typical audiovisual (A/V) system applications.
- List the qualifications of A/V system design professionals.
- Explain A/V system standards.
- Describe A/V system network integration.
- Explain how A/V system drawings and specifications are used for system installation.

Learner Resources

atplearningresources.com
access code 783821

AUDIOVISUAL SYSTEM APPLICATIONS

An *audiovisual (A/V) system* is a digital voice-data-video system that is a combination of sound and video elements. A/V systems consist of hardware that includes conference telephones, video cameras, interactive whiteboards, digital signage, computers, smartphones, tablets, and wireless connectivity. A/V systems can be installed for residential, commercial, and institutional applications. A/V systems can also integrate with other building systems, such as lighting, building management, and room-scheduling systems. A *room-scheduling system* is a digital calendar scheduling system used in commercial environments to allocate space and resources for upcoming events and meetings.

Residential applications for A/V systems include home theaters and other related devices such as high-definition television monitors, video projectors, video projector screens, and ceiling/wall-mount speakers. Historically, the A/V systems installed in residential applications were typically only for high-end and luxury residences due to the high cost of the systems. As A/V technology evolved, most applications have become more cost effective, resulting in effective and affordable A/V systems and products.

TECH FACT

Most audiovisual (A/V) system monitors and projectors can be installed with specially designed ceiling and wall mounts that hide any cabling from view.

Commercial applications for A/V systems include office spaces, hospitality venues, small health care facilities, theaters, broadcasting facilities, entertainment and sports venues, retail buildings, and public spaces related to transportation such as airport, railroad, and bus terminals. Commercial A/V systems provide advanced communication, collaboration between different systems, and the ability to share information over several digital platforms. **See Figure 10-1.**

For A/V system applications, the system requirements, equipment requirements, and costs are considered for the design, installation, and functionality of the system. For example, commercial retail applications require systems that attract customers by providing information or point-of-sale digital signage that appeal to the customers and enhance their shopping experience. Audio systems, such as background music, provide a pleasant shopping experience for customers. Paging systems provide additional information and the ability to communicate with customers and employees. Visual systems, such as digital signage, provide advertisement and information for select products in the store inventory.

Institutional applications include facilities such as educational campuses (schools), large office complexes, and hospitals. A/V systems for educational facilities are primarily designed to be a tool for instructors to present curricula, enhance class assignment collaboration, and test and grade students. A/V systems are used with digital displays, multimedia and interactive testing system platforms, student response systems (SRSs), and tablet and smartphone applications.

A *student response system (SRS)* is a digital educational platform that allows instructors to electronically pose questions and gather student responses during classroom (in-house or virtual) instruction. Student response systems are also commonly referred to as clickers, classroom response systems, personal response systems, or audience response systems. An SRS provides real-time assessment of student progress and comprehension of the material presented. Similar systems are used for office and hospital campuses to communicate with staff and visitors. Modern A/V systems can integrate with other building systems as part of smart building technology that utilizes the local area network (LAN).

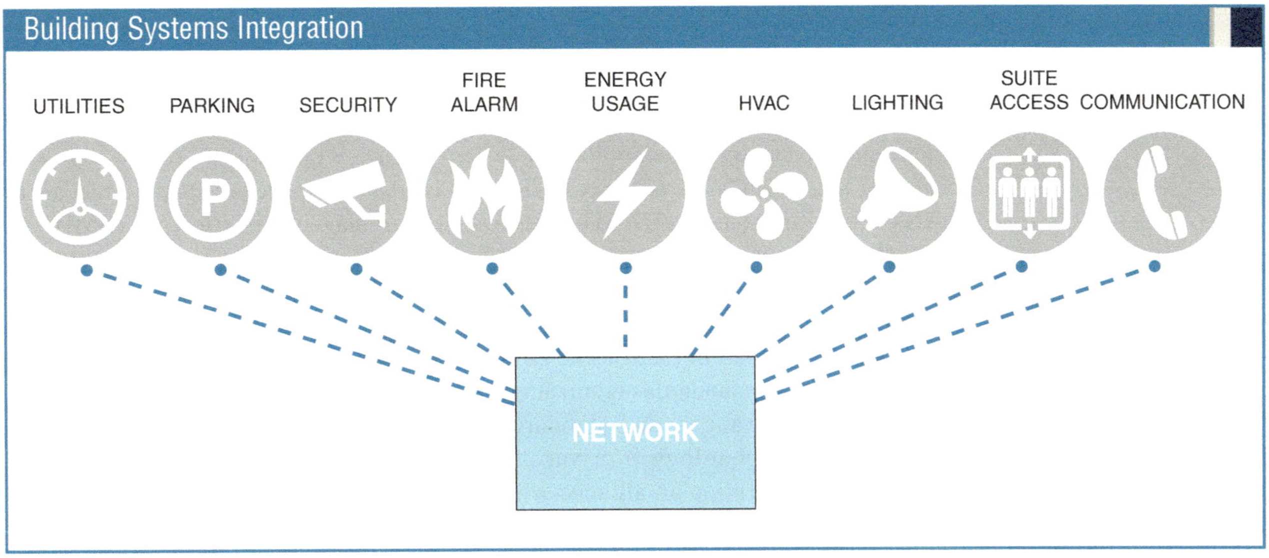

Figure 10-1. Modern A/V systems can integrate with other building systems as part of smart building technology that utilizes the LAN.

A/V SYSTEM DESIGN PROFESSIONALS

A/V system installations require skilled design and installation professionals who can provide standards-based designs that are installed and integrated with building and electrical codes. A/V applications, similar to other VDV applications, have evolved from being proprietary to standardized. Several industry associations, OEMs, and design professionals specialize in the design, collaboration, and coordination of construction specifications and drawings for A/V systems. These include professionals designated as information and communication technology (ICT) professionals with various levels and categories of design and installation certification. Certifications can be designated for individuals or for companies that specialize in specific installations.

OEMs of relevant A/V equipment, along with systems integrators and design consultants, may also be part of the A/V system planning and design team. Additionally, other design team members include electrical engineers, professional engineers, and RCDDs. Installation personnel can include factory-certified technicians, industry-certified or industry-licensed electricians, and VDV technicians. A/V system design professionals participate in the design process and in the development of the project construction documents. These professionals may also be consulted to review the accuracy of bid proposals, equipment submittals prior to installation, and the final commissioning of the project.

A/V SYSTEM STANDARDS

Standard-based designs for A/V products reduce costs over the usable life of the system and provide accommodation for future moves, adds, and changes (MACs). A/V products, equipment, and wiring are typically proprietary. Custom design and installations are completed by using best practice methods. Installed legacy A/V systems currently in use require that design professionals have experience in design and the ability to customize cables and equipment through various types of cables and connectors, which requires specialized skills to terminate, test, and troubleshoot. For example, RGB, RCA, BNC, VGA, mini-VGA, DVI, and DIN connectors are legacy interfaces that require additional expertise to install and troubleshoot.

As with other VDV systems, standardization has been implemented in parallel with the advancements in A/V technologies. *HDBaseT* is a firmware-based consumer electronics connectivity technology that enables a single standard CAT 5e or CAT 6 LAN cable with a standard RJ-45 connector to simultaneously carry uncompressed 10.2 Gbps high-definition video and audio, 100 Mbps of data, and 100 W of power over a distance of 100 m (328′). The *HDBaseT Alliance* is a standards organization that promotes, advances, and standardizes HDBaseT technology in a variety of industries by forming an alliance with manufacturers and other organizations.

In addition to the A/V industry, HDBaseT technology is also used in the medical, industrial, and residential consumer markets. For A/V systems, consumer electronics, and even industrial PCs, HDBaseT technology uses a category cable (CAT 5e or above) for up to 100 m (328′). For medical and government applications, fiber-optic cable is preferred over category cable because it can be used for longer distances.

HDBaseT technology has improved video distribution by providing increased transmission distance and better performance than legacy solutions. HDBaseT technology provides connection of uncompressed ultra-high definition (UHD) digital A/V, Ethernet, power, control signals, and USB-connected devices over a single cable. **See Figure 10-2.** The benefits of HDBaseT technology include the following:

- long-distance transmission—100 m (328′)
- signal convergence—audio and video, Ethernet, USB interfaces, and power over a single cable, which reduces infrastructure costs and the amount of cable pathways
- performance—4K and 8K UHD video
- simplicity—category (LAN) cable or UTP cables and standard RJ-45 interfaces
- standardization—HDBaseT is an industry standard, facilitating interoperability between different OEM products

HDBaseT applications for A/V systems include displays such as digital signage, video walls, and projectors. There are several factors A/V system designers must consider when coordinating the requirements for the specified system. These factors include the following:

- pathways and spaces where the A/V wiring is to be installed and subjected to electromagnetic interference (EMI)
- termination hardware, types, and locations of where the equipment is to be housed, including in-floor poke-thru devices, floor boxes, and wall-mounted control interfaces
- video display locations and mounting hardware, controls, and furniture associated with the system, such as lecterns and conference room tables

The four basic elements of an A/V system include the following:

- hardware (includes cables and connecting devices)
- software
- spaces where the equipment and devices are installed
- content that creates video, audio, and multimedia applications

Equipment incorporated into an A/V system design typically fulfills the requirements for the following:

- visual displays
- audio devices
- A/V controls
- hardware for the transmission of A/V signals including cable, connectors, and racks
- modular furniture
- lighting and controls
- room-scheduling/calendar software (for commercial office space installations)

Types of visual displays that may be specified include projectors, cameras, or video display monitors with or without interactive capabilities. Visual displays may also be specified with the following technology:

- electroluminescent display (ELD)
- light-emitting diode (LED) display
- liquid crystal display (LCD) with LED-backlight
- plasma display panel (PDP)

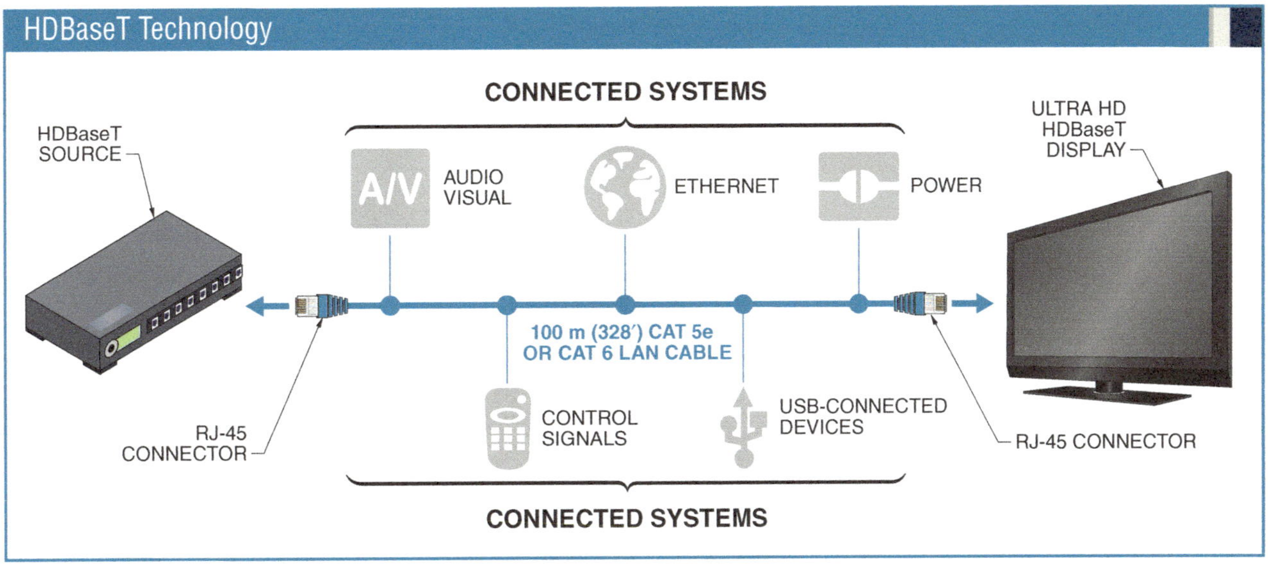

Figure 10-2. HDBaseT technology provides connection of uncompressed UHD digital A/V, Ethernet, power, control signals, and USB-connected devices over a single cable.

Audio equipment that may be specified for the design includes the following:

- microphones
- speakers
- amplifiers and sound mixers
- paging systems
- sound-masking devices
- smart boards and presentation tools

A/V control systems are required to integrate the various components for lighting, visual display screens, A/V recording devices, and HVAC equipment. The A/V control system may include the following:

- computers and software
- control modules
- switches and relays
- infrared (IR) handheld devices
- user interface touch panels
- voice-activated controls

The distribution and transmission of A/V signals can be accomplished by utilizing various types of cable types. **See Figure 10-3.** Common cable types used for A/V systems include the following:

- coaxial
- shielded twisted pair (STP) and unshielded twisted-pair (UTP)
- video graphics array (VGA)
- digital visual interface (DVI)
- high-definition multimedia interface (HDMI)
- fiber-optic cables for residential and commercial use
- RCA/composite
- component
- DisplayPort

A/V systems are commonly used in commercial applications, such as corporate auditoriums.

Figure 10-3. The distribution and transmission of A/V signals can be accomplished by using various types of cables.

In addition to the different cable types, the system requires the installation of compatible equipment rack termination interfaces and patch panels for video, audio, data, and optical transmissions. For A/V system hardware, various connector interfaces are used with both new and legacy A/V systems, including coaxial F type, RJ-45 insulation displacement (IDC), VGA, DVI, HDMI, RCA, and fiber-optic cables. **See Figure 10-4.** When modifying or adding to existing legacy A/V systems, an installer must install the same type or compatible types of connectors.

The A/V system design and drawings may specify a specific OEM's furniture and related components that are to be installed in the building. The furniture and related components that may be specified include the following:

- lecterns that provide control and interfaces
- tables and desks that accommodate A/V, network, and power access
- racks, credenzas, and carts
- visual display stands and wall-mount interfaces

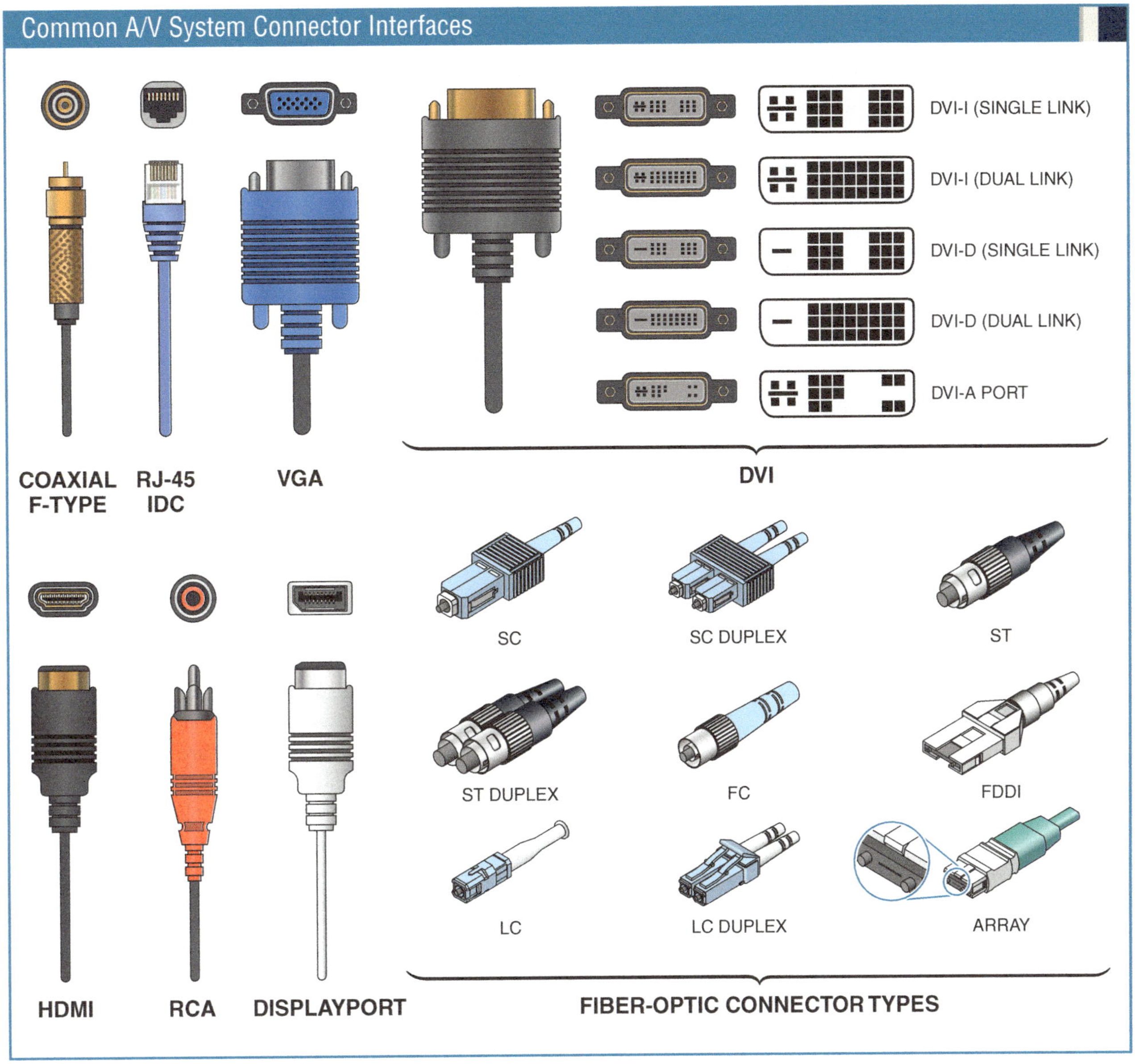

Figure 10-4. Various connector interfaces are used with both new and legacy A/V systems, including coaxial F type, RJ-45 IDC, VGA, DVI, HDMI, RCA, DisplayPort, and fiber-optic cables.

A/V System Network Integration

The merging of A/V and information technology (IT) has increased the IT network bandwidth requirements and the migration of the A/V system digital signal processing to the end user's network. A/V applications that reside on the end user's network can include the following:

- video on demand
- presentation and training systems
- videoconferencing
- digital signage
- presentation capture
- mobile sharing
- video streaming and archiving
- remote monitoring, management, and diagnostics
- lighting

Modern office technology requires separate cabling that is included as part of the network. The transmitting of audio and video signals across various mediums, utilizing a variety of pathways that are integrated with building systems during and after construction, can be complicated and require proper planning and coordination. A/V systems are controlled through hardware, software, and devices that are in compliance with A/V standards, building industry standards, and codes.

Audiovisual over internet protocol (AV/IP) is the transmission of audiovisual data over a network, such as a LAN, a wide area network (WAN), or the internet. As compared to traditional A/V applications, AV/IP refers to the distribution of audio, video, and control signals over the LAN using IP switching and configuration protocols. With AV/IP technology, traditional A/V infrastructures have been replaced with IP-based infrastructures.

TECH FACT

Ethernet devices are connected to external receivers through the use of an attachment unit interface (AUI) cable.

AV/IP covers applications such as internet-based live or on-demand video streaming, as well as video distribution infrastructures. Ethernet-based switched networks are the standard for A/V systems in commercial and residential applications and share the same network hardware as the other installed systems.

The IEEE 802.3 Ethernet standard is a collection of IEEE standards defining the physical layer and data link layer's media access control (MAC) of wired Ethernet and is generally a LAN technology with some WAN applications. Physical Ethernet connections are terminated between nodes and infrastructure devices (such as hubs, switches, and routers) by different copper or fiber-optic cables.

Power over Ethernet (PoE) is a networking feature that allows network cables to carry electrical power over an existing data connection with a single CAT 5e or CAT 6 cable. PoE allows for the installation of remote or outside equipment without having to connect to AC power, which allows power to be delivered to more areas without installing additional electrical infrastructure or needing power outlets at every point. Ethernet cable has a lower cost than a combination of VDV cables and electrical cables, and pathways for Ethernet cables are often already installed in existing buildings. PoE-based distribution systems are more cost-effective and efficient than traditional electrical distribution systems.

A primary benefit of PoE installations is the low cost of installing traditional VDV wiring. One twisted-pair cable delivers both data and power to devices, and in some cases existing copper from legacy phone systems can also be repurposed. Typically, PoE voltages are not greater than 60 V and do not require conduits or metal cladding. PoE networks and devices are controlled through the use of gateways. A *gateway* is a device that connects two or more dissimilar networks. Gateways can be hardwired or wireless. Wireless gateways transmit signals to a wireless relay controller to control network devices that are hardwired to the relay controller.

A common PoE application is lighting. PoE-controlled lighting is a low-power (48 V), high-performance LED lighting network technology consisting of fixtures and sensors that is managed remotely using intelligent software, such as that used with building management systems or room-scheduling systems. The lighting control module communicates data signals and power over CAT 5 or higher twisted-pair cable. Lighting controls allow end users to imitate the characteristics of daylight to promote health, productivity, and collaboration using a broad spectrum of visual frequencies.

PoE use in commercial and industrial applications is increasing as both the equipment costs and the installation costs for PoE systems decrease. Other PoE applications include motion and light sensors, motorized window shade and blind control, and A/V systems. **See Figure 10-5.** Data may be collected from motion sensors, which allows organizations to better monitor and control energy consumption.

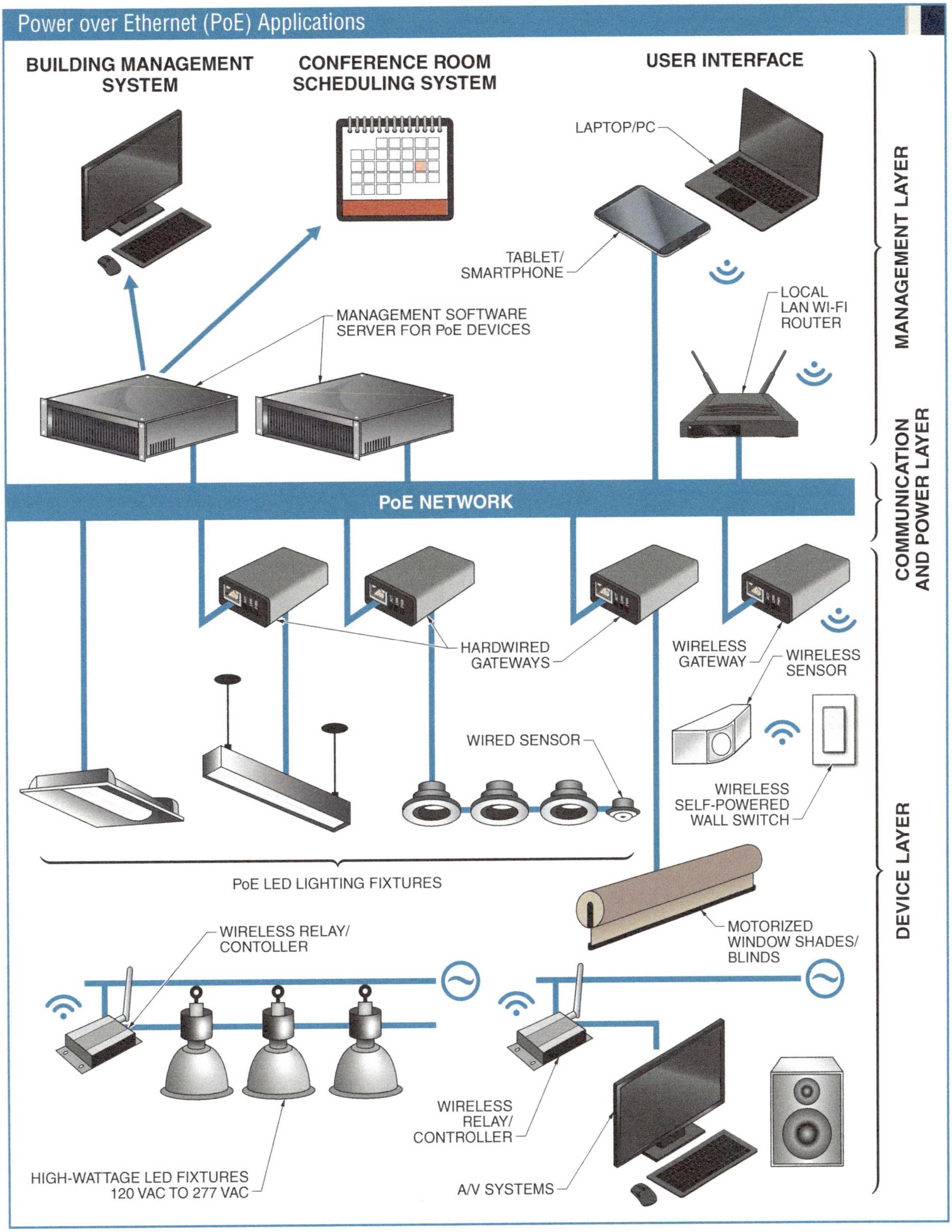

Figure 10-5. PoE applications include lighting systems, motion and light sensors, motorized window shade and blind control, and A/V systems.

A/V System Pathways and Spaces

A major requirement of any A/V design or installation is the proper specification of the pathways and spaces in which they are to reside. Adherence to building and electrical codes and industry standards for VDV wiring must be observed. For example, A/V premises distribution requires the installation of cable trays, conduits, floor boxes, poke-thru devices, and other pathways for wire and cable. The A/V wire and cables must be routed to avoid electromagnetic interference (EMI) and firestopped to applicable building codes.

The proper wires and cables must be specified for use in plenum and nonplenum spaces. For example, if nonplenum wires and cables are installed in plenum spaces, there could be a problem in the event of a fire, such as toxic gases released from the burning of the nonplenum insulation. While plenum-rated wires and cables are typically more expensive than nonplenum-rated wires and cables, they are considerably less toxic when exposed to fire.

The spaces utilized for A/V wiring are traditionally located in dedicated floor and wall spaces or in VDV equipment rooms. A/V wiring, including wiring for electrical and VDV systems, is commonly hardwired to furniture, conference room tables, and podiums or may be installed in wall-mounted faceplates and floor-mounted poke-thru devices. **See Figure 10-6.**

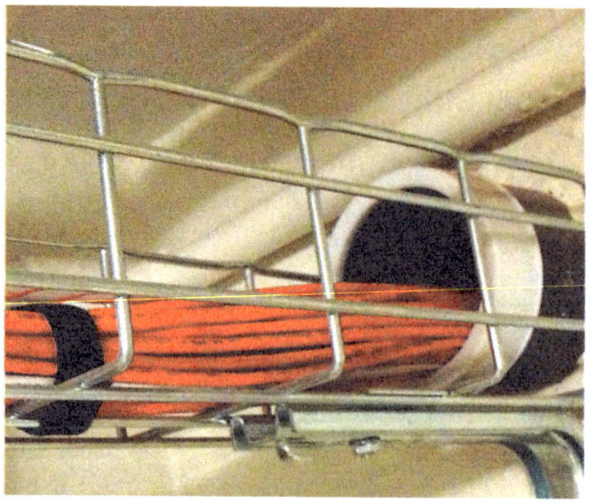

Common VDV cable pathways include cable trays.

A/V SYSTEM DRAWINGS AND SPECIFICATIONS

Much of the A/V design occurs in the design development phase of the project and is usually finalized by the construction documents phase. Coordinated drawings, specifications, and equipment lists are used for an efficient bidding process. A successful A/V project requires an experienced A/V design consultant to coordinate the A/V requirements with the design team. If A/V design elements are not properly captured, shown, and specified, the project could be exposed to costly and time-consuming change orders during the construction phase.

A/V System Drawings

The drawings for A/V infrastructure design typically show electrical, structural, mechanical, and architectural requirements. A/V infrastructure drawings must include power outlets; network outlets; floor, ceiling, wall, and table box locations; junction boxes; HVAC requirements; A/V equipment locations; and A/V equipment installation/mounting details. The A/V infrastructure drawings must be coordinated with the project architect and engineers and the A/V requirements must be incorporated into the other design disciplines' drawing package.

A/V infrastructure drawings contain an installation responsibilities schedule, which is a checklist of which trade or contractor is responsible for the different A/V system installations required to support the A/V design. The A/V system responsibilities schedule must be referenced during A/V system bidding and installation. **See Figure 10-7.**

A/V infrastructure drawings typically consist of the following:

- floor plans
- reflected ceiling plans
- wall elevations
- equipment details
- schedules
- risers
- power outlet locations
- network outlet locations
- floor, ceiling, wall, and table box locations
- junction box locations

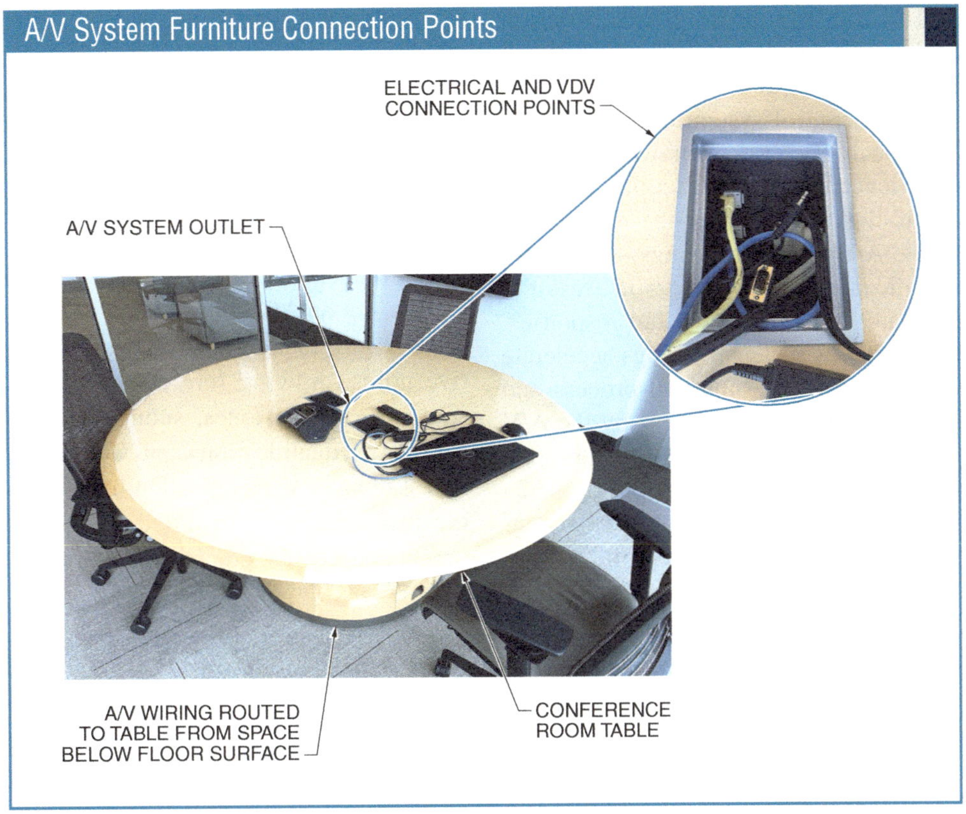

Figure 10-6. A/V wiring is commonly hardwired to furniture, such as conference room tables.

A/V System Responsibilities Schedules				
Device	**Electrical**	**Communications**	**A/V Vendor**	**General Contractor**
Conduit and backbox	X			
Wire and cable for A/V	X			
Telecommunications wire and cable		X		
Telecommunications device faceplate		X		
Device faceplates	X			
Projectors			X	
Displays			X	
Racks	X			
Switching equipment for A/V			X	
Amplifiers			X	
Wireless microphone system			X	
Digital signal processor			X	
Touch panels			X	
Projector screens	X			
Speakers	X			
Blocking				X

Figure 10-7. A/V infrastructure drawings contain a responsibilities schedule, which is a checklist of which trade or contractor is responsible for the different A/V system installations required to support the A/V design.

The drawings for the technical design of the A/V systems typically show signal flow, viewing angles, speaker coverage, and control system user interfaces. Technical A/V system drawings require experienced installers familiar with the symbology to interpret the scope. Accurate A/V system drawings, along with specifications or scopes of work, result in accurate and comparable bid responses. Detailed A/V system drawings accelerate the shop drawing submittal process and prevent design modifications by the A/V contractor. A/V system drawings also indicate how the different types of media interconnect with the equipment that controls the system. **See Figure 10-8.**

A/V System Specifications

A/V specifications can be written as descriptive, proprietary, performance-based, or referenced to specific standards. Most A/V specifications are written as proprietary, which lists specific acceptable OEMs and part numbers. A/V specifications are typically written in three parts.

Part 1 of an A/V specification includes general information such as administrative notes, drawing references, system summaries, milestones, submittals, site conditions, qualifications, and warranties. Part 2 of an A/V specification focuses on the products to be installed, including acceptable OEMs, equipment, materials,

A/V System Drawings

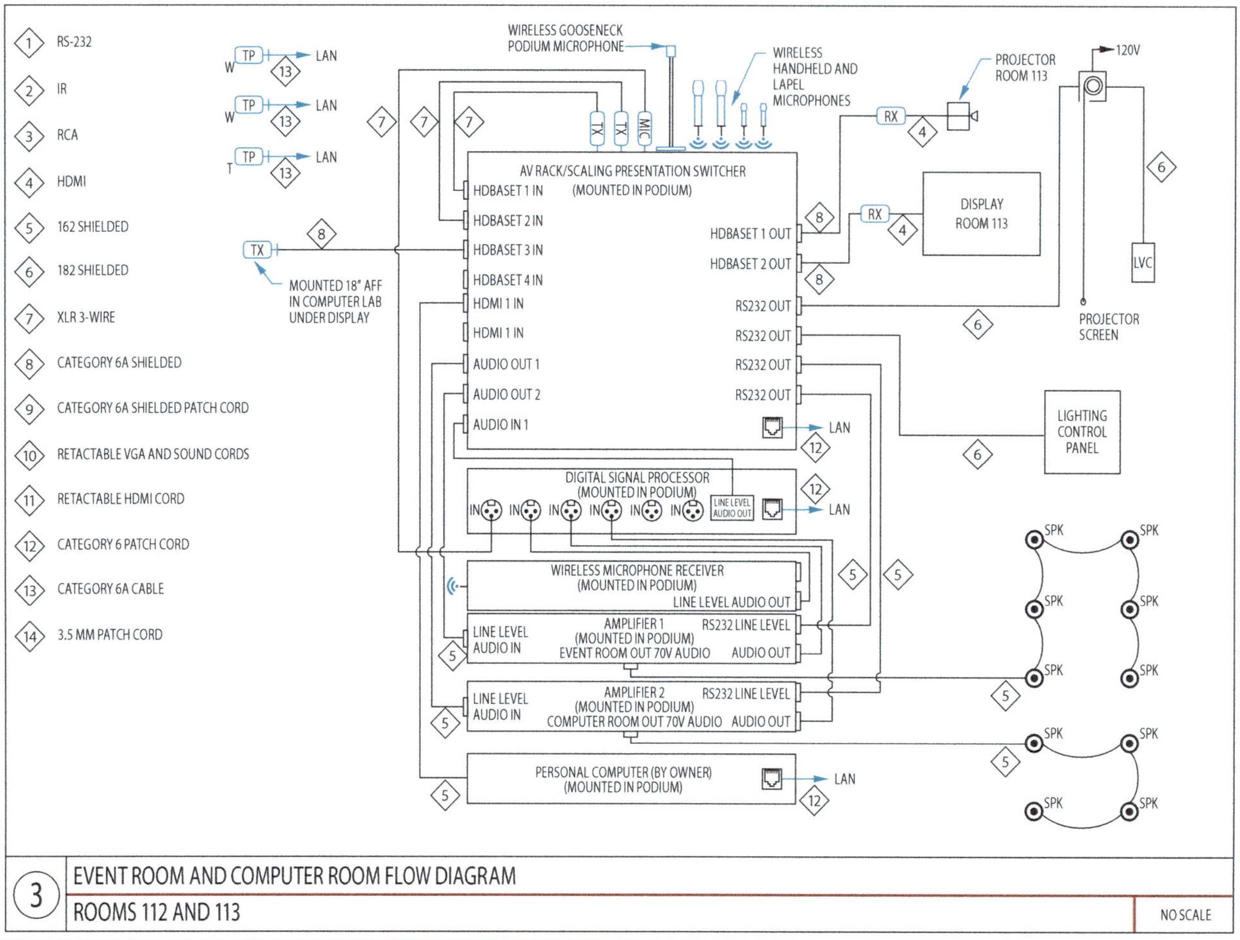

Figure 10-8. A/V system drawings indicate how the different types of media interconnect with the equipment that controls the system.

and substitutions. In certain specifications, due to the complexity of a system, a detailed equipment list may be an appendix or table on a detail drawing. Part 3 of an A/V specification includes execution details, such as installation and performance requirements, systems verification, testing requirements, and closeout activities.

Specifications may also be issued as a "scope of work" document for small projects that do not require full bid specifications. With small projects, the project specifications are sometimes included in the CAD drawings, in the form of detail drawings and notes.

A/V SYSTEM INSTALLATION

A/V equipment such as projection screens, projector mounts, floor boxes, and table boxes may be furnished and installed by different contractors on a project. Coordination of equipment installation between trades is required when planning the sequence of work. Often, several trades may be required to complete a specific installation. For example, the installation of motorized projection screens may require both structural work by the job-site carpenter for the mounting of the screen and electrical work by the job-site electrician for the operation of the equipment. If the installation is not properly coordinated, rework may be required, adding more cost for the installation of the equipment.

The submittal process is an essential part of the preinstallation planning in order for the final product to match the intent of the design. Preinstallation planning is the final opportunity for the owner, architect, and designer to approve the design before the contractor proceeds with procurement, construction, and installation. The time required for the submittal process depends on the size and complexity of the project. A/V submittal materials usually include shop drawings, the finalized equipment list, A/V furniture details, cut sheets, finish samples, product samples, and control system programming details. The typical practice is to not procure any equipment nor begin construction on the project until the submittal materials have been approved.

The procedure for resolving installation issues for A/V installations must follow the RFI process for the purpose of documenting the flow of information during the construction phase. During the work or as a contractor is working through the process of preparing for the installation, questions about the design intent can arise that may impact the schedule, design intent, or overall costs. The RFI response may be referenced in the future materials if a question concerning the same issue should arise. The RFI response process protects both the customer and the contractor.

Due to the complexity of A/V installations, most projects usually require a quality control (QC) program that includes input from the owner, architect, and designer. QC activities for an owner, architect, and/or designer may include a visit to the A/V contractor shop to observe testing of the A/V system after it has been fabricated and prior to shipment to the site for installation. A QC program can also include periodic project site visits while the various trades are installing the A/V-related infrastructure. The owner, architect, and designer may also perform final A/V room assessments after the rooms are constructed and before equipment is shipped to be installed in the rooms. Additionally, general site visits by the owner and A/V designer may occur periodically during the course of the installation. **See Figure 10-9.**

A/V SYSTEM VERIFICATION

Upon completion of the installation and prior to the owner accepting the system, a systems verification program is followed. A systems verification program consists of a checklist that inventories the installed systems. A final systems verification checklist must be clearly defined in the specifications or terms of the contract. It is essential that the systems verification program be clear and concise, so that additional time and significant costs are not incurred for the owner or the general contractor.

A/V Equipment QC Programs

Figure 10-9. A/V equipment QC programs include periodic site visits by the owner and A/V designer during the course of the installation.

During the systems verification program, the owner and designer confirms that the system(s) have been installed, configured, and tested as specified. A punch list is generated, with any deficiencies addressed along with any other outstanding items. Most OEMs of A/V equipment and systems provide installation and testing punch lists to complete the verification process.

After the specified systems have been fully tested and punch list items completed, client training can occur. Client training must only occur after the owner has approved that the system is complete. Starting the training process prior to final approval by the owner could result in longer training time requirements and negatively affect the overall sequence of the installation.

Typically, there are two groups who must receive A/V system training: the owner's technical A/V personnel and the end users who will use the system. The training for each group must be conducted by qualified personnel who are skilled in the operation of the system. Multiple training sessions may be required for both the nontechnical users and for the technical support staff that are to operate, service, and maintain the A/V systems.

Once the A/V equipment is installed and the owner's personnel training is complete, the project closeout for the A/V system can be initiated. The project closeout submittal package may consist of the as-built system drawings, control system programming files, equipment and system manuals, equipment inventory, MAC and IP address list, warranty statement, maintenance contract, keys, and spare parts—all of which are defined in the original bid specifications and project manual. The project owner and designer must review and approve the closeout submittal.

SUMMARY

A/V system applications are separate from the other VDV systems to be installed in both new and retrofit construction. A/V systems are designed by professionals who specialize in such systems. A/V systems must be designed and installed based on standards that apply to those systems. A/V system network integration is performed based on the specifications and drawings for the project. VDV installers must be familiar with the different A/V systems and any specialized installation practices required when performing installation, testing, and troubleshooting tasks.

CHAPTER 10

Audiovisual System Drawings and Specifications

Review

Name ____________________ Date ____________

True-False

T F 1. OEMs of relevant A/V equipment, along with systems integrators and design consultants, may also be part of the A/V system planning and design team.

T F 2. A/V standard-based designs increase costs over the usable life of the system and provide accommodation for future moves, adds, and changes.

T F 3. For medical and government applications, fiber-optic cable is preferred over category cable because it can be used for longer distances.

T F 4. Modern office technology does not require separate cabling that is included as part of the network.

T F 5. PoE allows for the installation of remote or outside equipment without having to connect to AC power.

T F 6. Much of the A/V design occurs in the design development phase of the project and is usually finalized by the construction documents phase.

T F 7. A/V systems do not integrate with other building systems, such as lighting, building management, and room-scheduling systems.

T F 8. The drawings for the technical design of the A/V systems typically do not show signal flow, viewing angles, speaker coverage, and control system user interfaces.

T F 9. Typical practice is to not procure any equipment nor begin construction on an A/V project until the submittal materials have been approved.

T F 10. Client training can occur before the owner has approved that the system is complete.

Multiple Choice

_______________ **11.** HDBaseT technology uses a category cable (CAT 5e or above) for up to ___′.
A. 328
B. 358
C. 378
D. 398

_______________ **12.** Typically, PoE voltages are not greater than ___ V.
A. 60
B. 75
C. 90
D. 115

_______________ **13.** A/V specifications are typically written in ___ parts.
A. two
B. three
C. four
D. six

_______________ **14.** ___-based switched networks are the standard for A/V systems in commercial and residential applications and share the same network hardware as the other installed systems.
A. A/C
B. D/C
C. Ethernet
D. Isolation transformer

_______________ **15.** The ___ system is not typically integrated with an A/V system.
A. building management
B. digital signage
C. fire alarm
D. lighting

Completion

_______________ **16.** A(n) ___ is a digital educational platform that allows instructors to electronically pose questions and gather student responses during classroom (in-house or virtual) instruction.

_______________ **17.** ___ is a firmware-based consumer electronics connectivity technology that enables a single standard CAT 5e or 6 LAN cable with a standard RJ-45 connector to simultaneously carry uncompressed 10.2 Gbps high-definition video and audio, 100 Mbps of data, and 100 W of power over a distance of 100 m.

_______________ **18.** ___ is the transmission of audiovisual data over a network such as a LAN, WAN, or the internet.

_______________ **19.** ___ is a networking feature that allows network cables to carry electrical power over an existing data connection with a single CAT 5e or CAT 6 cable.

_______________ **20.** A(n) ___ is a digital calendar scheduling system used in commercial environments to allocate space and resources for upcoming events and meetings.

_______________ **21.** A(n) ___ system is a digital voice-data-video system that is a combination of sound and video elements.

_______________ **22.** A(n) ___ is a device that connects two or more dissimilar networks.

_______________ **23.** A PoE lighting control module communicates data signals and power over CAT ___ or higher twisted-pair cable.

_______________ **24.** The procedure for resolving installation issues for A/V installations must follow the ___ process for the purpose of documenting the flow of information during the construction phase.

_______________ **25.** The ___ is a standards organization that promotes, advances, and standardizes HDBaseT technology in a variety of industries by forming an alliance with manufacturers and safety organizations.

Short Answer

26. List five types of A/V system hardware.

27. List four benefits of using HDBaseT technology.

28. List five parts of typical A/V infrastructure drawings.

29. List four types of visual display used in A/V systems.

30. List six items that are included in the project closeout submittal package.

Audiovisual System Drawings and Specifications

CHAPTER 10

Activity

Name ________________________ Date ______________

A/V SYSTEM

Instructions

Create a list of materials that would be included in a proposal by using the drawings and specifications to determine material types and quantities, room locations, and connecting hardware for the audiovisual (A/V) system. List any qualifications or exclusions required that may not be on any of the drawings. Categorize the work by system type.

Project Overview

The A/V system is to be installed in a commercial office building. The A/V system has the following requirements:

- The drawings and specification stipulate design-intent to develop the proposal.
- There are parts to the scope specified and not shown, as well as shown and not specified.
- The drawings and specification are a guideline to identify the materials and quantities required for a proposal and are not all inclusive.
- Identify the materials centered on the codes, standards, practices, and industry-accepted methodologies.

Drawings:

- AV0-01 – Audiovisual Legend and Schedule
- AV1-00 – Audiovisual Floor Plan
- AV1-01 – Audiovisual Floor Plan
- AV2-00 – Audiovisual Details
- AV2-01 – Audiovisual Details

Specification:

- Section 280400 – Audiovisual

Note: The drawings and the specification can be accessed in the online learner resources.

1. A/V equipment housings:

2. Location of VDV equipment rooms:

3. A/V media devices/Quantity:

4. Wire and cable/Length or quantity:

5. Cable pathways, support methods, and spaces:

6. Electrical hardware:

7. A/V and VDV connecting hardware/Quantity:

8. A/V equipment/Quantity:

9. A/V control components:

10. Integration required:

11. Items requiring coordination with owner, general contractor, and electrical contractor:

12. Missing scope items:

13. Questions/Qualifications/Exclusions:

VDV Rooms, Spaces, and Data Centers

CHAPTER

11

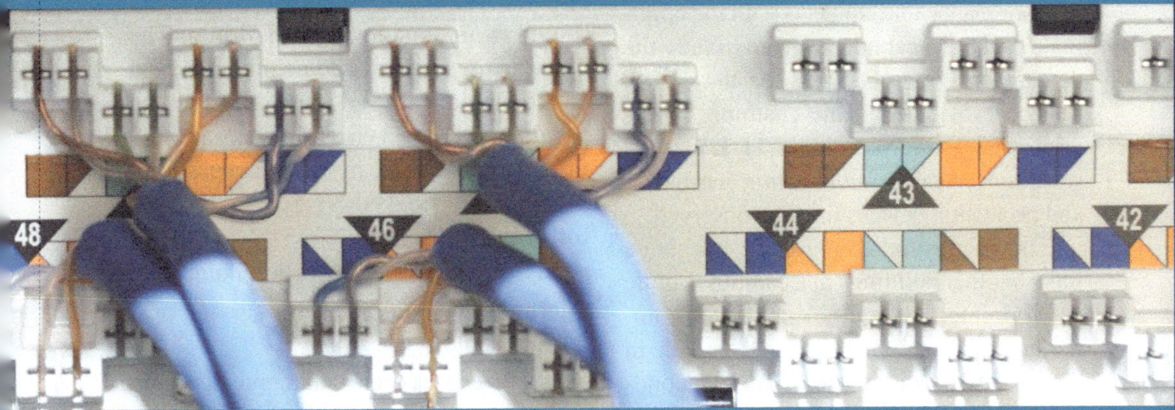

The foundation for any commercial, industrial, or educational facility is the numerous rooms, pathways, and building spaces that house VDV wiring and equipment. The systems that are commonly contained within these spaces provide normal and emergency operations that control the buildings services interconnecting the premises as well as connecting the premises to outside systems.

Industry standards have defined these building spaces as entrance facilities (EFs), equipment rooms (ERs), and telecommunications rooms (TRs). Data centers are centralized locations for a facility's computing equipment and require installation, monitoring, and maintenance by VDV technicians. These building spaces require the use of identification and documentation systems and rely on equipment that meets NEMA standardization. Prior to owner occupancy, building spaces such as ERs must be commissioned. Careful review of terminology and related acronyms must be conducted prior to any installation work.

Objectives

- Explain the differences between the entrance facilities, equipment rooms, telecommunications rooms, and data centers located within commercial buildings.
- Describe labeling and identification systems.
- List NEMA equipment configurations.
- Explain equipment room commissioning.

Learner Resources

atplearningresources.com
access code 783821

ENTRANCE FACILITIES

An *entrance facility (EF)* is the point where outside antenna, public, and private network service provider cables can enter a building and be consolidated. An EF is typically placed as close to the point at which the outdoor plant cable connects with the building's backbone cabling. EFs are usually the demarcation point between the third-party service provider and the customer's proprietary systems. **See Figure 11-1.** Careful selection of cables, connecting hardware, and protection devices must be observed.

The EF may also house the backbone links to other buildings in a campus. Public network interface equipment and telecommunications equipment may also reside in the EF. Some facilities may require secondary EFs for diversity and disaster recovery considerations. EFs are sometimes referred to as entrance rooms.

Local exchange carriers (LECs), such as the Regional Bell Operating Companies (RBOCs), CATV providers, and ISPs, provide the point of demarcation for their services in an EF. An EF houses the terminations of copper and fiber-optic cables that are owned by LECs and routed from outside of the building. The location must be dry and near the vertical backbone pathways. An EF must also be provisioned for services such as security, fire alarms, HVAC, lighting, access control, and electrical power. EFs include the following equipment:

- service provider demarcation point
- termination equipment for service provider cables
- interfaces between data-center cabling and interbuilding cabling (backbone, horizontal, and work area)
- pathways for entrance cables

Entrance Facilities (EFs)

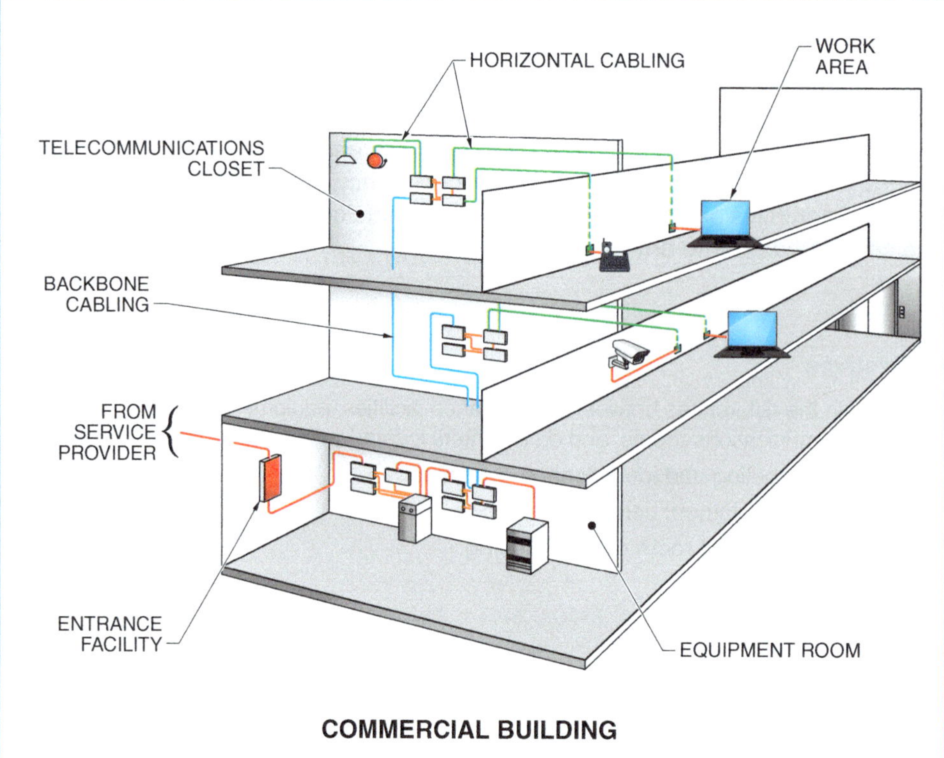

Figure 11-1. An EF is the point where outside antenna, public, and private network service provider cables can enter a building and be consolidated.

EQUIPMENT ROOMS

An *equipment room (ER)* is a centralized room or space for telecommunications equipment that serves the occupants of a building or multiple buildings within a campus environment. An ER is considered different from a telecommunications room because it serves a building or campus rather than a single floor. A telecommunications room is used to serve a single floor. The need for properly designed, installed, and maintained rooms to house the equipment for telecommunications systems is vital for fast, consistent, and reliable service as dependence on technology continues to increase, the systems become more advanced, and reliance on the network and internet strengthens.

The most common standards referenced for ERs include the ANSI/TIA-568, -569, and -942 standards, as well as the NEC®. NEC® Article 645 defines the requirements for information technology equipment. *Information technology (IT) equipment* is the hardware, software, and networks that are used by an organization to input, store, process, output, and disperse data and information. However, not all ERs with IT equipment fall under the requirements of NEC® Article 645.

TECH FACT

New to the 2020 NEC®, Article 800, General Requirements for Communications Systems, was created to reduce redundancy, while the material from the previous Article 800 from 2017 was renamed Article 805, Communications Circuits.

For example, a common mistake made by designers when designing an ER is to apply only NEC® Article 645 to ERs that are used solely for networking, telephone equipment, and building cable distribution. These types of ERs have additional information covered by other sections of the NEC® such as Articles 300, 770, and 800. Most often, the requirements of NEC® Article 645 apply to the equipment, power supply wiring, interconnecting wiring, and grounding of IT equipment in ERs and data centers. The other types of ERs for VDV systems have specific requirements, but not as advanced or as costly as the requirements for ERs and data centers with IT equipment.

A *main distribution frame (MDF)* is an apparatus that connects the telephone lines coming into the switching center from outside to the internal lines connected to the switching or transmission equipment inside the switching center. An MDF may also be the main location for backbone cabling; contain power protection equipment, uninterruptable power supplies (UPS), and LAN equipment; and function as a central testing point.

The ER must contain an adequate heating, ventilating, and air-conditioning (HVAC) system, preferably from a separate stand-alone system capable of operating on a continuous basis if required. If the ER will house LAN servers or similar IT equipment, the HVAC must be provided by a separate, stand-alone computer room air conditioning (CRAC) unit with back-up power. A *computer room air conditioning (CRAC) unit* is an HVAC device that monitors and maintains the temperature, air distribution, and humidity in a network room, space, or data center. CRAC units are used to replace the standard A/C units that were used in the past to provide cooling to such areas.

A detail drawing of an MDF shows backbone and horizontal cabling, connector blocks, telephone equipment, network equipment, and cable management and support devices. Several ER detail drawings may be required for complex installations. VDV technicians must consult ER detail drawings before installing any equipment.

ER detail drawings must specify the location and mounting method for VDV equipment. The drawings indicate where and how raceways are to be mounted in ceiling spaces and between floors. Elevation and plan views are used to identify the location of cable ladders, equipment racks, termination backboards, connector blocks, and raceways. **See Figure 11-2.** A legend is included to avoid clutter on the drawing.

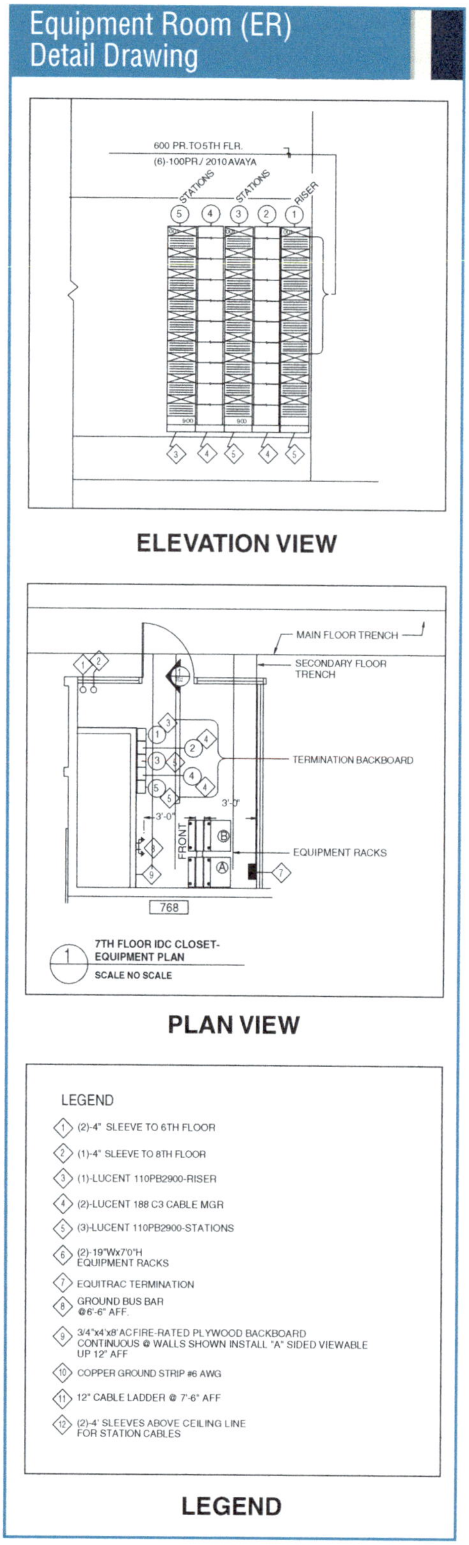

Figure 11-2. ER detail drawings include elevation views, plan views, and legends that show the locations of cable ladders, equipment racks, termination backboards, connector blocks, and raceways.

TELECOMMUNICATIONS ROOMS

A *telecommunications room (TR)* is an enclosed space that houses telecommunications equipment, cable terminations, and cross-connect cabling used to service work areas on the same floor of a building. *Note:* A TR is sometimes referred to as a telecommunications closet (TC). An *intermediate distribution frame (IDF)* is a metal rack designed to connect cables and is located in an equipment room or telecommunications room. The IDF in a TR is the critical point between the work area and the MDF in an ER, through backbone cabling and horizontal distribution cabling. **See Figure 11-3.**

TRs in a building house equipment, cable terminations, and cross-connects (connections between cables and other devices) that serve a specific office area and/or a specific floor. A TR is typically a floor-serving room for horizontal distribution cabling and is aligned vertically with the floors located above and below to provide pathways between rooms. Backbone pathways connect each TR to an ER.

It is preferable for the TRs to be stacked vertically to facilitate running backbone cables through them. In such cases, backbone pathways typically consist of sleeves run through the floor slab. Backbone pathways may also connect to other TRs or ERs. When a floor has more than one TR, they are often are interconnected by a backbone pathway. A TR should be provisioned for services such as HVAC, lighting, doorways, and normal, backup, and uninterrupted electrical power supplies. Each system must be sized for the equipment to be housed within the TR. As with ERs, TRs must have a stand-alone HVAC system.

A *telecommunications riser* is the vertical infrastructure that connects the tenant spaces of a building from the local exchange carrier presence in the entrance facility to each floor. A telecommunications riser typically extends vertically through a shaft or series of shafts, connecting to a TR on each floor. The telecommunications riser room serves as the central point of distribution for LEC services to tenants (or multiple tenants) in each floor of a building. Typically, no tenant equipment will reside in this room. Most telecommunications riser rooms support multiple LECs to provide competitive LEC services for VDV applications to building tenants.

DATA CENTERS

A *data center* is a centralized location where computing resources critical to an organization are maintained in a highly controlled physical environment. The main purpose of installing a data center is to provide secure and uninterrupted availability to computing, storage, networking, and communication resources, including VDV equipment. The increasing amount of VDV and IT applications require more pathways and equipment space that call for uninterrupted service and downtime.

Modern data centers include spaces within or adjacent to it for the technical support staff monitoring the systems, EFs for LECs, computers, servers, peripherals, power, applications (HVAC distribution equipment, lighting, etc.), network access, and databases. Modern data centers house a large number of diverse bandwidth-intensive devices, including servers, clustered storage systems, virtualization appliances, and back-up devices (battery power and diesel generators) that are interconnected by networking equipment. The three common types of data centers for commercial applications are the following:

- An *enterprise data center* is a private facility operated for the sole use of supporting a single organization. Enterprise data centers can be located on premises as part of a larger campus or business site. But in many cases, they are located off premises at a site chosen for connectivity, power, and security purposes.
- A *multitenant data center* is a data center operated by a third party for the purpose of serving multiple enterprise tenants. They are often referred to as colocation data centers. However, in some instances, multitenant data centers can operate a single-client data center on behalf of large clients that require more than one data center.
- A *service-provider data center* is an off-site data center that is leased by a telecommunications service provider to an organization. Clients rent or lease access to the provider's data center, accessing the network servers, data networking and storage, and other computing resources owned by the service provider.

Telecommunications Rooms (TRs)

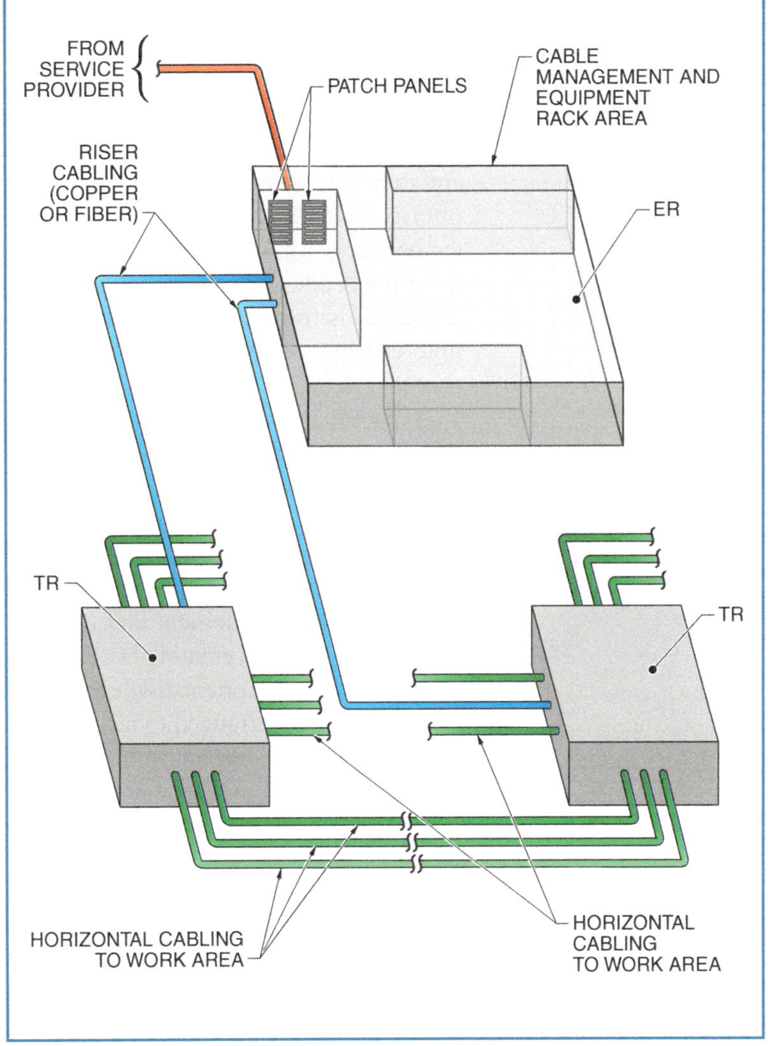

Figure 11-3. The primary function of a TR is to serve as a connection point between riser and horizontal cabling, while the primary function of an ER is to serve as a location for telecommunications equipment.

When a data center is being designed, coordination of data center aspects is required to determine the current needs of the organization at full capacity and the anticipated growth of the organization over the life of the facility. Equipment and systems in a data center that are affected by future growth and applied to the building drawings include the following:

- VDV spaces and systems
- equipment floor plans
- electrical distribution systems
- architectural plans
- HVAC systems (CRAC units)
- fire detection and suppression systems
- security systems (physical and electronic)
- lighting systems
- building management systems

The environmental factors that are taken into consideration for a data center design are similar to those for ERs and include interior humidity levels and operating temperatures. The electrical considerations for a data center may include redundant power feeds from the utility, UPSs, multiple circuits to equipment, and back-up power systems.

A significant aspect of a data center design is reducing the amount of heat generated by equipment and producing maximum airflow. The use of adequate cooling equipment in conjunction with a raised-floor system facilitates the movement of hot and cold air by arranging equipment and cabinets on alternate patterns, creating hot and cold aisles. With hot aisle orientation, equipment (server) racks are arranged to face to face, allowing outgoing hot air from the servers to be exhausted in opposite directions. *Note:* If server racks were oriented facing the same direction, the exhausted hot air would continue to build on each successive server rack, causing heat-related issues for the servers. **See Figure 11-4.**

Another method of controlling hot exhaust air is through the use of perforated floor tiles in the raised-floor system. Perforated floor tiles in the raised floor of the cold aisle allow cold air to be drawn into the face of the equipment. As the cold air is drawn into the face of the equipment, it is exhausted out the rear of the equipment to the hot aisle, where no perforated tiles are located, which cools the hot air present in that area.

Once equipment is routed from the ER to the data center, there are several different areas that are located within a data center. These areas are the main, horizontal, zone, and equipment distribution areas. **See Figure 11-5.** While some equipment is specific to a data center alone, there are other types of equipment that can be located in a data center and/or in other areas of a building. The different areas of a building include main distribution areas (MDAs), equipment distribution areas (EDAs), horizontal distribution areas (HDAs), and zone distribution areas (ZDAs).

As the central point of distribution, at least one MDA is required per data center. Specific equipment for an MDA includes the following:

- main cross-connect (MC)
- horizontal cross-connect (HC)
- core routers and switches for LANs

Specific equipment for an EDA includes the following:

- end equipment (servers, mainframes, back-up tape drives, etc.)
- horizontal cabling terminated into patch panels
- point-to-point cabling between equipment

In addition to a data center installation, at least one HDA is installed per floor. Specific equipment for an HDA includes the following:

- supports for cabling to work areas
- horizontal cross-connect (HC)
- EDA switches

A ZDA is the optional interconnection point within horizontal cabling and can serve a maximum of 288 connections to avoid cable congestion. A ZDA is located between the HDA and EDA, but allows for frequent reconfiguration and flexibility. A ZDA contains no active equipment other than DC power sources.

Data Center Hot Aisle Orientation

Figure 11-4. With a hot aisle orientation in a data center, server racks are arranged to face to face, allowing outgoing hot air from the servers to be exhausted in opposite directions.

Data Center Areas

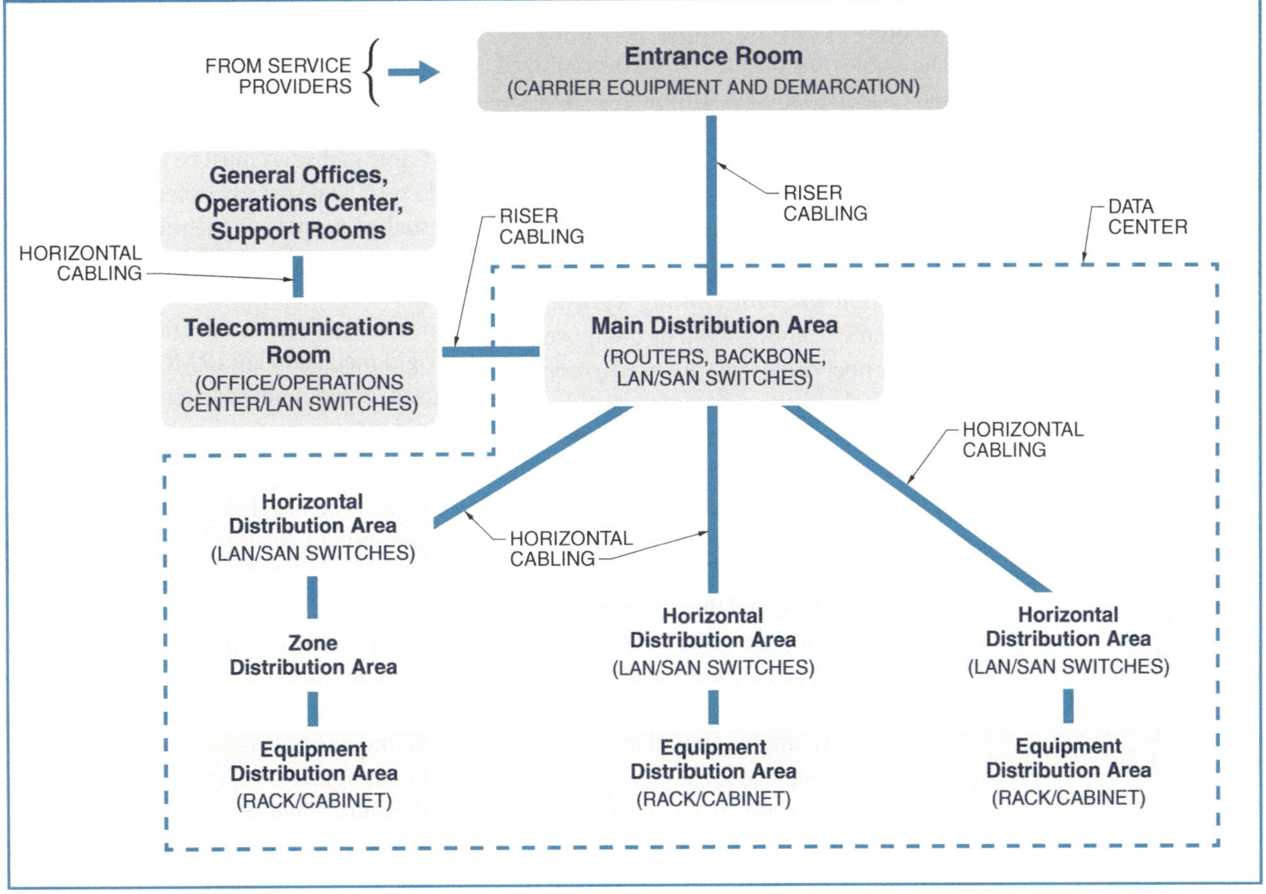

Figure 11-5. The different areas that are located within a data center include the main, horizontal, zone, and equipment distribution areas.

Data Center Cabling and Equipment

Category (copper) and fiber-optic cabling are the main types of network cabling used in data centers. The equipment interface is the primary means for determining which type of cabling is installed in a data center. The network data cabling may also be selected based upon the bandwidth requirements of the facility where the equipment is to be used. Cabling systems within a data center may be either VDV cabling or point-to-point systems.

A *voice-data-video (VDV) cabling system* is a telecommunications cabling system that is organized into a hierarchy of wiring termination and interconnection structures. VDV cabling systems use a standards-based design with predefined connection points and pathways. The cabling used in a VDV cabling system design is specified by the bandwidth requirements of the system and is tested and certified to ensure proper performance. The cables used must be well organized and properly labeled. Although a VDV cabling system may take longer to install and have a higher initial cost than a point-to-point cabling system, the operational cost is lower and the life cycle of the system is longer than that of a point-to-point system.

A *point-to-point cabling system* is a data center cabling system composed of jumper fiber cables that are used to connect one switch, server, or storage unit directly to another switch, server, or storage unit. Point-to-point cabling systems do not use predefined standards, connection points, or pathways. This type of cabling system can lead to cooling issues because the airflow is typically restricted due to the manner in which the cables are routed. Restricted airflow may also cause energy costs to be high. Also, managing system growth is difficult because there are no predefined pathways to run additional cabling.

Although a point-to-point cabling system may take less time to install and have a lower initial cost, the operational cost is higher and the life cycle is shorter as compared to a VDV cabling system. Whether the system installed is VDV or point-to-point cabling, the cable types for data centers may include the following:

- four-pair CAT 6 or higher cabling (shielded or unshielded)
- multimode fiber (62.5 μm or 50 μm)
- laser-optimized (LO) 50 μm multimode fiber
- single-mode fiber
- 75 Ω coaxial cable

Also, there are various options for each cable type, such as traditional field terminations and preterminated configurations. *Note:* Preterminated applications are commonly specified for data center cabling and require assembly by approved UL assembly facilities.

Data Center Cable Pathways. The cable pathways for data centers should adhere to ANSI/TIA-569, *Telecommunication Pathways and Spaces*. Per this standard, cabling must not be routed through public spaces unless enclosed in secure pathways. Access points such as maintenance holes, pull boxes, and splice boxes must be locked and monitored.

Cable pathways must be installed to provide sufficient space between electrical and twisted-pair cables. Branch electrical circuits must be installed in liquidtight flexible metal conduit (LFMC). Feeder electrical circuits to power distribution units must be installed in rigid metal conduit (RMC). If cable tray is used, a minimum spacing of 12″ must be provided between the installed cable trays.

Data Center Fiber-Optic Cable Raceway Systems. The installation and maintenance of fiber-optic cables is sometimes accomplished by utilizing an enclosed plastic raceway dedicated for fiber-optic cabling. Raceways used for fiber-optic cabling are often enclosed to prevent damage to cables and keep dust contamination to a minimum. Cable trays for fiber-optic cables also have fittings designed to maintain the proper cable bend radius. Fiber cable tray is commonly installed in conjunction with and supported by adjacent copper cable trays, such as ladder trays, for the purpose of providing an efficient pathway. **See Figure 11-6.**

Data Center Access Floors. When a data center is designed, separate aisles must be allocated for both power and telecommunications cabling. If separate aisles are not possible, such as with retrofit projects, then both horizontal and vertical separation must be provided. Cabling installed in under floor locations must be installed in ventilated cable trays and may be installed in multiple layers with a maximum depth of 6″.

Data center access floors are typically a system of floor tiles raised on pedestals. Plenum-rated cable is commonly specified for the space below the floor tiles. Although, depending on the design of the room and routing of the cable, the cable could be riser rated. **See Figure 11-7.** Self-contained data centers may utilize riser-rated components that meet codes and standards.

Data Center Cable Trays. When fiber-optic and category (copper) VDV cabling is installed in cable trays, the cable trays must be separated to improve cable management and minimize damage to small-diameter cables. If possible, cable trays may be installed in multiple layers, with trays containing fiber cabling installed above those that contain copper (category) cabling. Cable trays must be supported in accordance with standard ANSI/TIA-569 from a ceiling or raised floor.

Data Center Cabinets. A *cabinet* is an enclosure used for terminating telecommunications cables, wiring, and connection devices that has a hinged cover and is usually installed flush-mounted to a wall. *Note:* Cabinets are sometimes referred to as racks or enclosures. Data center cabinets are classified into the two following types:

- Equipment cabinets—Equipment cabinets are equipped with integrated power strips and air flow/cooling options for the cooling of high-density/high-powered devices.
- Ultra-high-density (UHD) cable termination cabinets—UHD cabinets have the capacity to hold the same amount of cables that would be installed in multiple standard equipment cabinets.

Fiber-Optic Cable Raceway Systems

Figure 11-6. Fiber cable tray is commonly installed in conjunction with and supported by adjacent copper cable trays, such as ladder tray, for the purpose of providing an efficient pathway.

Cabinets dedicated to switching equipment are typically wider than server or storage cabinets to accommodate the larger volume of cables and connecting (patch) cords. Cabinets can be configured to provide cable management within, cooling fans, heavy-duty shelving, dividers, side panels, and kits to attach multiple cabinets together in series.

Some data center applications may require zone cable management enclosures to provide connectivity to large or specialized equipment. These enclosures are plenum rated and may be used in environmental air spaces. Raised floor enclosures integrate into a standard 2′ × 2′ floor tile grid. Raised floor enclosures provide the ability to meet the following requirements:

- passive enclosure for voice and network wiring blocks
- passive enclosure for data network patch panels
- active enclosure for housing active equipment, including VDV connectivity

Data Center Access Floors

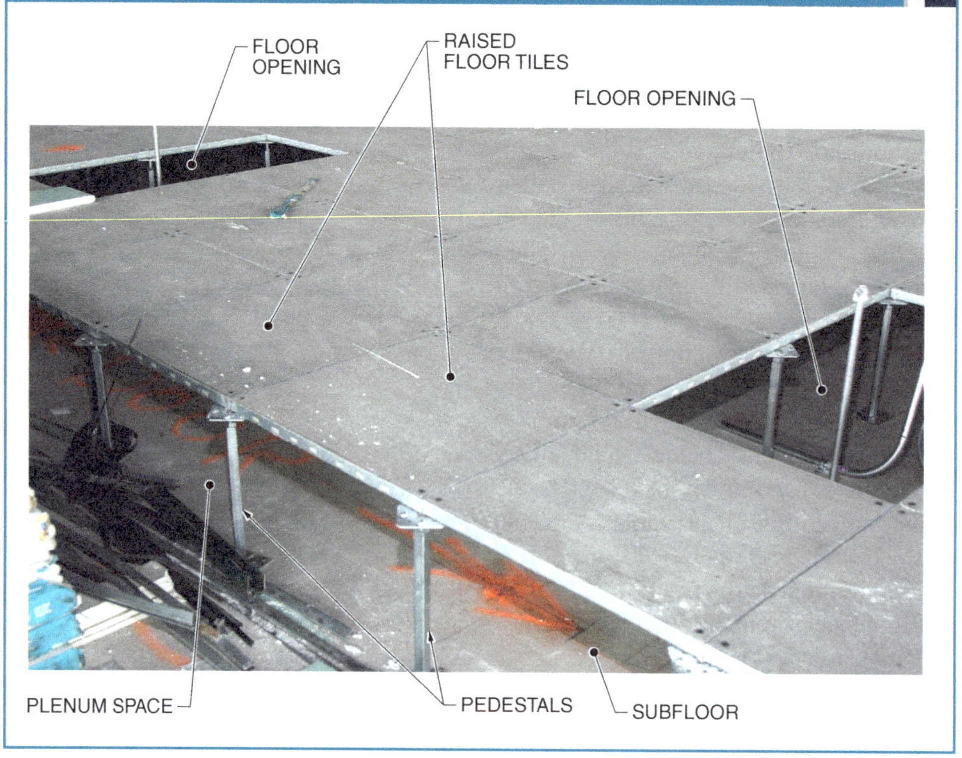

Figure 11-7. Data center access floors are typically a system of floor tiles raised on pedestals; plenum-rated cable is commonly specified for the space below the floor tiles.

Data Center Cable Management. Cable management is an essential part of a data center or any ER. Cable management is accomplished with cable managers. A *cable manager* is a device used to secure, organize, and manage cables in a data center or equipment room. Cable manager size is measured in rack units (RU), with 1 RU and 2 RU managers typically being the most common sizes. Horizontal cable managers allow neat, proper routing of the patch cables from and between equipment racks and protect them from damage.

However, there both advantages and disadvantages to horizontal cable managers. Advantages of horizontal cable managers include the ease of tracing cables for troubleshooting purposes as well as the ease of moving, adding, or changing cables. Another advantage is that cables can be secured at the front or rear of the rack. Disadvantages of horizontal cable managers include their large footprint, which requires extensive space, and their time-consuming installation.

Cable management can also be accomplished with vertical cable managers. Vertical cable managers are installed on the front or rear of the rack/cabinet and provide for the routing of cables and management of cable slack. Most vertical cable managers include features such as hinged doors and covers, which provide a neat appearance and hide excessive cable.

Vertical cable managers also have advantages and disadvantages. One advantage of vertical cable managers is reduced signal interference by keeping power and data cables separated. Another advantage of vertical cable managers is that they are cost effective and typically only one cable manager is required per cable rack. Vertical cable managers also have a small footprint, which conserves space. The disadvantages of vertical cable managers include their difficulty in tracing cables and difficulty in moving, adding, or changing cables.

Data Center Termination Hardware. As modern networking equipment is designed to be denser and port counts in the data center increase to several hundred ports, managing cable connected to these devices becomes critical to the operation of a facility. Copper and fiber termination hardware used for data center terminations must be both high quality and high density to maximize space in racks and cabinets. High-density copper patch panels, in 1 RU or 2 RU and ports of 24/48, many of which are angled, provide easy routing of interconnect cables to vertical cable managers. Copper patch panels for use in data centers are typically rated for CAT 5e through CAT 8 cables.

Using fiber-optic cable assemblies that have a single connector at one end of the cable and multiple duplex breakout cables at the opposite end of the same cable, such as with a multifiber push-on (MPO) connector, is a common method used for installing high-density fiber-optic cables. A *multifiber push-on (MPO) connector* is a connector for ribbon cables with a capacity of 8 to 72 fibers that is designed to provide multifiber connectivity in one connector to support high bandwidth and high-density cabling system applications.

MPO connectors have 12 times the density of standard connectors (SC), which make it ideal for high fiber-count data center and parallel-optical connections between servers. A *standard connector (SC)* is a fiber-optic cable connector that uses a push-pull latching mechanism. Female MPO connectors have holes in the ferrule to accept the alignment pins from the male connector. While MPO connectors can handle up to 72 fibers, most are factory terminated and commonly have either a 12 or 24 fiber array. **See Figure 11-8.**

Data Center Network Protocols. There are several different protocols used for data centers. For transmission between data centers, fiber channel is used. Fiber channel is a high-speed networking technology primarily used for transmitting data among data centers, computer servers, switches, and network storage at data rates of up to 128 Gbps. Network cable protocols for data centers include the following:

- 100Base-TX cable—100Base-TX cable is a CAT 5 UTP straight-through cable using two of the four available pairs. A 100Base-TX cable can support speeds up to 100 Mbps with a maximum length of 100 m.
- 1000Base-TX cable—1000Base-TX cable is a cable conforming to a physical layer standard created and managed by the TIA that enables transmitting data at speeds of 1000 Gbps on CAT 6 and CAT 7 cables with four pairs of wires and RJ-45 connectors and up to 100 m in length. This type of cable is not commonly installed due to the cost of CAT 6 and CAT 7 cabling and components.
- 10 GbE cable—10 gigabit Ethernet (GbE) cable is a group of computer networking technologies for transmitting Ethernet frames at a rate of 10 Gbps.
- 100 GbE cable—100 GbE fiber-optic cable can transmit at speeds of 100 Gbps and 40 GbE. The 100 Gbps speed is for long-distance switch-to-switch transmission, while the 40 GbE speed is designed for use within an organization between servers and the Ethernet switch.

Multifiber Push-on (MPO) Connectors

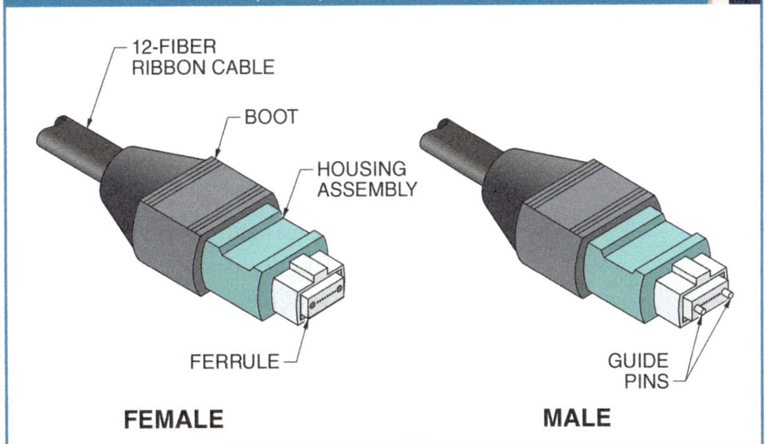

Figure 11-8. An MPO connector is a fiber-optic connector that has the same size as an SC connector but provides 12 times the density, making it ideal for high fiber count data center and parallel-optical connections between servers.

Data Center Cabling Applications

A data center cabling system is designed to span several generations of network electronics, such as servers and storage units, that are added or replaced more frequently than networking equipment such as Ethernet or fiber channel switches and core routers. Data center topologies can be centralized, partially centralized, and distributed.

In a centralized topology, all switches are grouped together in one area. The cabling infrastructure is used to connect switch ports to servers/storage units using the most cabling that is also the most flexible for data center management. The entire data center is managed as a single network entity. Switch port use is maximized and any port can connect to any server/storage unit.

In a partially centralized (in-row) topology, each row of server/storage cabinets has a set of switches in a cabinet at the end of the row. Less cabling is used than with centralized topology. A partially centralized topology has moderate flexibility for management because each row is managed as a separate network entity. Some switchports may remain unused, as ports are dedicated to the row.

In a distributed (in-cabinet/top of rack) topology, each server/storage cabinet is equipped with a switch and uses the least amount of cabling. This type of topology is the least flexible for data center management because each cabinet is managed as a separate network entity. Many switch ports may remain unused, as ports are dedicated to the cabinet.

A general design recommendation for cabling systems is to manage active device connectivity using passive (does not require power to operate) cabling components. For example, a centralized copper or fiber-optic patch panel with patch cables or cords can terminate cables connected to cabinets populated with switches, servers, or storage arrays. **See Figure 11-9.** An adjacent patch panel can do the same for switch cabinets. Connecting any server or storage array to any switch port in the data center would consist of manipulating a copper or fiber-optic patch cord. In addition to ease of use, other advantages to this design recommendation include the following:

- Isolation of network electronics—After initial cabling is routed to the cabinets, there is no reason to touch the equipment to make or change connections. The central patch panels can even be in another area or room of the data center.
- Investment protection—Changing cabinet equipment would not require any changes to the cabling system. The existing cabling can connect to the port(s) in the new device and be managed, as normal, from the central patch panels.

The general design recommendation for cabling systems also includes the following requirements:

- Horizontal cable (copper or fiber optic) from a switch cabinet in an HDA to an equipment cabinet in an EDA cannot be longer than 90 m (295′). A total of 10 m (33′) is allocated for the modular cords used at both ends, resulting in a port-to-port maximum channel length of 100 m (328′).
- If a ZDA is used with copper cabling between an HDA and an EDA, the ZDA must be located at least 15 m (49′) from the HDA to minimize interference caused by multiple connection points in close proximity. The maximum channel length from the HDA to EDA is 100 m (328′).
- Copper backbone cabling channels from a core switch in the MDA to a second-tier switch cabinet in an HDA must not exceed 100 m (328′). The HDA must be located at least 15 m (49′) from the MDA to minimize interference caused by multiple connection points in close proximity.

Unlike copper cabling, allowable lengths for fiber-optic backbone cabling channels vary by application: 2 Gbps copper vs. 4 Gbps fiber channel or 10 GbE copper vs. 10 GbE fiber channel. In general, 150 m to 300 m (492′ to 984′) is specified by standards when using laser-optimized (LO) 50 µm multimode fiber (OM3). *Laser-optimized (LO) multimode fiber* is a high-capacity multimode fiber designed for short-wave 10 Gb optical transmission. Longer cable runs are possible for specific applications by using LO multimode fiber at 500 m (1640′) for 10 GbE or single-mode fiber at 10 km (6.2 mi) for fiber channel.

A *transceiver* is an electronic device that can send and receive signals on the Ethernet media system. Transceivers can be either a separate device or may be built into an Ethernet port. Adapter and switch vendors may offer multiple fiber-optic transceiver options with varying distance limits for their products. If HDAs are omitted and cabling is routed from the MDA directly to an EDA, a maximum channel length of 300 m (984′) is allowed for fiber-optic cabling and 100 m (328′) is allowed for copper cabling.

A *storage area network (SAN)* is a dedicated network that provides access to servers, local area networks, and shared storage devices. The primary function of a SAN is to make storage devices accessible to servers. Sharing storage simplifies storage administration and adds flexibility since cables and storage devices do not have to be physically moved to shift storage from one server to another. A SAN also enables effective disaster-recovery processes. **See Figure 11-10.**

Copper Distribution Cabling. The two types of networks associated with data center installations are Ethernet for server networking and fiber channel for storage networking. While it is possible to use copper cabling for fiber channel connections, the distance limitations of 33 m (108′) at 1 Gbps and 17 m (56′) at 2 Gbps make it impractical to consider copper as a suitable medium for SANs.

Previously, data center standards recommended CAT 6 cabling runs. However, innovations in design and manufacturing have improved and the current industry recommendation is to install augmented category 6 (CAT 6a) cabling, which has improved immunity to crosstalk and double the bandwidth (500 MHz vs. 250 MHz) for short cabling runs (up to 100 m) compared to standard CAT 6 cabling. Twisted-pair copper cabling is suitable for server networking, where 100 m (328′) port-to-port (channel) cabling runs are possible for the full range of Ethernet data rates, including 10 GbE and 10GBase-T.

Passive Cabling Components

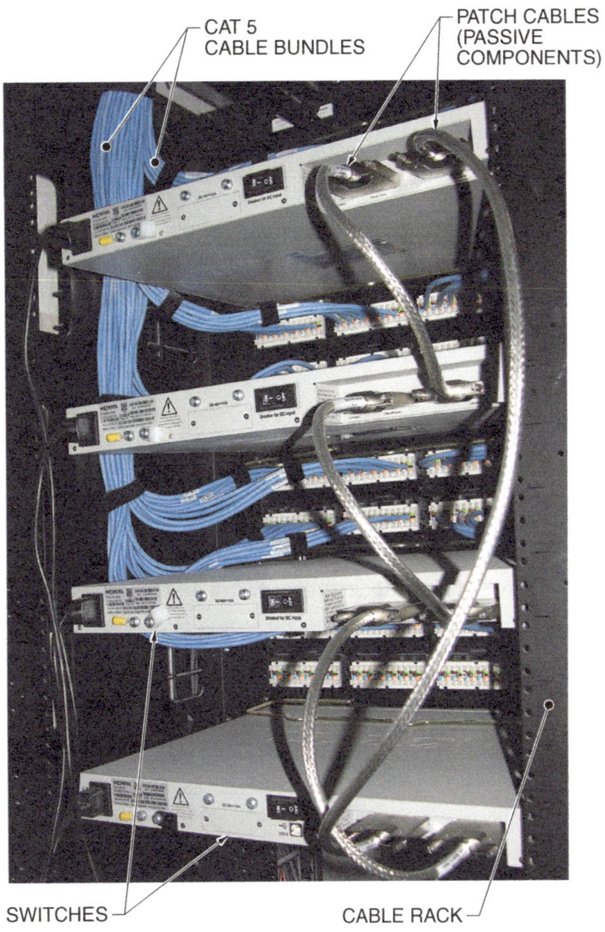

Figure 11-9. A copper patch cable can terminate cables connected to cabinets populated with switches.

Installed CAT 6 cabling infrastructure can support the operation of 10GBase-T only up to 55 m (180′) under certain conditions. CAT 6a cabling supports the operation of 10GBase-T over a maximum of 100 m (328′) of twisted-pair copper cabling with up to four connectors, as well as the low-power "short-reach mode" version of 10GBase-T over a maximum of 30 m (98′) of twisted-pair copper cabling with up to two connectors.

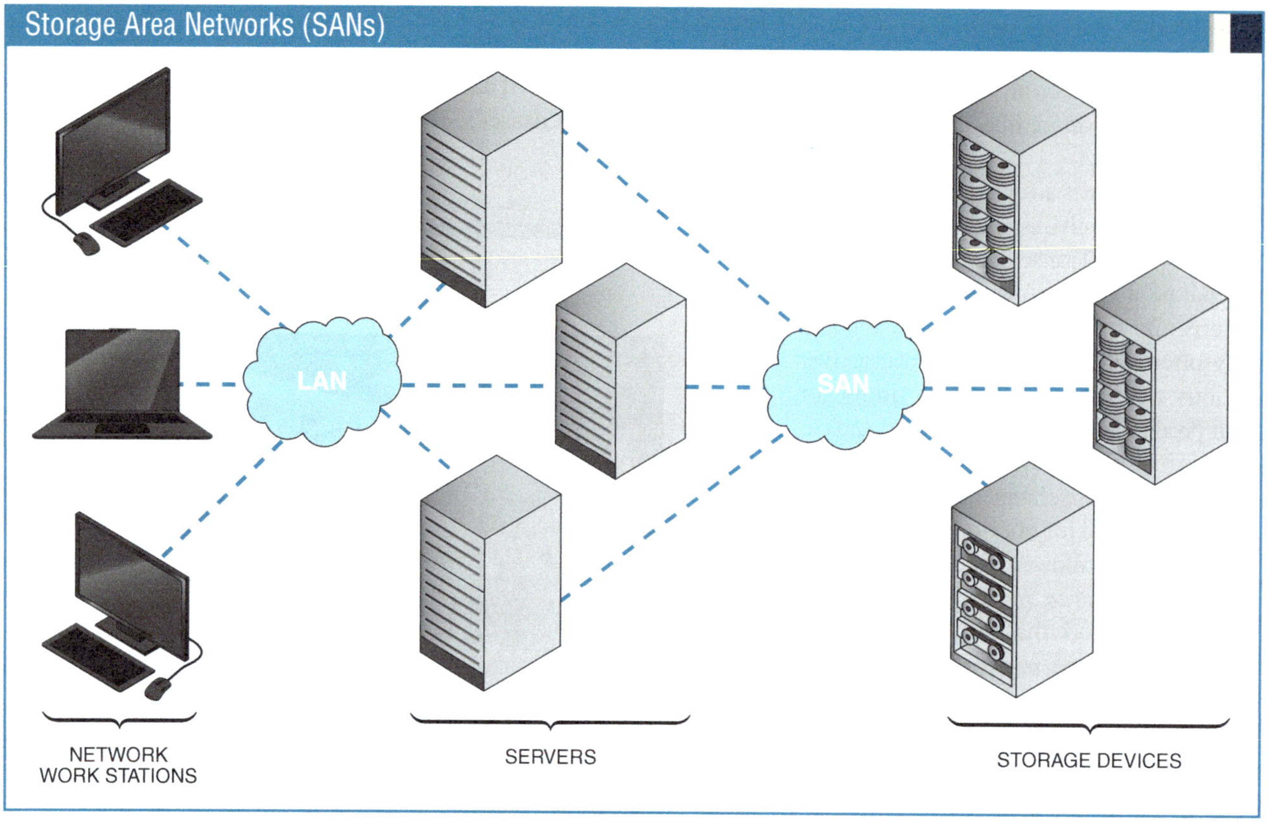

Figure 11-10. A SAN is a dedicated network that provides access to servers, LANs, and shared storage devices.

While many OEMs of cable produce CAT 7 cabling and associated hardware, currently ANSI/TIA does not recognize CAT 7 cable as an applicable standard. CAT 8 is designed only for data centers where distances between switches and servers are short. CAT 8 cabling is not intended for general commercial building cabling.

Fiber-Optic Distribution Cabling. Single-mode (SM) and multimode (MM) fiber-optic cabling can be used exclusively or together with copper cabling in a data center. The advantages of fiber-optic cabling include immunity to electromagnetic interference (EMI), flexibility, smaller cable and patch cord diameters, and allowable spans exceeding 100 m (328′). The principal disadvantage of fiber-optic cabling is the higher cost of provisioning devices with fiber-optic interfaces rather than copper interfaces.

Multiple types of fiber-optic cabling are available for installation; however, LO 50 μm MM fiber works best for maximum flexibility and transmission of data. Fiber-optic cables are identified by the cable jacket color. **See Figure 11-11.** For the installation of fiber-optic equipment, the two following options are generally considered for data centers:

- Factory-made, preconnectorized assemblies for plug-and-play connectivity are used in cases where rapid connectivity is needed or there is a lack of technicians on the project that have expertise in fiber installation. Plug-and-play modules provide a quick-connecting interface with plastic body cassettes, which are used for collecting individual fibers into a multifiber cable, between multiple fiber-optic connectors on a communications path (trunk) and the LC duplex, SC duplex, or MPO jumper cables. An *LC connector* is a small form factor connector that is half the size of a standard connector. These assemblies can then connect directly to the electronic devices. **See Figure 11-12.**

- Field-installable, factory-terminated, and polished connectors are used for mounting on individual fibers. In some environments, such as military or financial institutions, security is an additional concern since the fiber infrastructure usually transports data between sites or groups of servers. In such cases, secure connectors and transparent enclosures can be used to minimize the risk of data theft through physical access to fiber connection points.

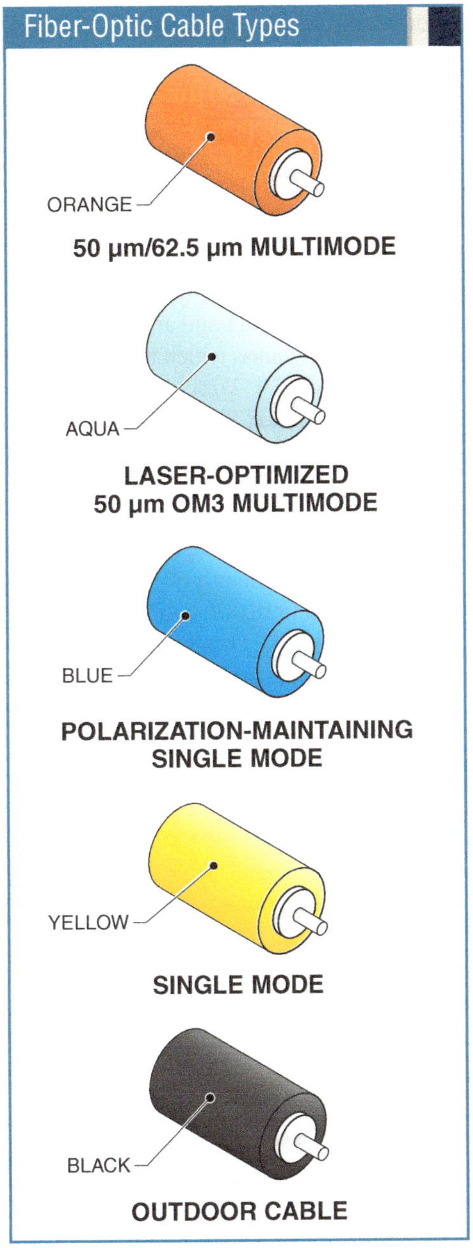

Figure 11-11. Fiber-optic cables are identified by the cable jacket color.

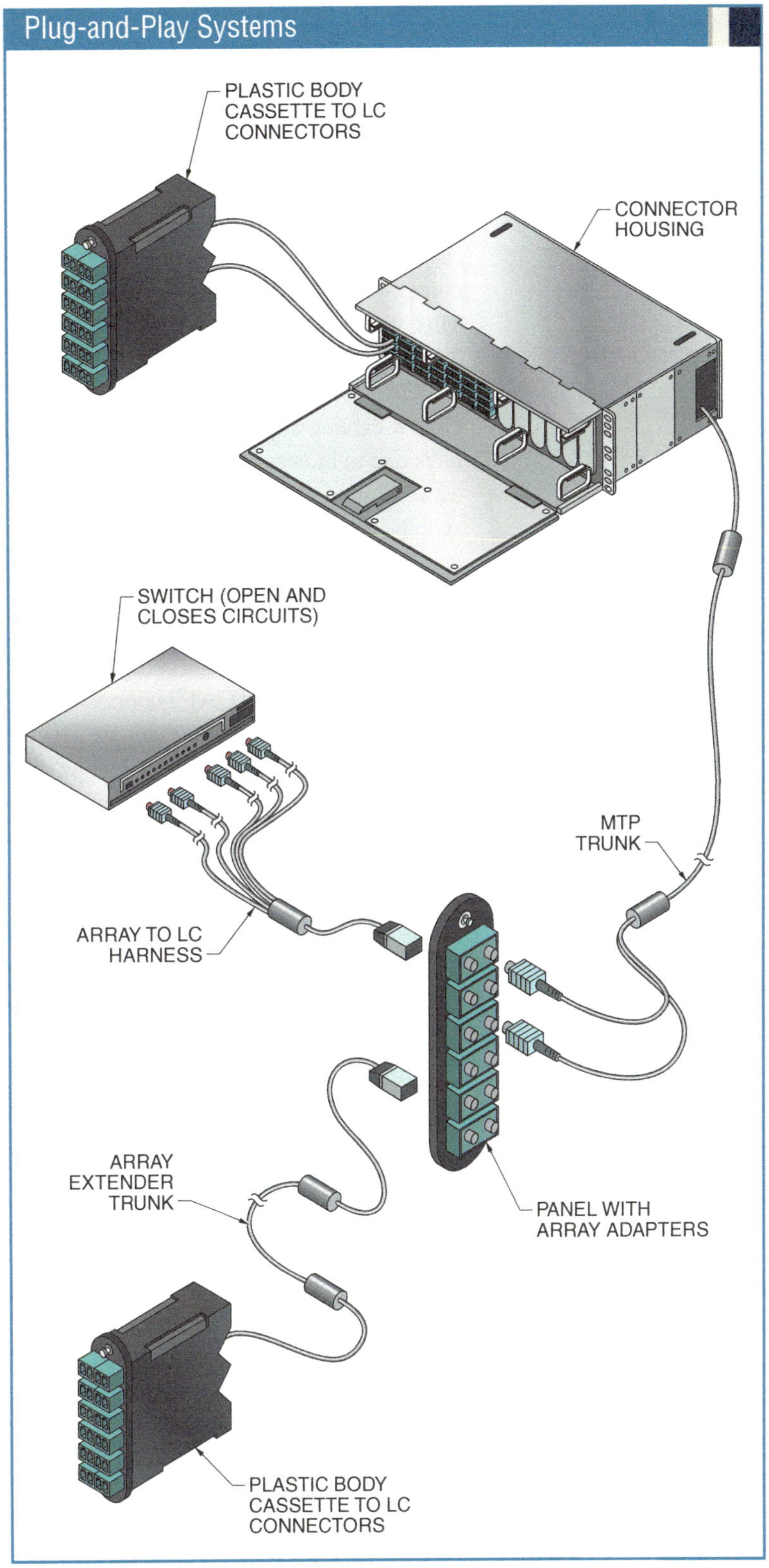

Figure 11-12. Plug-and-play modules provide a quick-connecting interface with plastic body cassettes between multiple fiber-optic connectors on a communications path (trunk) and the LC duplex, SC duplex, or MPO jumper cables. These then connect directly to the electronic devices.

Few data center installations are 100% copper or 100% fiber-optic cabling. The optimal installation for most data centers is a combination of copper and fiber-optic cabling in different parts of the network infrastructure. The use of fiber-optic media converters adds flexibility by interconnecting different cabling formats and extending the reach of copper-based Ethernet equipment over SM/MM fiber links spanning considerably longer distances. While future upgrades to the existing CAT 6 and CAT 7 cabling will require CAT 8 cabling supporting 25/40 Gbps data rates due to increased capacity requirements over short-reach connections at the server, switch, and top of rack levels, data center operators can then aggregate that traffic over larger capacity SM/MM fiber backbones for core interconnect and cross-campus links.

Most facilities with data centers rely on a mixture of CAT 6 and CAT 7 copper cabling that supports 10 Gbps bandwidth over 100 m and higher data rates of up to 40 Gbps over shorter distances. However, the evolution of those copper cabling specifications is fundamental to meeting the requirements of not only cloud service providers but also larger enterprises and LECs for their use or delivery of private-, hybrid-, or cloud-hosted applications and services. Data centers typically have MM fiber-optic cables used in their installations. Identified by core and cladding diameters and the amount of speed that data can be transmitted, multimode cable types are designated OM1 through to OM5 by the International Electrotechnical Commission (IEC) as follows:

- OM1 cables are identified with an orange-colored jacket and have a 62.5 μm core. They support 10 GbE applications at 33 m but are usually used for 100 MbE applications. OM1 cables are commonly used in premises applications.
- OM2 cables are identified with an orange-colored jacket and have a 50 μm core. These cables are standardized to support 10 GbE applications at 82 m but are usually used for 1 GbE applications. Both OM1 and OM2 cables are used in short-haul networks, LANs, and private networks and work with LED transmitters that send hundreds of light modes down the fiber, but they will not support future requirements for higher speeds.
- OM3 cables are identified with an aqua-colored jacket and a 50 μm core. OM3 cables can run 10 GbE applications at 300 m, though they have been improved to work with 40 GbE and 100 GbE applications if using an MPO connector. OM3 cables use fewer light modes, which enables higher speeds, and have 1.5 dB connector loss at all speeds.
- OM4 cables are identified with an aqua-colored jacket and 50 μm core. They are optimized to support 10 GbE at 550 m and 100 GbE at 150 m using MPO connectors. OM4 cables are usually used in high-speed networks, data centers, financial centers, and corporate campuses.
- OM5 cables are identified with a lime-green-colored jacket and are the first type of cable approved as wideband multimode fiber (WBMMF). *Wideband multimode fiber (WBMMF)* is a fiber-optic cable that can transmit up to four times the amount of data that could be transmitted on a single multimode fiber-optic cable. The optical strands are designed to allow the simultaneous transmission of four different wavelengths of light, quadrupling the throughput. This type of cable is designed to specifically handle high-speed data center applications using two fibers to transmit from 40 Gbps to 100 Gbps. OM5 cable offers bandwidth of 28 GHz. **See Figure 11-13.** *Note:* Fiber-optic cable jackets are marked with the glass core size.

Most fiber-optic cables are colored orange for multimode and yellow for single mode.

Multimode Cable Types				
Type	**Max. Reach***	**Core/Cladding Dia.†**	**Jacket Color**	**Applications**
OM1	220	62.5/125	Orange	Low-bandwidth premises installations
OM2	220	50/125	Orange	Low-bandwidth LANs; private networks
OM3	100	50/125	Aqua	Data centers
OM4	150	50/125	Aqua	High-speed data center networks; financial networks; corporate campuses
OM5	400	50/125	Lime green	High-speed data center networks

* in m
† in μm

Figure 11-13. Multimode cable types are designated by the IEC as OM1 through OM5.

LABELING AND IDENTIFICATION SYSTEMS

A major aspect of any VDV installation is the labeling and identification scheme used, specifically on the cables and connector ports. While there are industry standards and recommendations, there is no specific standard that addresses labeling and identification systems for VDV and data center hardware. Nonetheless, the identification of the system is a key part of the process. Effective labeling promotes better communication and eliminates confusion when trying to locate or connect a component or port.

Depending on the size and type of premises, a labeling and identification system may be applied to equipment in rack, cabinet, or floor tile locations or for patch panels, individual ports, and individual cables. With some labeling and identification systems, a color-code system is used for the visual identification of specific components, such as cables, connectors, and connector ports. **See Figure 11-14.**

Labeling and Identification Systems		
Color	**Type**	**Application***
Aqua	OM3 Fiber	LAN/SAN device-to-device
Yellow	Single Mode Fiber	LAN/SAN device-to-device over long distance
Orange	OM1 or OM2 Fiber	LAN/SAN device-to-device
Blue	Copper	LAN device-to-device
Green	Copper	KVM host to KVM switch, KVM switch to LAN switch,KVM switch to KVM switch
Yellow	Copper	Serial host to terminal server, terminal server to LAN switch
White	Copper	Power strip to LAN switch

* connections made through patch panels

Figure 11-14. With some labeling and identification systems, a color-code system is used for the visual identification of specific components, such as cables, connectors, and connector ports.

NEMA CONFIGURATIONS

A design aspect of VDV rooms, spaces, and data centers is the electrical distribution to power the equipment. The standardization of power distribution and plug strips allows for their proper selection and use. The *National Electrical Manufacturers Association (NEMA)* is a national organization that provides standardization for electrical equipment. NEMA assigns designations to the various electrical configurations used in the electrical industry. The purpose of different types of electrical configurations is to prevent the wrong combinations of electrical system connections and the potentially dangerous conditions created by an incorrect installation. For example, many of the electrical receptacles located in VDV rooms, spaces, and data centers require a variety of straight-blade nonlocking and locking wiring devices. **See Appendix.**

Most NEMA connectors are identified by an alphanumeric code consisting of a letter and/or a numeral preceding a hyphen and a numeral and/or a letter following the hyphen. The letters at the beginning and end of the code indicate whether the connector is a locking type and whether it is a plug (male connector) or a corresponding receptacle (female connector). The following are the two basic classifications of NEMA connectors:

- Curved blade, locking—Connectors with numbers prefixed by "L" are curved-blade, twist-locking connectors. Twist-locking connectors are used for heavy industrial and commercial equipment, where increased protection against accidental disconnection is required.
- Straight blade, nonlocking—Connectors without an "L" as a number prefix are straight blade and nonlocking. The numeral preceding the hyphen indicates the type's particular configuration, such as the number of poles (current-carrying terminals) and wires connected to it, the voltage limitations, and whether the connector is for use with single- or three-phase power.

A connector with a ground terminal is described as having more wires than poles, such as a two-pole, three-wire connector or a four-pole, five-wire connector. Examples of connectors without a ground terminal (nongrounding) devices may be two-pole, two-wire connector or a three-pole, three-wire connector, etc. The numeral following the hyphen is the current rating of the device in amperes. This number is followed by the letter "R" to indicate a receptacle or "P" to indicate a plug. **See Figure 11-15.**

EQUIPMENT ROOM COMMISSIONING

Commissioning is the process of planning, documenting, scheduling, testing, adjusting, verifying, and training in order to operate a facility as a fully functional system per the owner's project requirements. ERs must go through a separate (from other parts of a building) commissioning process prior to building occupancy and operation of related equipment. Commissioning occurs for both new and retrofit construction projects.

A critical part of the commissioning and warranty process is the testing of data center cabling media. A *one-cord test method* is a fiber-optic cable test method that allows for testing of a fiber optic link from end-to-end and includes losses from all connectors. Including the connector losses is required as the fiber link becomes shorter, since connectors are the major contributor to the overall loss, not the fiber itself. A *two-cord test method* is a fiber optic cable test method used when the connector on the power meter is different from the connector on the fiber link to be tested. The two-cord test method involves using a hybrid test cord to allow the connectors to match.

The best practice for testing multimode fiber includes using the one-cord test method. While the two-cord test method may seem easier, referencing both test cords can result in optimistic results and may provide negative loss results. Only the one-cord reference includes the loss of the connections at both ends of the channel for the highest accuracy. Many cabling OEMs reject results carried out with the two-cord reference, which could prevent acquiring a warranty.

An *optical time domain reflectometer (OTDR)* is a test instrument used to measure fiber-optic cable attenuation (signal loss). If an OTDR is used to test fiber in the ER of MDA, the measurement of the loss of specific connections, such as individual connectors and splices, as well as overall link loss, depends on the direction from where the measurement is taken. Bidirectional testing that averages the results from measurements taken at both ends of the link is required to complete this type of testing. While bidirectional testing may have once been time consuming, there are OTDRs available that simplify the process by using a loop at the remote end that allows for testing from one end and automatically reports the bidirectional average for that link.

Copper testing in a data center focuses primarily on testing for the presence of alien crosstalk. *Alien crosstalk* is the crosstalk noise experienced by two or more adjacent twisted-pair cables. Alien crosstalk is the key performance parameter that can affect the ability of 10 Gb HDBase-T to operate. Since 10 Gb HDBase-T is the typical application used for switch-to-server links in the data center, testing for alien crosstalk is often required to warranty a system. When a sample size is specified for alien crosstalk testing, the best practice is to test an equal amount of short, medium, and long disturbed links.

Another copper testing consideration is shield integrity. Shielded copper cabling is in data centers more than any other space. Shielded cabling offers far superior noise immunity and security compared to unshielded cabling. When installed properly, there is no alien crosstalk present on a shielded cabling system. For example, measured to 2 GHz and specified for 2-connector channels at 30 m, CAT 8 cabling is intended as a data center application and is specifically targeted for data center edge applications in copper server-to-switch connections where the emerging 25 Gbps and 40 Gbps speeds are used.

NEMA Connector and Receptacle Codes

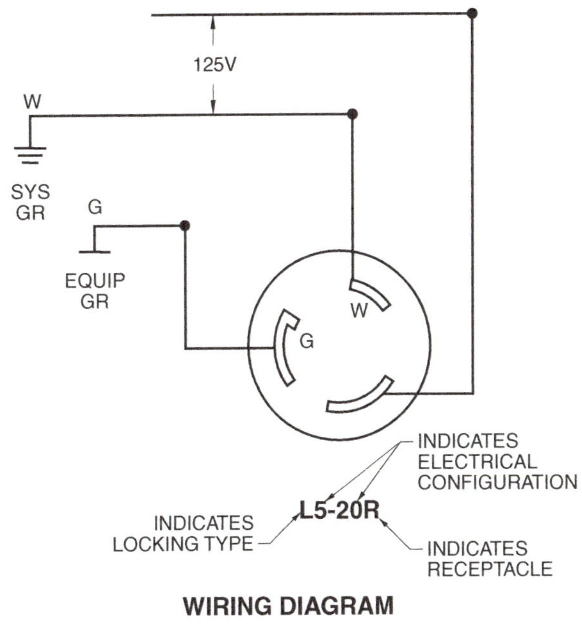

Figure 11-15. Most NEMA connectors and receptacles are identified by an alphanumeric code consisting of a letter and/or a numeral preceding a hyphen and a numeral and/or a letter following the hyphen.

SUMMARY

VDV prints must be referenced when installing, troubleshooting, and maintaining equipment in areas where such equipment is installed such as rooms, spaces, and data centers. While VDV equipment is installed in EFs, ERs, and telecommunications rooms, the most critical area for equipment installation is the data center. Cabling in a data center is typically a combination of copper and fiber-optic cables. Cable organization is accomplished with VDV labeling and identification systems. Prior to ER commissioning, all equipment is configured to conform to NEMA specifications.

CHAPTER 11

Review

VDV Rooms, Spaces, and Data Centers

Name ______________________ Date ______________

True-False

T F **1.** Entrance facilities (EFs) are sometimes referred to as entrance rooms.

T F **2.** All entrance rooms with IT equipment fall under the requirements of NEC® 645.

T F **3.** A telecommunications room (TR) is sometimes referred to as a telecommunications facility (TF).

T F **4.** With a hot aisle orientation, equipment (server) racks are arranged to face to face.

T F **5.** The cable pathways for data centers should adhere to ANSI/TIA-569, Telecommunication Pathways and Spaces.

T F **6.** Fiber channel is a high-speed networking technology primarily used for transmitting data among work areas.

T F **7.** Data center installations are either 100% copper or 100% fiber-optic cabling.

T F **8.** OM2 cables are identified with an orange-colored jacket and have a 50 μm core.

T F **9.** OM5 cables have an aqua-colored jacket.

T F **10.** There is no specific standard that addresses labeling and identification systems for VDV and data center hardware.

Multiple Choice

______________ **11.** When a floor has more than one TR, they are often are interconnected by a(n) ___.

A. backbone pathway
B. EMT system
C. patch cord system
D. wireless access point (WAP)

______________ **12.** The ___ is the primary means for determining which type of cabling is installed in a data center.

A. cable tray specifications
B. data center location
C. equipment interface
D. material cost

__________ **13.** Branch electrical circuits must be installed in ___.
- A. electrical metallic tubing (EMT)
- B. fiber channel
- C. liquidtight flexible metal conduit (LFMC)
- D. rigid metallic conduit (RMC)

__________ **14.** The size of cable managers is measured in ___.
- A. cord fill (CF)
- B. cubic centimeters (cm^3)
- C. rack units (RU)
- D. tile units (TU)

__________ **15.** A multifiber push-on (MPO) connector is a connector for ribbon cables with a capacity of eight to ___ fibers.
- A. 24
- B. 48
- C. 56
- D. 72

Completion

__________ **16.** A(n) ___ is the point where outside antenna, public, and private network service provider cables can enter a building and be consolidated.

__________ **17.** A(n) ___ is a centralized room or space for telecommunications equipment that serves the occupants of a building or multiple buildings within a campus environment.

__________ **18.** A(n) ___ is an apparatus that connects the telephone lines coming into the switching center from outside to the internal lines connected to the switching or transmission equipment inside the switching center.

__________ **19.** The cabling used in a VDV cabling system design is specified by the ___ requirements of the system and is tested and certified to ensure proper performance.

__________ **20.** If cable tray is used as a cable pathway in a data center, a minimum spacing of ___″ must be provided between the installed cable trays.

__________ **21.** A(n) ___ is a device used to secure, organize, and manage cables in a data center or ER.

__________ **22.** A(n) ___ is an electronic device that can send and receive signals on the Ethernet media system.

__________ **23.** ___ is the crosstalk noise experienced by two or more adjacent twisted-pair cables.

__________ **24.** The ___ is a national organization that provides standardization for electrical equipment.

__________ **25.** ___ is the process of planning, documenting, scheduling, testing, adjusting, verifying, and training in order to operate a facility as a fully functional system per the owner's project requirements.

Short Answer

26. Explain the main purpose for installing a data center.

27. Define the three different types of data centers.

28. List the four different areas of a building that are not considered a data center.

29. List the different network cable protocols for data centers.

30. Explain the advantages of using fiber-optic cabling.

VDV Rooms, Spaces, and Data Centers

CHAPTER 11

Activity

Name______________________________ Date ______________

DATA CENTER

Instructions

Create a list of materials that would be included in a proposal by using the drawing and original equipment manufacturer (OEM) catalog to determine cable, connecting hardware, and rack types and quantities for the data center.

Project Overview

The project is a commercial office building data center. The data center has the following requirements for the copper and fiber-optic systems:

- All cable shall be UL listed CMP/OFNP type and meet applicable industry standards.
- Cable and connector requirements are CAT 6a, 12-strand, LO 50 μm MM (OM4) cable (terminated LC connectors and coupler panels).
- Connecting hardware shall be manufactured by Leviton® and components must be selected from the 2019 North America Leviton Network Solutions Catalog.
- Cabinets are existing and have standard TIA/EIA rack units available for mounting termination hardware.
- Provide 1U wire managers between every two fiber panels and each copper panel at the network racks.
- Install angled 110-type patch panels, 48 port, for network racks.
- Install angled 110-type patch panels, 24 port, for zone boxes.
- Provide CAT 6a terminations.
- Install Leviton Atlas-X1® Mission-Critical components for copper.
- Install six CAT 6a cables from network rack #3 to the left side of each "S" cabinet (#1 through #42).
- Install twelve CAT 6a cables from network rack #3 to each "Z" raised floor enclosure.
- Install twelve CAT 6a cables from network rack #6 to each "Z" raised floor enclosure.
- Install Vertigo LC adapters for "S" racks.
- Install UHDX fiber-optic panels for network racks and zone boxes.
- Install MPO connectors at network racks and LC-type interface at "S" and zone boxes.
- Install Opt-X® HDX (half duplex) for fiber-optic systems.
- Install 12-strand cable from network rack #4 to the left side of each "S" cabinet (#1 through #42).
- Install 12-strand cable from network rack #5 to the left side of each "S" cabinet (#1 through #42).
- Install 12-strand cable from network rack #4 to each "Z" raised floor enclosure.
- Install 12-strand cable from network rack # 5 to each "Z" raised floor enclosure.

Drawing:

- T0.3 – Typical Telecommunications Rooms

OEM catalog:

- 2019 North America Leviton® Network Solutions

Note: The drawing and the catalog can be accessed in the online learner resources.

1. Cables/Length:

2. Connecting hardware/Quantity:

3. Rack elevations/Quantity:

CHAPTER

12

Outside Plant and Premises Distribution Systems

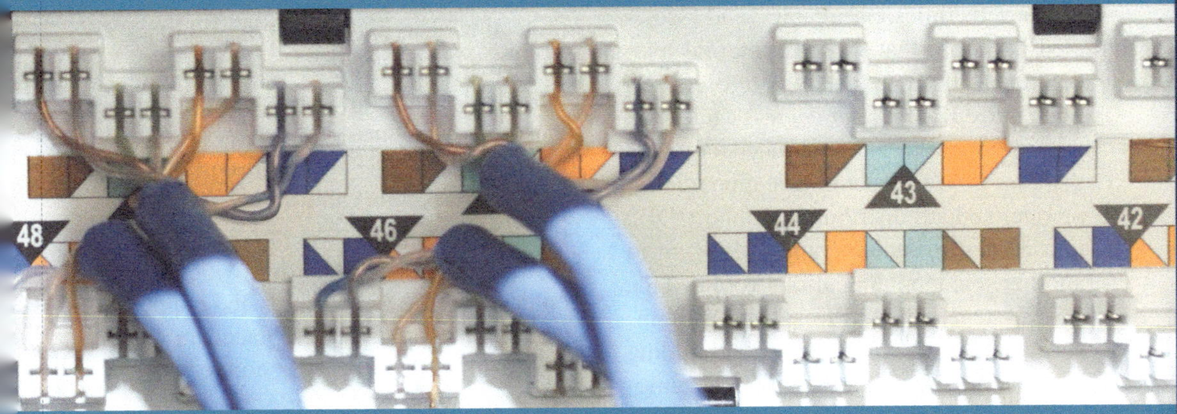

Before any cabling and connections can be installed inside a building, the cables must be routed to the building from the outside. Outside cabling is referred to as outside plant (OSP) and is routed to separate buildings. Outside cabling provides the system that distributes cable from the outside of a building to the inside of the building, to its risers, and eventually to the end users of VDV systems. Both copper and fiber-optic cabling, along with the associated hardware, are used for outside, riser, and premises distribution systems. For added protection, multiple routes for riser cables can be installed between buildings or within the same building.

Objectives

- Describe outside plant (OSP) distribution systems.
- Describe premises distribution systems.
- List the different types of cables and hardware used for interbuilding and intrabuilding backbones.
- Explain how drawings and site plans provide information for determining riser requirements.

Learner Resources

atplearningresources.com
access code 783821

OUTSIDE PLANT DISTRIBUTION SYSTEMS

An *outside plant (OSP) distribution system* is a telecommunications infrastructure designed for installation outside any structure. OSP systems consist of pathways, spaces, entrance facilities, riser cables, and associated hardware that provide external telecommunications connectivity to premises. **See Figure 12-1.** A *riser cable* is a telecommunications cable and associated hardware that runs from a telecommunications closet, or intermediate distribution frame (IDF), to a main distribution frame (MDF) or to another telecommunications room (TR). OSP systems are used to support systems such as VDV, electronic security, building automation, fire, life, and safety systems.

The elements of OSP systems include the following:

- OSP pathways—OSP pathways are the conduits that interlink OSP spaces and end points within a premises. Pathways can be direct-buried or underground infrastructure that utilizes tunnels, conduit systems that vary by application and use and may be direct buried or encased in concrete, or aerial pathways that utilize public and private utility poles. Each pathway type has advantages and disadvantages. For example, aerial pathways are not aesthetically pleasing and are subject to damage from vehicular traffic, falling tree limbs, high winds, ice loading, and other environmental factors. However, an underground conduit system encased in concrete is impervious to such hazards and can provide better overall protection.
- OSP spaces—OSP spaces include utility holes, handholes, pedestals, cabinets, and vaults. Utility holes are used as access points for pulling and splicing cables. Handholes are smaller than utility holes, are used for pulling cables, and do not typically serve as splice points. Pedestals provide access to small splices, interconnects, and cables. Cabinets are used as cross-connects for aerial and direct-buried applications. Vaults provide environmentally protected spaces either above or below grade and can accommodate equipment electronics.
- OSP entrance facilities (EFs)—OSP EFs serve as the entry point into a premises for the riser equipment. EFs interconnect the building riser to other building risers. The EF is where the copper system receives its primary fused overcurrent protection for hazardous voltages and lightning protection. The EF is also referred to as the service entrance (SE).
- OSP riser—The OSP riser is the pathway that provides connectivity between premises. *Note:* The OSP riser is also referred to as the backbone cable and often has a red jacket.
- OSP hardware—OSP hardware includes outdoor interface enclosures and copper and/or fiber-optic cables placed in the OSP pathways to provide connectivity between premises.

OSP Interbuilding Risers

The interbuilding riser is the portion of the OSP cabling infrastructure that includes cabling between building EFs. An *interbuilding riser system* is a system of telecommunications cables that connect one building to another.

OSP cables are typically gel-filled or air-core cables. A *gel-filled cable* is a cable that has a gel-type substance inside its jacket that encases its conductors. The water-resistant gel provides protection from underground moisture and water. An *air-core cable* is a cable in which the conductors are surrounded by an air space within the jacket. Air-core cables can be installed underground or in aerial locations between poles.

In addition to gel-filled and air-core multipair cabling, the most common cables used in OSP and interbuilding risers for utility and private applications include the following:

- Bell-type gel-filled and air-core multipair cable
- 100 Ω twisted-pair cable
- 75 Ω coaxial cable
- single-mode fiber-optic cable
- multimode fiber-optic cable

Figure 12-1. OSP systems consist of pathways, spaces, entrance facilities, riser cables, and associated hardware that provide external telecommunications connectivity to premises.

A *Bell cable* is a VDV cable specified for outdoor use by private companies and Regional Bell Operating Companies (RBOCs). Bell cable designs specify bare or uncoated shields for most products, but they allow the use of coated steel in certain armored designs. The metallic shield of a cable, when properly grounded at each end, effectively cancels the effect of power-induced noise. When used in conjunction with surge protection devices, the metallic shield helps to protect the cable, associated terminals, and customer equipment from damage caused by voltage surges, electromagnetic interference (EMI), and lightning strikes.

Interbuilding riser requirements are application dependent. Factors that are considered in the design and placement of OSP cables and the interbuilding riser include the following:

- existing pathways
- costs associated with installing new pathways
- types and quantities of cable hardware required for a specific application
- distance between premises
- expected life of the riser cabling
- size and occupancy population of the premises

OSP copper cabling can be limited by distance and application, but OSP fiber-optic cabling provides increased flexibility. This is because fiber-optic cabling can be installed in longer runs for advanced network application requirements. *Note:* Different cable types are frequently placed together in the same OSP pathways.

OSP Interbuilding Riser Cable Types

Hardware associated with the OSP cables must be either protected from the elements or constructed with materials suitable for exposure to the elements. Installation of the OSP often requires construction of significant physical infrastructure, such as handholes and underground vaults. A *handhole* is a buried nonmetallic box, in which the lid is flush with the surface of the ground, that provides access, housing, and protection for cables.

OSP Interbuilding Copper Cabling. Traditional gel-filled and air-core multipair OSP cables have a 300 V working voltage capability when used in communication circuit applications. With OSP cables, the voltage capability is purposefully omitted from the cable jacket and replaced with a symbol of a telephone handset to indicate that it is a telecommunications cable and not a high-voltage electric power cable. The traditional 300 V working capability applies to OSP copper cables manufactured in compliance to specifications, including PE-39, PE-89, and PE-86.

Polyethylene (PE) is a thermoplastic material with electrical insulating properties. PE is used as cable jackets, piping, foam, or sheets. One of the differences in electrical properties between solid PE and PE foam is the dielectric constant. The dielectric constant of PE foam is lower than solid PE. Solid PE is typically used as copper conductor insulation. Due to the moisture resistance of PE, it is typically used in wet and damp environments. Overall, PE insulation is rigid, but its rigidity can vary based on whether it is high density (HDPE) or low density (LDPE). Common industrial uses of PE-insulated wire and cable include petroleum product refineries, electronics wiring, and networking and data systems wiring.

An *aluminum-polyethylene (ALPETH) cable* is a cable covered with separate aluminum and polyethylene jackets. Shielded ALPETH multipair cables are constructed with a PE-foam-skin-insulated single jacket intended for direct-burial or aerial applications. A water-blocking compound completely coats each insulated twisted-pair conductor and fills the air space between conductors. Corrugated, bare, 8 mil aluminum tape is applied around the PE core wrap and provides environmental protection for direct-burial and underground conduit applications. **See Figure 12-2.**

ALPETH cables are fabricated with a PE jacket that provides a tough, flexible, protective covering that can withstand exposure to sunlight, atmospheric temperatures, ground chemicals, and stresses expected in standard installations. ALPETH cables are designed for outdoor use only, are rated as general-purpose type (Type CM), and are not allowed to be used in riser (intrabuilding backbone) or plenum applications. Riser cable, so named because it "rises" between the floors of a multistory building, is also called intrabuilding riser cable. An *intrabuilding riser system* is telecommunications cables that are part of the building subsystem that connect one telecommunications room (TR) to other communication rooms within the same building.

Ratings and grounding requirements of all types of telecommunications cables are covered in NEC® Article 800. Typically, there is a 50′ limitation when installing Type CM cables in a premises. *Note:* NEC® Articles 770.48 (fiber optic) and 805.47 (copper) state that unlisted nonconductive and conductive communication and fiber-optic cables are permitted to extend from the point of entrance beyond 50′ when they are continuously enclosed in ridged metal conduit (RMC) or intermediate metal conduit (IMC) to the point of termination.

Multipair interbuilding copper OSP cable has a maximum bandwidth of 16 MHz, is rated as CAT 3, and supports analog voice and 10BaseT data applications within standard distance limitations. Additional types of copper cables installed in OSP applications consist of 75 Ω coaxial (RG-6 and RG-11), CAT 5e (100 Mhz), CAT 6 (250 Mhz), and CAT 6a (500 MhZ) cables that are constructed with armored protection and gel-filled water-blocking agents. These types of cables are designed to be used with pair counts of 4 pair to 25 pair in direct-buried, aerial, or in-conduit applications. Multipair interbuilding copper cables extend connectivity to the OSP to meet the requirements of the NEC®.

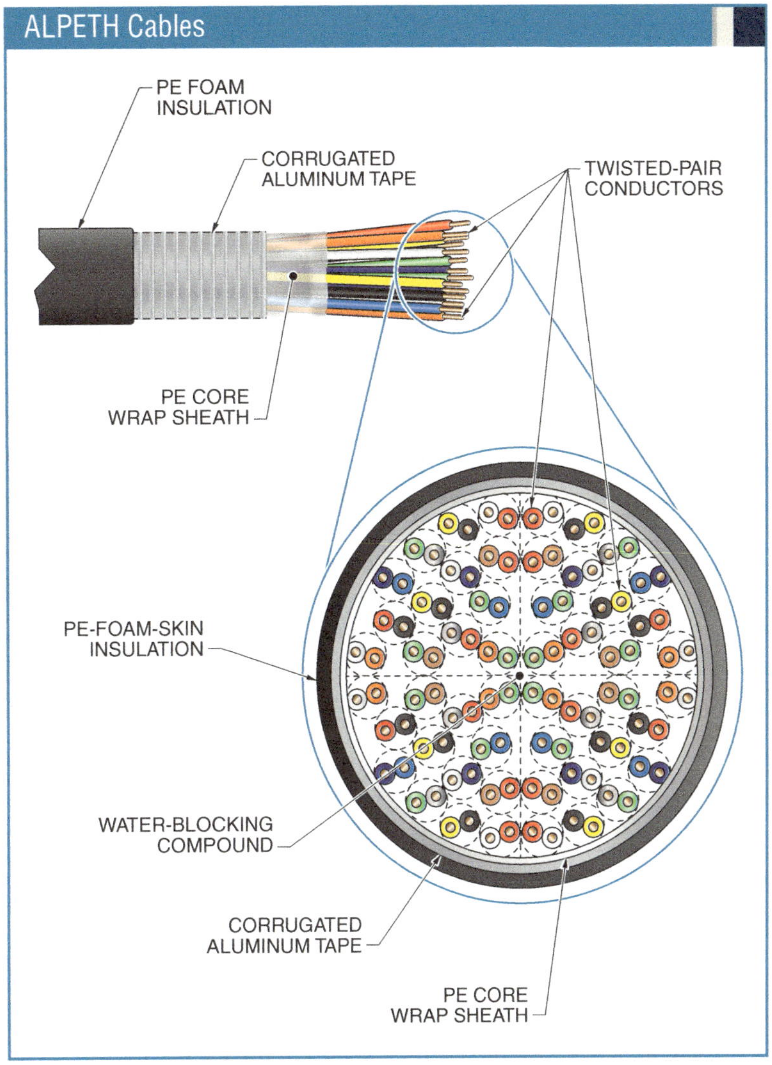

Figure 12-2. An aluminum-polyethylene (ALPETH) cable is a cable covered with separate aluminum and polyethylene jackets.

OSP Interbuilding Fiber-Optic Cabling. Fiber-optic cable is the most commonly specified and installed type of cable used in OSP applications due to its ability to transmit large amounts of data at long distances. Applications that fiber-optic cabling supports include VDV, audio, security, and fire alarm systems. Because they cannot be spliced as easily as copper cables, fiber-optic cables provide added security and less susceptibility to EMI and lightning. Fiber-optic cables are lightweight, have durable construction and flexibility, and are more economical to install.

The two types of fiber-optic cable used in OSP applications are single-mode and multimode cables. Each type is constructed of a core, cladding, and buffer. A *core* is the smallest section of a glass fiber strand and is used to carry light waves. The core is the center part of the fiber. *Cladding* is a layer of glass fused to a glass fiber-optic core to aid light transmission and to act as protection for the core. A *buffer* is a fiber-optic cable layer that provides protection for optical fiber within a cable. The buffer (or coating) is a hard-plastic coating on the outside of the fiber that protects the glass from moisture and physical damage. The buffer is stripped when terminating or splicing the fiber.

The two types of buffering for fiber-optic cable are loose tube and tight. A *loose-tube buffer* is a fiber-optic cable construction in which the glass is carried loosely in a buffer that has a diameter greater than the diameter of the fiber. A loose-tube buffer is used primarily in outdoor applications, typically in armored and gel-filled cables for protection from water, and has limited applications within buildings due to its cable rating. A *tight buffer* is a fiber-optic cable construction where each glass fiber is buffered tightly by a protective thermoplastic coating to a diameter of 900 μm. Tight-buffered cable is suitable for both indoor and outdoor applications (depending on the rating) and is dry-constructed.

Other fiber-optic cable constructions include ribbon fiber and blown fiber. A *ribbon fiber* is a type of fiber-optic cable in which multiple fibers are clad in a single, flat, ribbon-like cable. Ribbon fiber provides the ability to splice multiple fibers simultaneously. The cables are constructed in a ribbon, or flat, manner and are positioned next to each other. Standard strand counts are 12 per ribbon. The cost of these cables offsets the costs associated with splicing multiple individual fiber strands.

Blown fiber is a type of fiber-optic installation method in which fiber-optic cables are blown through preinstalled conduit or tubes using air pressure to float the cable through the conduit or tube. Blown fiber uses a system of microtubes installed in a point-to-point distribution that is composed of tube distribution units (enclosures) and connected by tube couplers. Upon completion of the tube system, technicians use air compressors and cable-blowing machines to install point-to-point fibers (in various strand counts and fiber types) between locations at great distances. The integrity of the fiber-optic cable is tested with an OTDR at each end of the installation to verify that there are no issues with the performance of each fiber. **See Figure 12-3.**

Blown fiber has a lower rate of failure because the tubes are empty when installed in place. Because the fiber is positioned with air pressure and no pulling or tugging is required, installation damage is also minimized. While the upfront costs of installing blown fiber are high, future upgrades can be easily performed with minimal cost since the cabling infrastructure is already in place.

OSP fiber cables are constructed with a durable jacket that is available in various materials, from PE (Type CM) to riser/plenum materials for riser applications (Type CMR) and plenum applications (Type CMP). OSP fiber cables have internal strength members to assist in pulling the cable or aerial installation.

Fiber-optic cables are available in several designs with various jacketing options. In certain specifications, a nonarmored cable can be referred to as a "duct cable." Duct cables can be armored (metallic) or nonmetallic. Armored cables have a steel armor layer added under the outer cable jacket. The armor is usually added to increase the protection against rodents for a direct-buried cable. However, the armor also serves as an extra layer of protection against other factors, such as rocky soil. Nonmetallic, dielectric versions of duct cable incorporate a nonmetallic central member and are suitable for duct or conduit placement. A dielectric cable has no metallic or conductive components, such as a metallic central member, metallic strength member(s), armor, or copper wires. Dielectric duct cables are designed for duct, tunnel, or aerial installations.

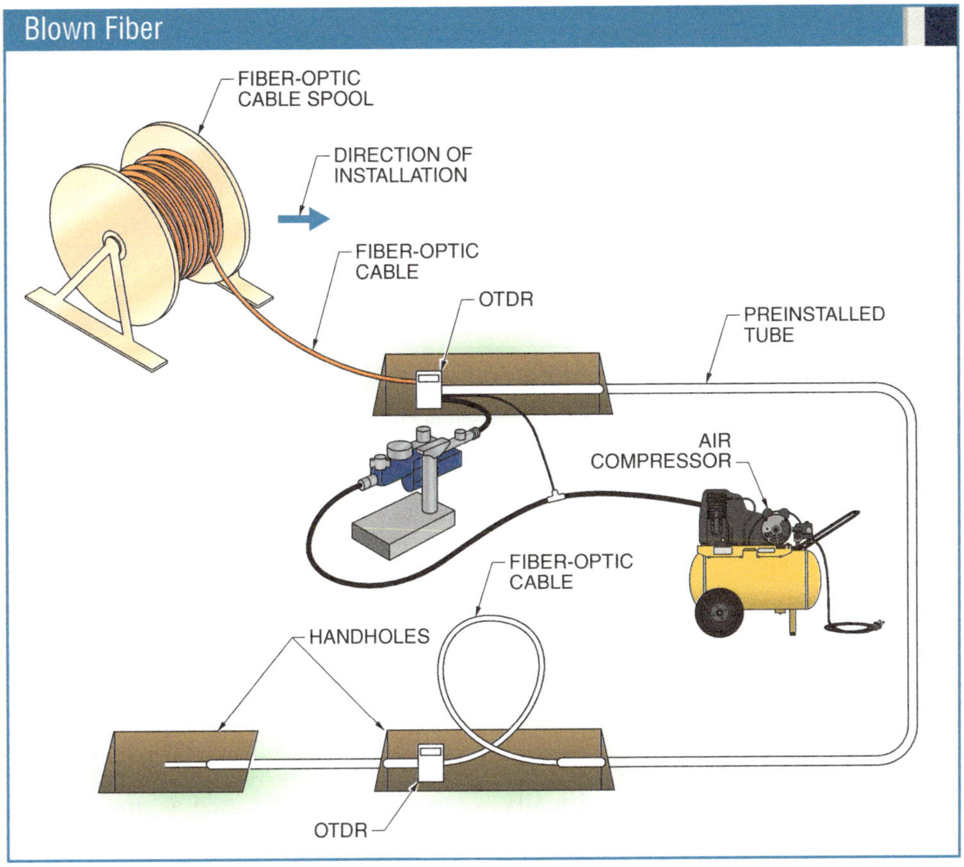

Figure 12-3. Blown fiber is a type of fiber-optic installation method in which fiber-optic cables are blown through preinstalled conduit or tubes using air pressure to float the cable through the conduit or tube.

Aerial cables typically have the same cable construction as duct cables. Self-supporting aerial cables are typically duct cables with modifications to their design to simplify the aerial installation. Self-supporting cables are designed to be installed without the installation of a preinstalled messenger wire. If properly installed, self-supporting cables can be installed in less time than lashing a conventional duct cable to a metallic messenger wire.

All-dielectric fiber-optic cables are recommended for aerial applications since fiber-optic cables are not as susceptible to lightning strikes, are not subject to induced voltages, and are not required to be grounded like cables with metallic components. All-dielectric self-supporting cables have a duct cable core with a layer of strength members that allows installation without the use of a separate messenger wire. Typically, there are length limitations depending upon location of installation and special mounting hardware may be required. Because these cables are dielectric, no grounding is required.

Indoor/outdoor cables that can be installed in both indoor and outdoor locations and that are water-blocking and UV resistant are available. Indoor/outdoor OSP fiber cable jackets are composed of a flame-retardant material that allows the cables to pass the NEC® flame test requirements for indoor installation and carry a cable flame rating (such as riser/plenum rated). The cable ratings associated with indoor/outdoor OSP fiber cable applications are listed in NEC® Article 770. The cable markings are listed in Table 770.179 and the permitted cable substitutions are listed in Table 770.154(b). **See Figure 12-4.**

Fiber-Optic Cable Marking Types and Substitutions		
Cable Marking	**Cable Type**	**Permitted Substitutions**
OFNP	Nonconductive optical fiber plenum cable	None
OFCP	Conductive optical fiber plenum cable	OFNP
OFNR	Nonconductive optical fiber riser cable	OFNP
OFNG	Nonconductive optical fiber general-purpose cable	OFNP; OFNR
OFCG	Conductive optical fiber general-purpose cable	OFNP; OFCP; OFNR; OFCR; OFNG; OFN
OFN	Nonconductive optical fiber general-purpose cable	OFNP; OFNR
OFC	Conductive optical fiber general-purpose cable	OFNP; OFCP; OFNR; OFCR; OFNG; OFN

Figure 12-4. Fiber-optic cable types and markings are listed in NEC® Table 770.179 and permitted cable substitutions are listed in Table 770.154(b).

OSP fiber-optic cables can be installed in several different types of pathways, including the following:

- high-density polyethylene (HDPE) pipes—HDPE pipes are specifically designed for extended, leakproof performance and are a common choice for fiber applications. Heat-fused joints provide for the reliable coupling of pipe sections. Corrugated HDPE is also available for ductwork in tight spaces.
- flexible fabric innerduct systems—Flexible fabric innerduct systems include a flexible fabric material that is easier to use than corrugated innerduct piping. Flexible fabric can be installed in tight spaces while retaining its cable protection properties. **See Figure 12-5.**
- plenum/riser-rated innerduct systems—Plenum/riser-rated innerduct is a flexible, nonmetallic material used for cable pulls of 1000′ or less. The nonmetallic flexibility properties of plenum/riser-rated innerduct reduce surface contact of fiber cables that are pulled through the ducts and can be specified in different wall thicknesses, coil cable lengths, straight cable lengths, and colors.
- ribbed innerduct systems—Ribbed innerduct provides less friction when pulling cables and is stronger than nonribbed innerduct. Ribbed innerduct is built for in-ground installation and provides protection from moisture, groundwater, and rodents.
- smooth-wall innerduct systems—Smooth-wall innerduct is used as underground raceway and can be installed through trenching and directional drilling.

OSP Interbuilding Connecting Hardware. Multipair copper cables can be spliced in utility holes or terminated in the premises SE to provide cross-connects to premises riser distribution systems. An *insulation displacement connector (IDC)* is a type of telecommunications connector in which the wire is punched down into a metal holder that strips the insulation on the wire and allows electrical contact to be made. Multipair copper cables spliced in the OSP use high-density, modular IDCs that are crimped together with a special crimp tool, providing a gastight connection in 25 pair increments. The connecting hardware meets CAT 3 and CAT 5 performance specifications. The modular splice assemblies are housed in watertight and airtight enclosures to protect the system from the elements and provide easy re-entry for service and maintenance.

Multipair copper cables that enter a premises must be protected from overcurrent and lightning surges commonly encountered with OSP installations. A *building entrance terminal* is a cable entrance point where an input cable between buildings is terminated. **See Figure 12-6.** The multipair copper cables are terminated on building entrance terminals that provide various termination interfaces, such as input/output splice modules or direct IDC terminations. Between the input/output splice modules, solid-state fuses are installed for each pair that is terminated. The armored sheath of the cable must also be grounded by using a shield bond connector and extended to the building entrance terminal.

Fiber-optic interbuilding connecting hardware includes fiber-optic splices, including fusion and mechanical splices. Fiber-optic interbuilding connecting hardware is intended for use in a variety of environments, such as in utility holes, utility vaults, aerial applications, and open-trench applications. Splicing is performed to join individual fibers or multiple fibers such as ribbon fibers.

A *fusion splice* is a permanent splice accomplished by the application of heat with a temperature high enough to melt the ends of two sections of glass fiber. Fusion splices are accomplished with a fusion splicer. A *fusion splicer* is a device that uses a high-voltage electrical arc between two electrodes to heat and melt the ends of glass fibers to form a continuous strand. Since a splice is similar to a welded joint, it is possible to get a stable fiber-optic connection. Because a fusion splice creates a continuous strand with no optical loss, fusion splicing is recommended for fiber-optic connections in the OSP.

A *mechanical splice* is a fiber-optic splice in which the ends of two different fibers are mechanically connected. A mechanical splice has a gripping mechanism to prevent fiber separation and maintain fiber alignment and includes index-matching material. An *index-matching material* is a material in liquid, paste, film, or gel form with a refractive index nearly equal to the core index of a fiber that is used to reduce reflections from a fiber end face. Because mechanical splices depend on a physical contact between two cleaved fiber ends, mechanical splices are typically more sensitive to large variations in environment temperature, which can affect performance.

OSP SE and Route Diversity. Cable routing diversity is often integrated into OSP designs as part of disaster recovery plans and to avoid a single point of failure. Each OSP pathway is susceptible to weather damage, deliberate acts of damage, and accidents. For example, underground cables and pathways can be accidently damaged during digging operations. Aerial installations using utility poles can be destroyed or damaged by weather or vehicular collisions. Local exchange carrier (LEC) central offices are considered prime targets for acts of terror. End users often require multiple LECs routed from different directions to the property and connecting to multiple LEC central offices.

Flexible Fabric Innerduct Systems

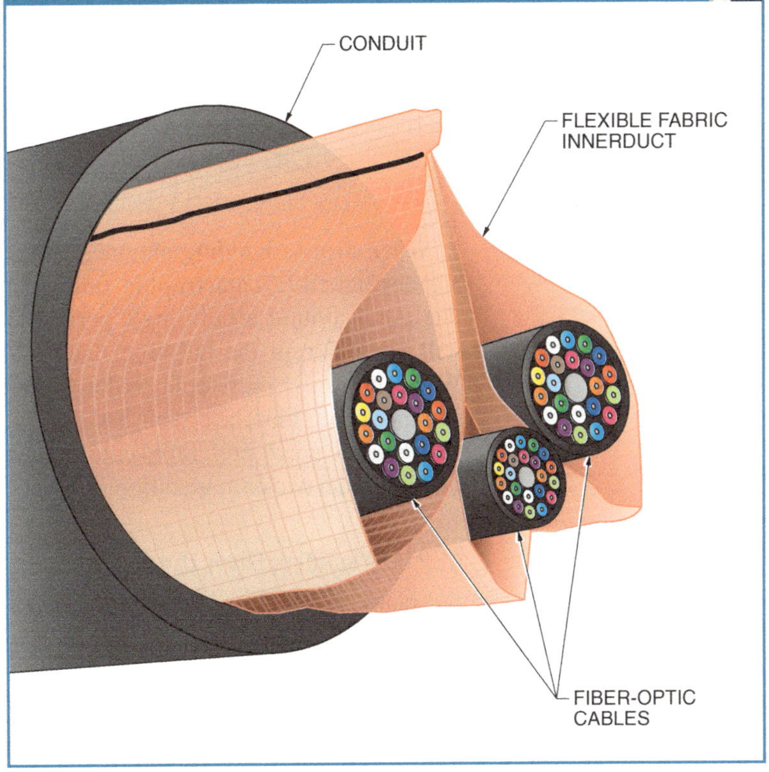

Figure 12-5. Flexible fabric innerduct systems include a flexible fabric material that is easier to use than corrugated innerduct piping.

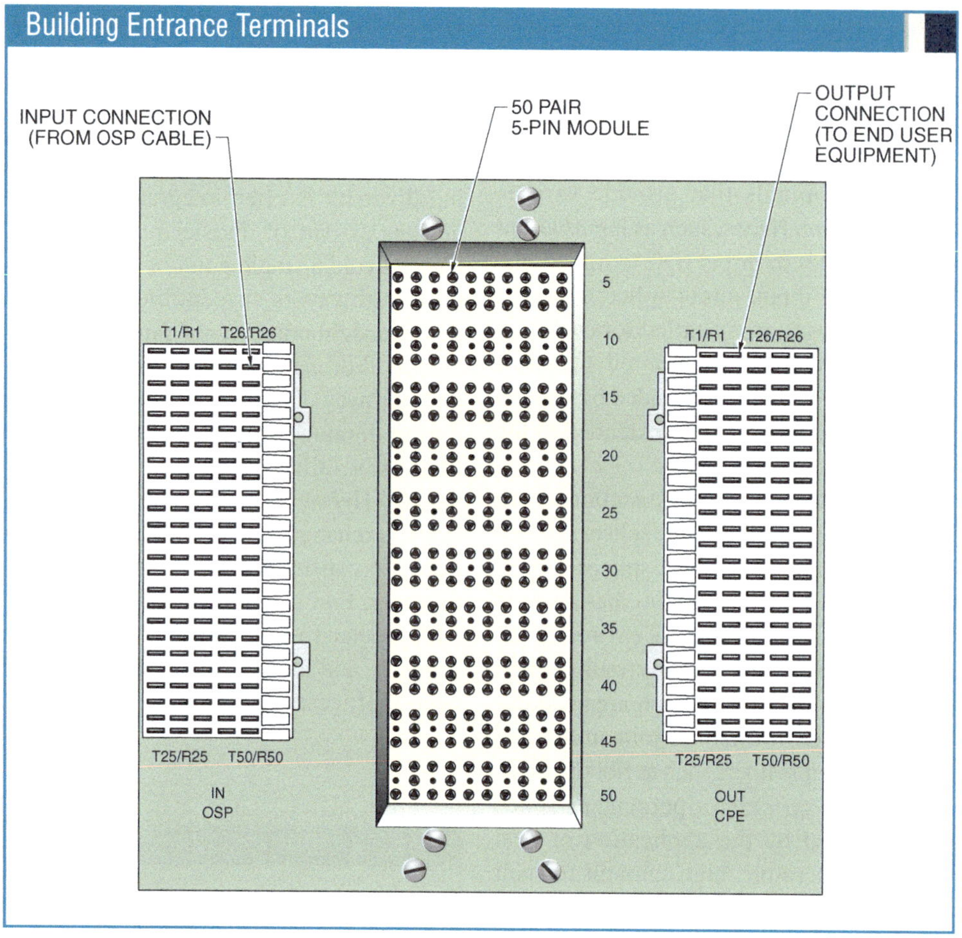

Figure 12-6. A building entrance terminal is a cable entrance point where an input cable between buildings is terminated.

When diverse building entrance points are developed, a catastrophic failure at one point around a building's perimeter will not interrupt the building's service. When entrance point diversity is developed, entrance points must be established a specified distance apart, preferably entering the building from two or more different streets.

OSP Interbuilding Riser Drawings

Site plans or civil plans (sometimes referred to as plot plans) are used for interpreting OSP design and installation requirements. *Note:* Site plan scales are not typical of interior floor plan scales. Site plan scales can vary from 1″ equals 10′ to 1″ equals 20′ or no scale at all. Additional items typically shown on a site or civil plan include built features such as roadways, sidewalks, parking lots, lighting, and landscaping.

When site plans are used, whether to scale or not, all dimensions must be field-verified for accuracy. Accuracy can be verified by measuring and comparing items located on the plan. Site plans often use a variety of different symbols and abbreviations. In addition to OSP locations, other utilities, such as gas, water, sewer, and electrical systems, can often be located on site plans. In cases where multiple utilities are shown on a single drawing, the drawing can become cluttered and difficult to read. Care must be practiced when referencing plans that show multiple trade disciplines, in order to properly understand what is required for each system.

When an estimate is being formulated or OSP work is performed, the existing and new conduit distribution system must be carefully reviewed to determine the quantity, size, type, and routing of each conduit. The distance between access points, buildings, and property lines must also be considered. The estimator and technician must understand not only the VDV site plan information but also the other utilities on the site plan to avoid costly mistakes during an installation.

PREMISES DISTRIBUTION SYSTEMS

A *premises* is a space occupied by an authorized user in a building on continuous or contiguous property that is not separated by public road or highway. The requirements for multitenant, single-occupant, or mixed-use premises used in commercial buildings are an extension of the services provided by the OSP interbuilding backbone. Although with today's reliance on network convergence and IP, additional services are typically required to be delivered throughout a premises.

The telecommunications engineers for a building must manage horizontal and vertical cable distribution, whether for multiple tenants or a single tenant. For example, intelligent building networks use intelligent premises equipment to manage building systems, such as physical security, mechanical, electrical, hydraulic, vertical transportation (elevators and escalators), car park, tenant improvement, building service, fire and life safety, communication, and sustainability systems. **See Figure 12-7.**

Commercial buildings have specified technology requirements that are met through the vertical and horizontal riser distribution system connecting EFs, equipment rooms (ERs), and TRs. A building's telecommunications riser is typically designed to provide a vertically aligned shaft, or a series of shafts, connected to rooms on each floor. LECs access these shafts and closets to extend services to end users.

Telecommunications engineers that are staffed by the building management must verify that risers are properly organized and systematically labeled. A properly installed riser system can lead to revenue for the building owner by providing multiple-carrier access and a potential increase in property value. A poorly organized riser system can lead to dangerous situations, such as owner's liability and building code violations. For example, code violations, such as the installation of nonrated cable and improper fire and smoke protection, can result in damage to property and harm to personnel. Premises riser systems can be intrabuilding or interbuilding types.

Premises Riser Systems

A premises riser system is composed of vertical and horizontal pathways consisting of a series of conduits, sleeves, slots, and cable trays connected to various types of dedicated ERs by predetermined shafts within a building. Pathways and spaces require physical supports for cable routed through and within them, firestop systems, and proper grounding. Physical protection and security must be maintained to avoid disruptions in service. **See Figure 12-8.**

Premises Riser System Cable Types

Intrabuilding riser systems have multiple media types to support a range of applications that include LEC services, public safety systems, and tenant network distribution. The fiber-optic or copper cables that carry VDV systems into a building are available in a range of sizes and types. High-density category (copper) cabling and fiber-optic cabling connect the building's central TR to ERs and are distributed typically to one ER per floor.

The entire riser system consists of the actual cables arranged in intermediate and main cross-connects. Also included are the patch cords between switches or hubs, the mechanical terminations, and any cable runs that terminate or originate with the EFs. The riser system includes the riser cables, intermediate and main cross-connects, mechanical terminations, and patch cords or jumpers used for riser-to-riser cross-connections. The riser cabling system must be isolated from electrical lines to minimize EMI.

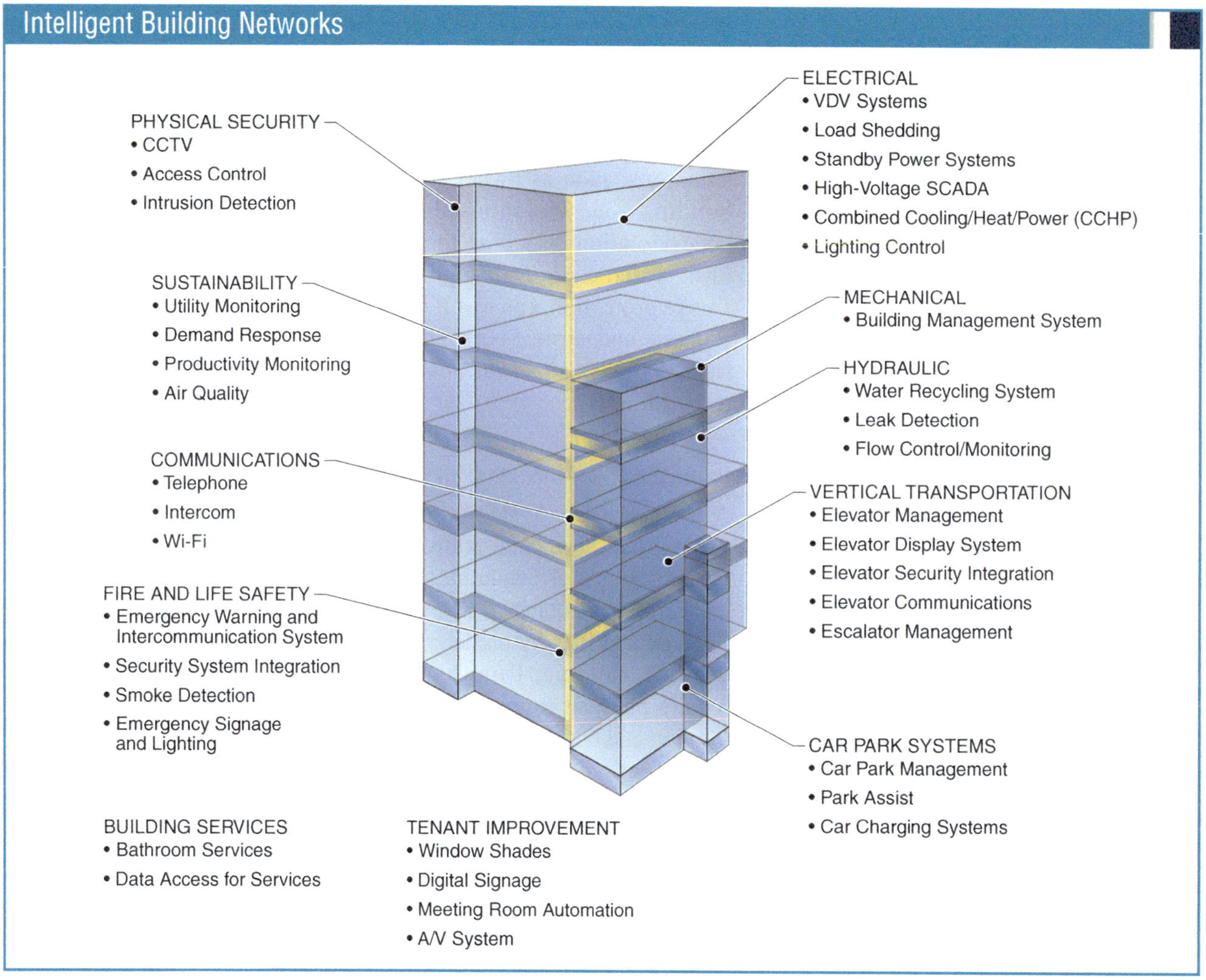

Figure 12-7. Intelligent building networks include systems such as physical security, mechanical, electrical, hydraulic, vertical transportation (elevators and escalators), car park, tenant improvement, building service, fire and life safety, communication, and sustainability systems.

Premises Intrabuilding Copper Cabling. For low-speed applications in commercial building riser distribution, a copper cable called an ARMM cable is commonly installed. An *ARMM cable* is a color-coded riser cable that consists of bare copper insulated with foam polyethylene (PE) and a polyvinyl-chloride (PVC) jacket. ARMM cables contain individual conductor pairs, up to 2400 pair, and are color-coded to match standard OSP cable designs. An ARMM cable is typically installed in conduit or vertical shafts that are rated for riser applications. ARMM cables are intended for vertical and horizontal distribution in commercial buildings and meet CAT 3 specifications. ARMM cables can be used for all applications with the exception of plenum space installations. ARMM cables have a fire-retardant ALVYN sheath and are CMR rated, in accordance with the NEC®.

TECH FACT

ARMM cables used for intrabuilding riser applications typically consist of telecommunications cables with multipair, 24 AWG, solid-copper, insulated conductors assembled into twisted pairs.

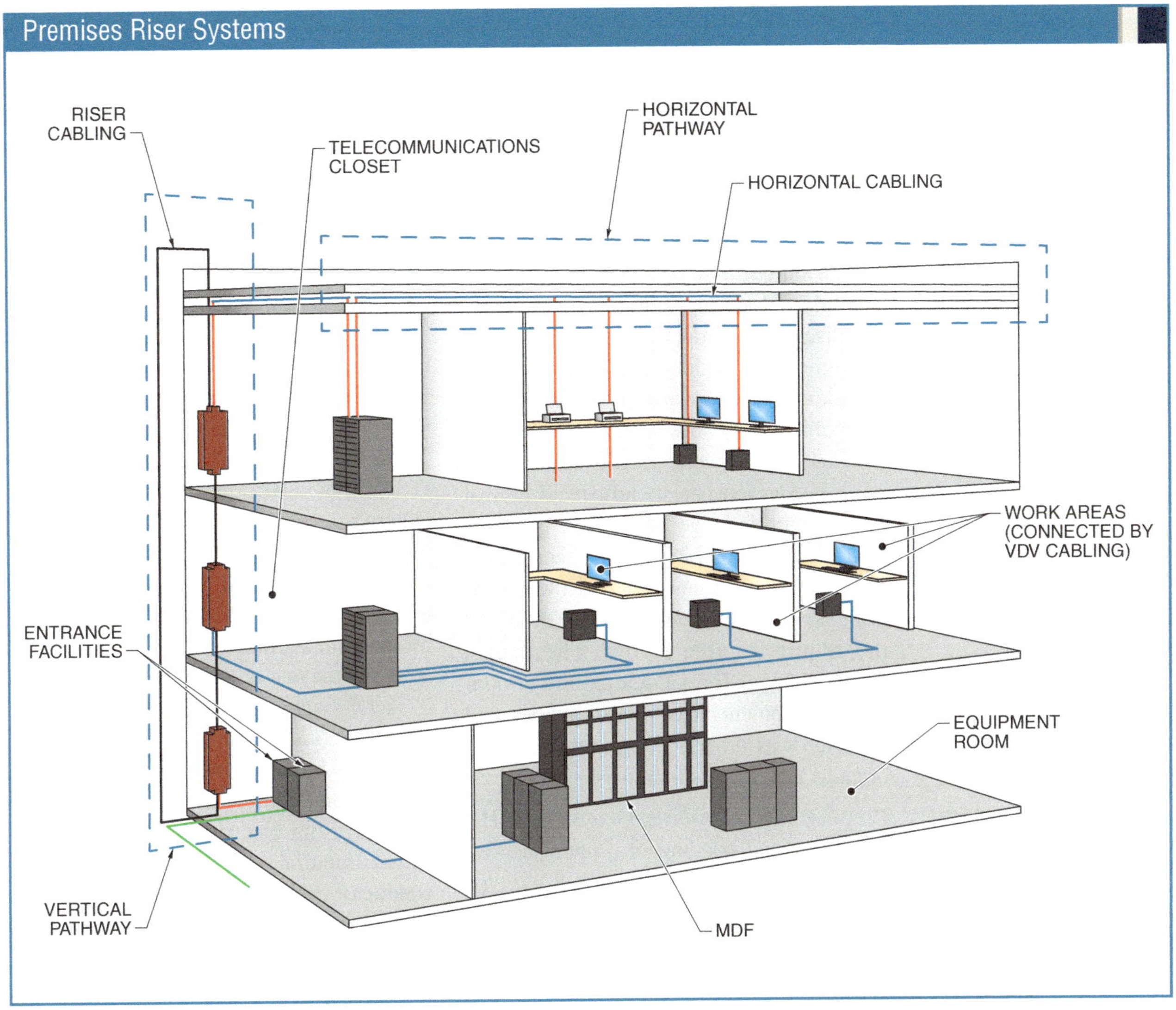

Figure 12-8. A premises riser system consists of vertical and horizontal pathways that are connected to, and in between, various types of dedicated equipment rooms by predetermined shafts within a building.

An *aluminum-polyvinyl-chloride (ALVYN) sheath* is an aluminum-polyvinylchloride (PVC) cable sheath used for riser cable where a flame-retardant sheath is required. An ALVYN sheath consists of 8 mil thick aluminum tape applied along and bonded to a gray or black PVC outer jacket. **See Figure 12-9.** Additional applications for ALVYN sheaths include 4 Mbps token ring LAN systems, analog voice systems, and 10Base-T Ethernet. The shielded design of an ALVYN sheath provides protection from EMI and radio-wave interference.

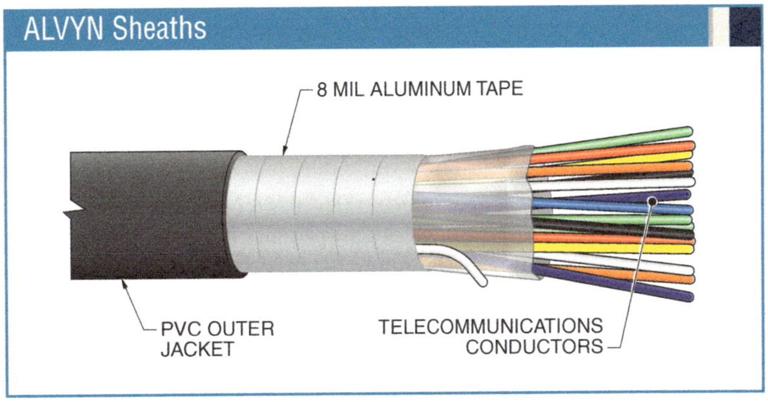

Figure 12-9. An ALVYN sheath consists of 8 mil thick aluminum tape applied along and bonded to a gray or black PVC outer jacket.

Cable and satellite television service providers utilize 75 Ω coaxial (RG-6 and RG-11) for riser distribution. Category cables from CAT 5e (100 MHz) to CAT 6 (250 MHz) and CAT 6A (500 MHz) can also be installed in intrabuilding riser systems.

Intrabuilding copper cabling for premises is also used for distributed antenna systems. A *distributed antenna system (DAS)* is a network of low- and high-powered antennas distributed vertically and horizontally throughout a building. DAS applications include emergency responder radio coverage and public safety networks. A *public safety network* is a network used by emergency services to prevent or respond to incidents that can harm or endanger persons or property. A DAS uses a combination of Ethernet and fiber-optic cables constructed in a variety of designs, depending on the application and the required cable rating. The cables and equipment for a DAS require similar space, equipment racks, power supplies, and enclosures as other VDV systems and are often located in ERs together.

Premises Intrabuilding Grounding. The grounding and bonding of a premises system begins at the EF. The number and type of components used in a building depend on the size and design of the building. The current standard, ANSI/TIA-607-D, *Generic Telecommunications Bonding and Grounding (Earthing) for Customer Premises,* does not require short metal pathways, such as floor sleeves or metal cable supports, to be bonded.

A *primary bonding busbar (PBB)* is a busbar that serves as the central connection point between the grounding electrode system for a building electrical service and the grounding and bonding system for telecommunications. A *telecommunications bonding conductor (TBC)* is a conductor that bonds a primary bonding busbar to the grounding electrode system for a building electrical service. A PBB is connected to the grounding electrode system for a building electrical service by the TBC. PBBs are listed copper or copper alloy busbars that are ¼″ thick by 4″ wide and are available in various lengths. A PBB has drilled holes for mounting and terminating listed lugs.

A *secondary bonding busbar (SBB)* is a busbar located in the telecommunications room (TR) or equipment room (ER) of a building that connects to the primary bonding busbar (PBB). **See Figure 12-10.** Some buildings have TRs or ERs in addition to an EF. ANSI/TIA-607-D requires that each of these rooms have an SBB.

An SBB is connected to a PBB by the telecommunications bonding backbone (TBB). SBBs are listed copper or copper alloy busbars that are ¼″ thick by 2″ wide and available in various lengths. An SBB also has drilled holes for mounting and terminating listed lugs. If a TR or ER has an electrical panel that serves IT loads, the SBB must be mounted near it. Otherwise, the SBB must be mounted near the terminations for riser cabling. An SBB must be insulated from its support by a listed insulator. The recommended minimum size for the insulator is 2″ in height, which allows access to the rear of the SBB.

A *telecommunications bonding backbone (TBB)* is a cable that is connected to a primary bonding busbar in the telecommunications entrance facility, to an earth ground in the electrical entrance facility, and to structural steel on each floor. Bonding conductor cabling must be colored green or labeled appropriately. Telecommunications equipment, frames, cabinets, and voltage protectors are typically grounded to a TBB.

A TBB is a type of surge protector and is installed to establish an electrically conductive path for telecommunications, the metallic parts of which are permanently joined, to ensure electrical continuity, the ability to conduct safely any current likely to be imposed, and the ability to limit dangerous electric potentials. **See Figure 12-11.** The cable ratings associated with intrabuilding cable applications include communications plenum (Type CMP) and communications riser (Type CMR).

Note: NEC® Article 800.113 covers installation of wires, cables, cable-routing assemblies, and communications raceways. NEC® Article 800.179 covers plenum, riser, general-purpose, and limited-use cables. The NEC® ratings, articles, and tables used for OSP interbuilding copper cables are also used for fiber-optic cables.

Secondary Bonding Busbars (SBBs)

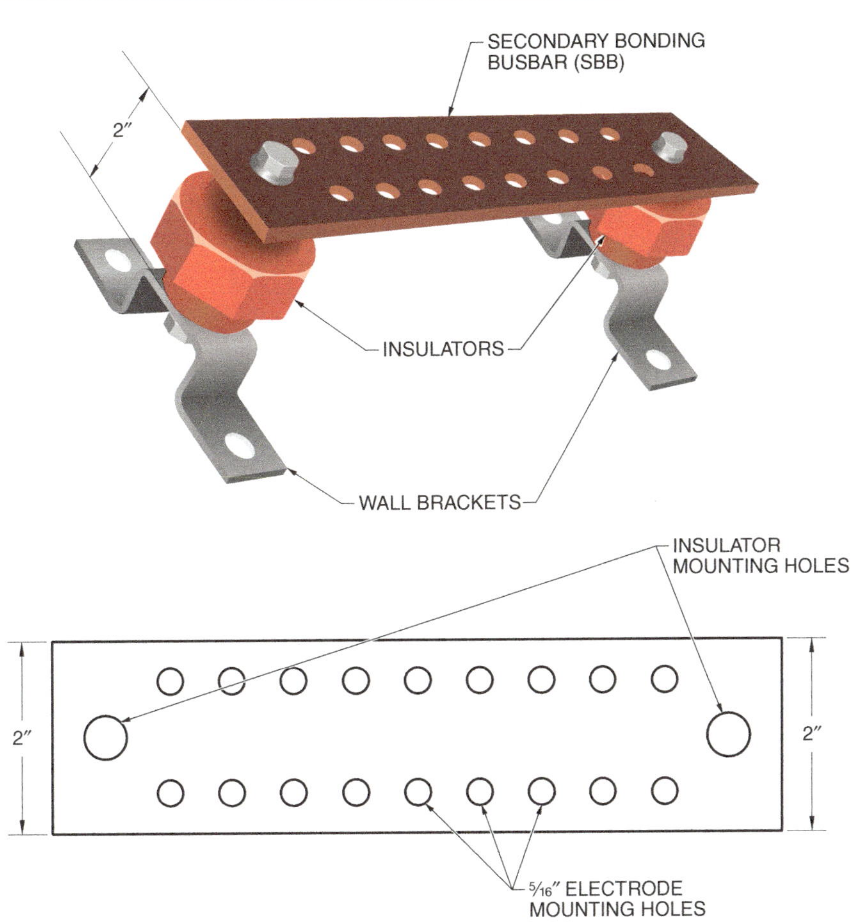

Figure 12-10. An SBB is a busbar located in the TR or ER of a building and connects to the PBB.

Premises Intrabuilding Fiber-Optic Cabling. Fiber-optic cables are the most commonly specified type of cables used for intrabuilding installations. In some installations, depending on the building construction, fiber-optic cables installed in the OSP transition from the OSP to the interbuilding riser. This is accomplished either by fusion-splicing the fiber-optic cables from the OSP to the riser or by installing them together in RMC. **See Figure 12-12.** Another option is to specify customized cables with ratings that allow them to be installed in both environments.

Premises Intrabuilding Riser Connecting Hardware. The connecting hardware for copper and fiber-optic intrabuilding riser cabling consists of equipment and wall-mounted racks to accommodate high-density interfaces. Located adjacent to work area outlets (horizontal cabling), interfaces must be compatible to cross-connect and interconnect the riser to the horizontal cabling. Copper and fiber-optic interfaces consist of VDV hardware types such as 110 (IDC), RJ-45 modular patch panels, patch cords, cross-connect wire, coaxial, and fiber-optic connectors. All terminating hardware must comply with industry-recognized building cabling standards, such as ANSI/TIA-569, *Telecommunications Pathway and Spaces,* and ANSI/TIA-570, *Residential Telecommunications Infrastructure.*

Telecommunications Bonding Backbones

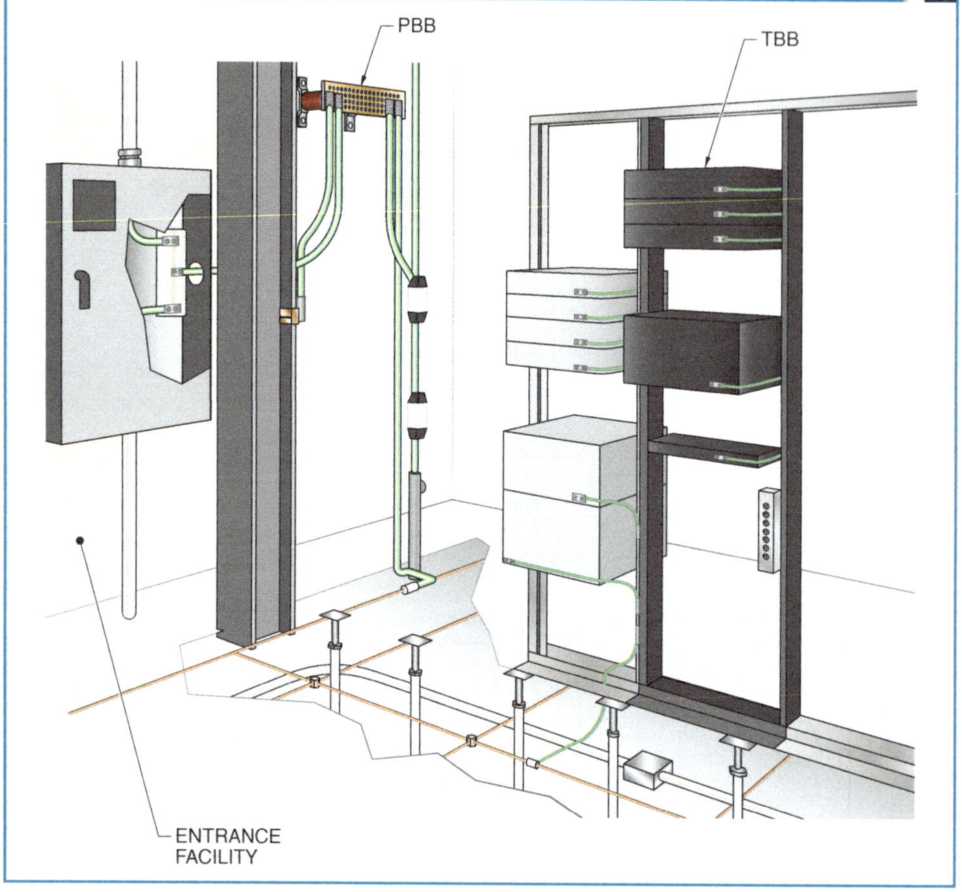

Figure 12-11. A TBB is a type of surge protector and is installed to establish an electrically conductive path for metallic telecommunications parts to limit dangerous electric potentials.

Fusion Splicers

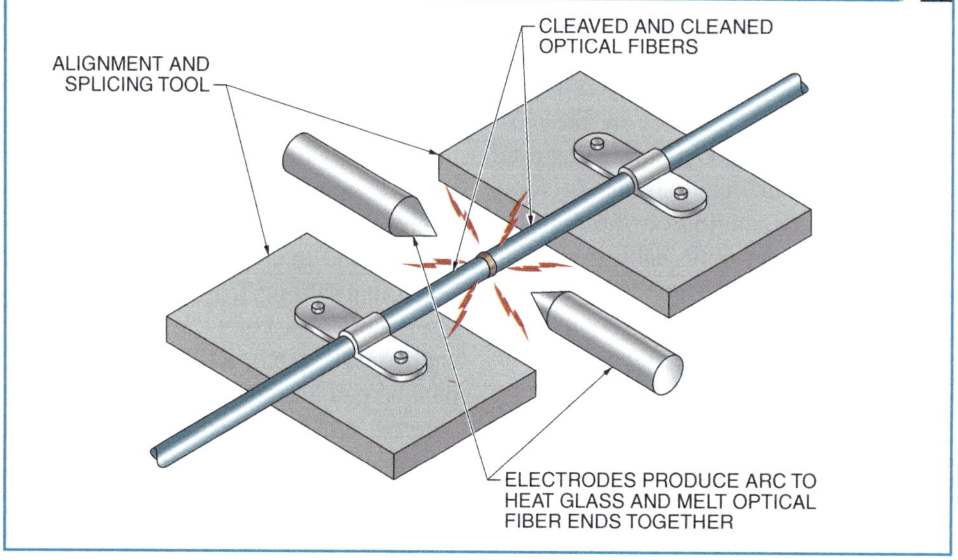

Figure 12-12. A fusion splicer creates a fiber-optic splice by applying heat to the ends of two different fibers in order to form one continuous strand.

Premises Intrabuilding Route Diversity. As with the intrabuilding riser distribution, developing more than one pathway can provide protection in the event that the route of the pathway is damaged or otherwise compromised. Additionally, diverse pathways must be designed to maintain the longest distance between them if possible and should never be routed through the same room or space. Commercial buildings, depending on size or specific use, may have multiple risers installed for route diversity and redundancy purposes. **See Figure 12-13.**

Fusion splicers are used to make solid connections with no signal loss between separate glass fibers.

Multiple Riser Installations

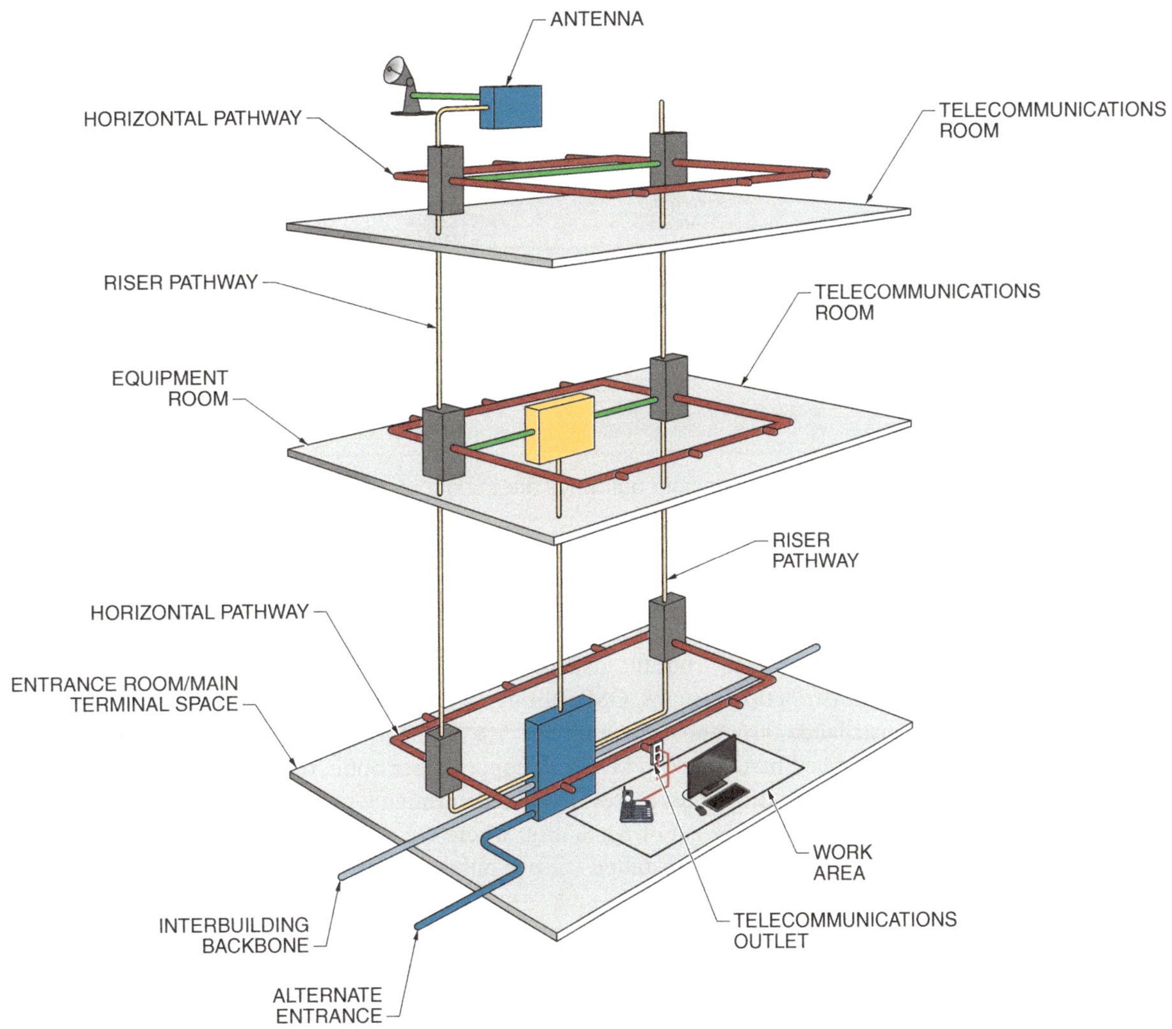

Figure 12-13. Commercial buildings, depending on size or specific use, may have multiple risers installed for route diversity and redundancy purposes.

Premises Interbuilding Drawings

When riser pathways are to be routed to multiple buildings, interbuilding drawings are used. The most common drawings that depict interbuilding riser pathways, routing, type, and quantity are shown on building diagrams. A *building diagram* is a simple view of the services and media types installed from one point to another point, from one building to another building. Building diagrams are typically not drawn to scale. In order to determine the distances between points, scaled floorplans must be referenced to calculate vertical and horizontal distances. The project specifications also provide additional details, such as specific part numbers for all cables, connectors, termination hardware, firestops, and grounding equipment. **See Figure 12-14.**

Interbuilding Riser Pathways

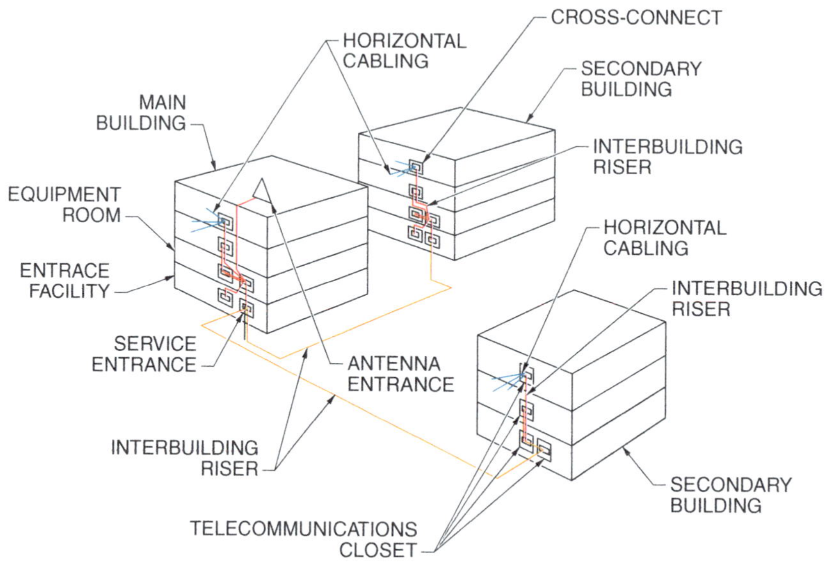

Figure 12-14. When riser pathways are to be routed to multiple buildings, interbuilding drawings are used.

SUMMARY

Cables routed from the outside of the building are known as outside plant (OSP) distribution systems. OSP cabling is connected to the building riser system and then distributed throughout the premises via horizontal cabling. Cabling systems and their associated hardware can be either copper, fiber optic, or a combination of both types. Project drawings must be reviewed to determine which type of cabling system is to be installed. In the event of a situation where the cable route is compromised, additional cable routes can be installed for use either between different buildings or within the same building.

CHAPTER 12

Outside Plant and Premises Distribution Systems

Review

Name ______________________________ Date ______________

True-False

T F **1.** An OSP riser is also referred to as the backbone cable and often has a red jacket.

T F **2.** Different cable types are never placed together in the same OSP pathways.

T F **3.** Due to the moisture resistance of PE, it is typically used in wet and damp environments.

T F **4.** ALPETH cables are designed for indoor use only.

T F **5.** Copper cable is the most commonly specified and installed type of cable used in OSP applications due to its ability to transmit large amounts of data at long distances.

T F **6.** Aerial cables typically have the same cable construction as duct cables.

T F **7.** When site plans are used, whether to scale or not, only the EF dimensions must be field-verified for accuracy.

T F **8.** Bonding conductor cabling must be colored white or labeled appropriately.

T F **9.** Fiber-optic cables are the most commonly specified type of cables used for intrabuilding installations.

T F **10.** The NEC® ratings, articles, and tables used for OSP interbuilding copper cables are also used for fiber-optic cables.

Multiple Choice

_______________ **11.** ___ are used as access points for splicing cables.

A. Handholes
B. Pedestals
C. Utility holes
D. Vaults

_______________ **12.** Traditional gel-filled and air-core multipair OSP cables have a ___ V working voltage capability when used in communication circuit applications.

A. 300
B. 500
C. 600
D. 1200

_______________ **13.** Typically, there is a ___′ limitation when installing Type CM cables in a premises.

A. 50
B. 75
C. 100
D. 112

_______________ **14.** A(n) ___ is a space occupied by an authorized user in a building on continuous or contiguous property that is not separated by public road or highway.

A. entrance facility (EF)
B. pathway
C. premises
D. work area

_______________ **15.** ARMM cables contain individual conductor pairs, up to ___ pair.

A. 2400
B. 1000
C. 2000
D. 2500

Completion

_______________ **16.** A(n) ___ is a telecommunications cable and associated hardware that runs from a telecommunications closet, or intermediate distribution frame (IDF), to a main distribution frame (MDF) or to another telecommunications room (TR).

_______________ **17.** The ___ is where the copper system receives its primary fused overcurrent protection for hazardous voltages and lightning protection.

_______________ **18.** Multipair interbuilding copper OSP cable has a maximum bandwidth of ___ MHz.

_______________ **19.** A(n) ___ is the smallest section of a glass fiber strand and is used to carry light waves.

_______________ **20.** The cable ratings associated with indoor/outdoor OSP fiber cable applications are listed NEC® Article ___.

_______________ **21.** A(n) ___ is a type of telecommunications connector in which the wire is punched down into a metal holder that strips the insulation on the wire and allows electrical contact to be made.

_______________ **22.** A(n) ___ sheath is an aluminum-polyvinyl chloride (PVC) cable sheath used for riser cable where a flame-retardant sheath is required.

_______________ **23.** The recommended minimum size for the insulator on an SBB is ___″ in height, which allows access to the rear of the SBB.

_______________ **24.** A(n) ___ is a cable that is connected to a primary bonding busbar in the telecommunications entrance facility, to an earth ground in the electrical entrance facility, and to structural steel on each floor.

_______________ **25.** A(n) ___ is a cable in which the conductors are surrounded by an air space within the jacket.

Short Answer

26. List five types of OSP spaces.

27. Explain why gel-filled cables are often used for OSP applications.

28. Describe ribbon fiber and how it is used in a VDV system.

29. List five different pathways where OSP fiber-optic cable is installed.

30. Explain why a fusion splice is recommended for fiber-optic connections in an OSP installation.

Outside Plant and Premises Distribution Systems

CHAPTER 12

Activity

Name ______________________________ Date ______________

OUTSIDE PLANT DISTRIBUTION SYSTEM

Instructions

Create a list of materials that would be included in a proposal by using the drawings and specifications to determine pathways, cable, connecting hardware, and rack types and quantities for an outside plant (OSP) distribution system.

Project Overview

The project is for an OSP distribution system to be used for a single-occupancy commercial office facility. The OSP distribution system has the following requirements:

- The office facility primary riser consists of 72 strands of single-mode optical fiber and originates in the existing building 300 basement MDF equipment room.
- The office facility secondary riser consists of 72 strands of single-mode optical fiber and originates in the existing building 300 basement MDF equipment room.
- The guard shack primary riser consists of 12 strands of single-mode optical fiber and 25-pair CAT 3 copper cable originating in the existing building 200 lower level MDF equipment room.
- The guard shack secondary riser consists of 12 strands of single-mode optical fiber cable originating in the existing building 200 lower level MDF equipment room.

Drawings:

- 200ATP-T-00-00 – Telecom Legend and Specification Sheet
- 200ATP-T-00-01 – Office Lower Level Plan
- 200ATP-T-00-02 – Lab Lower Level Plan
- 200ATP-T-01-01 – Office Level 1 Plan
- 200ATP-T-01-02 – Lab Level 1 Plan
- 200ATP-T-02-01 – Office Level 2 Plan
- 200ATP-T-02-02 – Lab Level 2 Plan
- 200ATP-T-03-01 – Office Level 3 Plan
- 200ATP-T-03-02 – Lab Level 3 Plan
- 200ATP-T-04-01 – Telecom Details
- 200ATP-T-05-01 – Telecom Grounding and Pathway Details
- 200ATP-T-06-01 – Telecom Device Mounting Details
- 200ATP-T-07-01 – Telecom Riser Diagrams and Guard Shack Details
- 200ATP-T-08-01 – Telecom Site Plan and Details
- 200ATP-T-09-01 – Telecom Rack Elevations and Fiber Details

Specification:

- ATP B200 Structured Cabling Specification – Section 271000

Note: The drawings and specification can be accessed in the online learner resources.

OSP interbuilding system:

1. Pathways:

2. Cables/Length:

3. Connecting hardware/Quantity:

Appendix

ANSI/TIA Telecommunications Standards

- Generic Telecommunications Cabling for Customer Premises (ANSI/TIA-568.0-D)
- Commercial Building Telecommunications Infrastructure Standard (ANSI/TIA-568.1-E)
- Balanced Twisted-Pair Telecommunications Cabling and Components Standard (ANSI/TIA-568.2-D)
- Optical Fiber Cabling and Components Standard (ANSI/TIA-568.3-D)
- Broadband Coaxial Cabling and Components Standard (ANSI/TIA-568.4-D)
- Telecommunications Pathways and Spaces (ANSI/TIA-569-E)
- Residential Telecommunications Infrastructure Standard (ANSI/TIA-570-D)
- Administration Standard for Telecommunications Infrastructure (ANSI/TIA-606-C)
- Generic Telecommunications Bonding and Grounding (Earthing) for Customer Premises (ANSI/TIA-607-D)
- Customer-Owned Outside Plant Telecommunications Infrastructure Standard (ANSI/TIA-758-B)
- Structured Cabling Infrastructure Standard for Intelligent Building Systems (ANSI/TIA-862-B)
- Telecommunications Infrastructure Standard for Data Centers (ANSI/TIA-942-B)
- Telecommunications Infrastructure Standard for Industrial Premises (ANSI/TIA-1005-A)
- Healthcare Facility Telecommunications Infrastructure Standard (ANSI/TIA-1179-A)
- Telecommunications Infrastructure Standard for Educational Facilities (ANSI/TIA-4966)

Bid Invitation—Commercial Building Renovation Project

AMERICAN TECHNICAL PUBLISHERS
Career and Technical Training Materials

Bid Invitation

You are reminded to bid only on items shown on the prints and detailed in the specification. In the event the owner should request additional work or deletion of work, they shall do so in writing during and after the bid phase.

If you feel additional work not specifically outlined on the drawings or in the specification is required, list those items separately from the bid price. All additional items will be considered separately from the bid and any changes will be negotiated with the owner after the bid opening.

The owner has the right to reject any bids at their sole discretion. Competitive bids will be descoped and reviewed by your instructor and peers for accuracy and technical merit.

Provide a complete and detailed bill of materials depicting the following:

- Material Quantity and type with manufacturer part number
- Quantity of wire & cable by type and in feet/meters
- Quantity and type of termination & connecting hardware
- Quantity and type of all raceways and Server Room equipment
- Quantity and type of all necessary hardware obviously necessary for a complete installation

The drawings and specifications associated with this chapter are as follows:

Owner Specification Dated February 17, Pages 1-9.

Owner Drawing February 17
T-001 Tel Data Legend & Notes
T-103 Tel Data 3rd Floor Pathway Plan
T-203 Tel Data 3rd Floor Coordination Plan
T-303 Tel Data 3rd Floor Existing Plan
T-403 Tel Data 3rd Floor Plan
T-503 Tel Data 3rd Floor Details

Upon award, a Notification to Proceed will be issued.

Thank you for your consideration of this Invitation to Bid.

Your bid breakdown should include the following items:

Tel/Data - Structured Cabling

Item		Labor		Material	
		Hrs	Labor Costs	Quantity	Mat. Costs
1.0	**Phase 1**				
0.0	Level 3 - Telecom Horizontal Cabling, Demolition				
0.1	Level 3 - Telecom Horizontal Cabling, Connecting Hardware, Supports				
	Subtotal	0			
2.0	**Phase 2**				
2.1	Level 3 - Telecom Horizontal Cabling, Demolition				
2.2	Level 3 - Telecom Horizontal Cabling, Connecting Hardware, Supports				
2.3	Level 3 - Copper & Fiber Backbone Distribution				
2.4	Level 3 - IDF Room Demolition & Equipment Removal				
	Subtotal	0			
3.0	**IDF Fitout: racks, runways, wire management, power strips, APC UPS, copper/fiber backbone and copper/fiber cords.**				
3.1	Level 3 - IDF Rooms Fitout				
3.2	Copper Patch Cords				
3.3	Fiber Optic Patch Cords				
	Subtotal	0			
4.0	**Other**				
4.1	Wireless Access Points (install, patch, & test; units provided by others)				
4.2	Premium Time Phase 1 Tenant Space Level 2				
4.3	Premium Time Phase 2 Tenant Space Level 2				
4.4	Premium Time Phase 2 Phase 1 Cable Relocation				
4.5	As-Built Drawings/Project Closeout Documentation				
4.6	Communication Testing, Adjusting and Identification				
4.7	Clean-Up to general location (disposal by Construction Manager)				
	Subtotal	0			
5.0	**General Conditions - Field**				
5.1	Project Administration				
5.3	Mobilization/Demobilization				
	Subtotal	0			
6.0	**Sales Tax**				
7.0	**Permits and Fees**				
8.0	**Warranty**				
	TOTAL Labor Hours and Pricing	0			

Unit Prices as specified.

9.0	Unit Pricing	Add	Deduct
9.1	Typical 3-Cat 6 Comm Outlet (150' avg)		
9.2	Typical 1-Cat 6 Comm Outlet (150' avg)		
9.3	Typical 3-Cat 6 Comm Outlet (250' avg)		
9.4	Typical 1-Cat 6 Comm Outlet (250' avg)		
9.5	Cat 6 Nine Foot (9') Connecting Cord		
9.6	Cat 6 Seven Foot (7') Connecting Cord		
9.7	Cat 6 Five Foot (5') Connecting Cord		
9.8	Cat 6 Three Foot (3') Connecting Cord		
9.9	1 meter LC-LC Duplex single mode cord		
10.0	3 meter LC-LC Duplex single mode cord		
10.1	Cat 6 48 port High Density Patch Panel		
10.3	Wire Manager 2U		
10.2	Labor Rate; Straight Time		
10.3	Labor Rate; Time and Half		

Common Industry Acronyms for Cable Construction			
Industry Acronyms	**ISO/IEC 11801 Name**	**Cable Shielding**	**Pair Shielding**
UTP	U/UTP	none	none
STP, ScTP	–	none	foil
FTP, STP, ScTP	–	foil	none
STP, ScTP	S/UTP	braiding	none
SFTP, S-FTP, STP	SF/UTP	braiding, foil	none
FFTP	F/FTP	foil	foil
SSTP, SFTP, STP	S/FTP	braiding	foil
SSTP, SFTP	SF/FTP	braiding,foil	foil

The code before the slash designates the shielding for the cable itself, while the code after the slash determines the shielding for the individual pairs:

U = Unshielded
F = Foil shielded
S = Braided shielding (outer layer only)
Sc = Screened
TP = Twisted Pair

Copper Minimum Bend Radius	
Cable Type	**Bend Radius**
4 Pair UTP	4 × cable diameter
4 Pair ScTP	8 × cable diameter
Backbone	10 × cable diameter
Patch cords	6 × cable diameter

Determining Loss Budgets*			
Component	**Qty / Unit**	**Unit Loss from Specification Multiplier**	**Loss†**
Cable/loss	3 / km	3.0 dB / km	9
Connector loss	2 / connectors	0.75 dB / connector	1.5
Splice loss	1 / splice	0.3 dB / splice	0.3
Total loss			10.8

* 3 km, 50 µm system with 2 connector pairs and a splice at 850 nm
Note: Cable loss will vary with operating wavelengths of 850 nm, 1300 nm, 1310 nm, and 1550 nm
† in dB

Electrical/Electronic Abbreviations/Acronyms. . .			
Abbr/ Acronym	**Meaning**	**Abbr/ Acronym**	**Meaning**
A	Ammeter; Ampere; Anode; Armature	DPST	Double-Pole, Single-Throw
AC	Alternating Current	DS	Drum Switch
AC/DC	Alternating Current; Direct Current	DT	Double-Throw
A/D	Analog to Digital	DVM	Digital Voltmeter
AF	Audio Frequency	EMF	Electromotive Force
AFC	Automatic Frequency Control	F	Fahrenheit; Fast; Forward; Fuse; Farad
Ag	Silver	FET	Field-Effect Transistor
ALM	Alarm	FF	Flip-Flop
AM	Ammeter; Amplitude Modulation	FLC	Full-Load Current
AM/FM	Amplitude Modulation; Frequency Modulation	FLS	Flow Switch
ARM.	Armature	FLT	Full-Load Torque
Au	Gold	FM	Frequency Modulation
AU	Automatic	FREQ	Frequency
AVC	Automatic Volume Control	FS	Float Switch
AWG	American Wire Gauge	FTS	Foot Switch
BAT.	Battery (electric)	FU	Fuse
BCD	Binary-Coded Decimal	FWD	Forward
BJT	Bipolar Junction Transistor	G	Gate; Giga; Green; Conductance
BK	Black	GEN	Generator
BL	Blue	GRD	Ground
BR	Brake Relay; Brown	GY	Gray
C	Celsius; Capacitance; Capacitor, Coulomb	H	Henry; High Side of Transformer; Magnetic Flux
CAP.	Capacitor	HF	High Frequency
CB	Circuit Breaker; Citizens Band	HP	Horsepower
CC	Common-Collector; Configuration	Hz	Hertz
CCW	Counterclockwise	I	Current
CE	Common-Emitter Configuration	IC	Integrated Circuit
CEMF	Counter-Electromotive Force	INT	Intermediate; Interrupt
CKT	Circuit	IOL	Instantaneous Overload
cmil	Circular Mil	IR	Infrared
CONT	Continuous; Control	ITB	Inverse Time Breaker
CPS	Cycles Per Second	ITCB	Instantaneous Trip Circuit Breaker
CPU	Central Processing Unit	J	Joule
CR	Control Relay	JB	Junction Box
CRM	Control Relay Master	JFET	Junction Field-Effect Transistor
CT	Current Transformer	K	Kilo; Cathode
CW	Clockwise	kWh	Kilowatt-Hour
D	Diameter; Diode; Down	L	Line; Load; Coil; Inductance
D/A	Digital to Analog	LB-FT	Pounds Per Foot
DB	Dynamic Braking Contactor; Relay	LB-IN.	Pounds Per Inch
DC	Direct Current	LC	Inductance-Capacitance
DIO	Diode	LCD	Liquid Crystal Display
DISC.	Disconnect Switch	LCR	Inductance-Capacitance-Resistance
DMM	Digital Multimeter	LED	Light-Emitting Diode
DP	Double-Pole	LRC	Locked Rotor Current
DPDT	Double-Pole, Double-Throw	LS	Limit Switch

. . .Electrical/Electronic Abbreviations/Acronyms			
Abbr/ Acronym	**Meaning**	**Abbr/ Acronym**	**Meaning**
LT	Lamp	RF	Radio Frequency
M	Motor; Motor Starter; Motor Starter Contacts	RH	Rheostat
MAX.	Maximum	rms	Root-Mean-Square
MB	Magnetic Brake	ROM	Read-Only Memory
MCS	Motor Circuit Switch	rpm	Revolutions Per Minute
MEM	Memory	RPS	Revolutions Per Second
MED	Medium	S	Series; Slow; South; Switch; Second; Siemen
MIN	Minimum	SCR	Silicon-Controlled Rectifier
MMF	Magnetomotive Force	SEC	Secondary
MN	Manual	SF	Service Factor
MOS	Metal-Oxide Semiconductor	1 PH; 1ϕ	Single-Phase
MOSFET	Metal-Oxide Semiconductor Field-Effect Transistor	SOC	Socket
MTR	Motor	SOL	Solenoid
N; NEG	North; Negative; Number of Turns	SP	Single-Pole
NC	Normally Closed	SPDT	SIngle-Pole, Double-Throw
NEUT	Neutral	SPST	Single-Pole, Single-Throw
NO	Normally Open	SS	Selector Switch
NPN	Negative-Positive-Negative	SSW	Safety Switch
NTDF	Nontime-Delay Fuse	SW	Switch
O	Orange	T	Tera; Terminal; Torque; Transformer
OCPD	Overcurrent Protection Device	TB	Terminal Board
OHM	Ohmmeter	3 PH; 3ϕ	Three-Phase
OL	Overload Relay	TD	Time-Delay
OZ/IN.	Ounces Per Inch	TDF	Time-Delay Fuse
P	Peak; Positive; Power; Power Consumed	TEMP	Temperature
PB	Pushbutton	THS	Thermostat Switch
PCB	Printed Circuit Board	TR	Time-Delay Relay
PH	Phase	TTL	Transistor-Transistor Logic
PLS	Plugging Switch	U	Up
PNP	Positive-Negative-Positive	UCL	Unclamp
POS	Positive	UHF	Ultrahigh Frequency
POT.	Potentiometer	UJT	Unijunction Transistor
P-P	Peak-to-Peak	UV	Ultraviolet; Undervoltage
PRI	Primary Switch	V	Violet; Volt
PS	Pressure Switch	VA	Voltampere
PSI	Pounds Per Square Inch	VAC	Volts Alternating Current
PUT	Pull-Up Torque	VDC	Volts Direct Current
Q	Transistor; Quality Factor	VHF	Very High Frequency
R	Radius; Red; Resistance; Reverse	VLF	Very Low Frequency
RAM	Random-Access Memory	VOM	Volt-Ohm-Milliammeter
RC	Resistance-Capacitance	W	Watt; White
RCL	Resistance-Inductance-Capacitance	w/	With
REC	Rectifier	X	Low Side of Transformer
RES	Resistor	Y	Yellow
REV	Reverse	Z	Impedance

Industry and Standards Organizations . . .	
American National Standards Institute (ANSI) www.ansi.org	Coordinates and encourages activities in national standards department; identifies industrial and public needs for standards; acts as national coordinator and clearinghouse for consensus standards
Building Industry Consulting Service International (BICSI) www.BICSI.org	A professional telecommunications standards and certification organization that provides professional credentials such as RCDD, RTPM, DCDC, and OSP designer
Canadian Standards Association (CSA) www.csa.ca	Tests equipment and products to verify conformance to Canadian codes and standards
Construction Specifications Institute (CSI) www.csiresources.org	An organization that develops standardized construction specifications and publishes the MasterFormat, which is a standardized classification for construction information, such as detailed construction project cost estimates or construction specifications
Factory Mutual (FM) www.fmglobal.com	Tests equipment and products to verify conformance to national codes and standards
The Institute of Electrical and Electronics Engineers (IEEE) www.ieee.org	Provides guidance in all electrical and electronic systems including aerospace systems, computers and telecommunications, biomedical engineering, electric power, and consumer electronics
International Association of Electrical Inspectors (IAEI) www.iaei.org	Focuses on interpretation of the National Electrical Code® and teaches safe installation and use of electricity
International Code Council (ICC) www.iccsafe.org	Administers the International Building Code®, including building materials, building systems, energy efficiency, fire protection systems, and structural design
InterNational Electrical Testing Association (NETA) www.netaworld.org	Defines standards by which electrical equipment is deemed safe and reliable; also provides training for and certification of electrical testing technicians
International Electrotechnical Commission (IEC) www.iec.ch	Provides standards for most international installations and some domestic installations
International Organization for Standardization (ISO) www.iso.org	A developer of voluntary international standards; facilitates world trade by providing common standards between nations
Mine Safety and Health Administration (MSHA) www.msha.gov	Develops and enforces safety and health rules for mines and for certain equipment that may be used in other locations

. . .Industry and Standards Organizations	
National Electrical Installation Standards (NEIS) **National Electrical Contractors Association (NECA)** www.neca-neis.org	Standards developed by NECA in partnership with other industry organizations; all NEIS are submitted for approval to ANSI
National Electrical Manufacturers Association (NEMA) www.nema.org	Provides standards for manufacturers of electrical equipment
National Fire Protection Association (NFPA) www.nfpa.org	Provides guidance in assessing hazards of products of combustion; publishes the National Electrical Code®; develops hazardous materials information
National Institute for Occupational Safety and Health (NIOSH) www.cdc.gov/niosh	Acts in conjunction with OSHA to develop recommended exposure limits for hazardous substances or conditions located in the workplace; recommends preventive measures to reduce or eliminate adverse health and safety effects
Occupational Safety and Health Administration (OSHA) www.osha.gov	Develops and enforces safety and health rules for most public and private workplaces; provides training, outreach, education, and assistance
Telecommunications Industry Association (TIA) www.tiaonline.org	An international trade group that represents several hundred telecommunications companies and is accredited by ANSI to develop voluntary, consensus-based industry standards for a wide variety of information and communication technologies (ICT) products
Underwriters Laboratories Inc. (UL) www.ul.com	Tests equipment and products to verify conformance to national codes and standards

General Pathway Requirements

1. Install conduit runs in the most direct route possible with no more than two 90 degree bends between pull points or pull boxes.
2. 90 degree conduit bodies or Type LB fittings shall not be used for telecommunications cabling installations.
3. Continuous conduit sections shall not be greater than 100 linear feet in length without an adequately sized pull box.
4. Conduits shall be bonded to ground on one or both ends in accordance with national and local requirements.
5. Telecommunications conduits shall be installed to withstand the environment in which they will be exposed and in accordance with ANSI/NFPA-70, the NEC®.
6. The total distance of conduit runs for telecommunications horizontal cable use shall not exceed 150 feet.
7. Do not install conduits over or adjacent to boilers, incinerators, or hot water or steam lines.
8. The minimum bend radius of any portion of the conduit pathway shall be six (6) times the internal diameter of the conduit for all conduits 2-inches in diameter or less.
9. The minimum bend radius of any portion of the conduit pathway shall be ten (10) times the internal diameter of the conduit for all conduit with an internal diameter greater than 2-inches.

LEED Rating System Descriptions. . .

LEED for Building Design and Construction (BD+C). Buildings that are new construction or major renovation. At least 60% of the project's *gross floor area* must be *complete* by the time of certification (except for LEED BD+C: Core and Shell). Must include the entire building's gross floor area in the project.

- **LEED BD+C: New Construction and Major Renovation.** New construction or major renovation of buildings that do not primarily serve K–12 educational, retail, data centers, warehouses and distribution centers, hospitality, or healthcare uses. High-rise residential buildings four stories or more can use new construction or multifamily midrise.
- **LEED BD+C: Core and Shell Development.** Buildings that are new construction or major renovation for the exterior shell and core mechanical, electrical, and plumbing units, but not a complete interior fit-out. LEED BD+C: Core and Shell is the appropriate rating system to use if more than 40% of the gross floor area is incomplete at the time of certification.
- **LEED BD+C: Schools.** Buildings made up of core and ancillary learning spaces on K–12 school grounds. LEED BD+C: Schools may optionally be used for higher education and non-academic buildings on school campuses.
- **LEED BD+C: Retail.** Buildings used to conduct the retail sale of consumer product goods. Includes both direct customer service areas (showroom) and preparation or storage areas that support customer service.
- **LEED BD+C: Data Centers.** Buildings specifically designed and equipped to meet the needs of high density computing equipment such as server racks, used for data storage and processing. LEED BD+C: Data Centers only addresses whole building data centers (greater than 60%).
- **LEED BD+C: Warehouses and Distribution Centers.** Buildings used to store goods, manufactured products, merchandise, raw materials, or personal belongings, such as self-storage.
- **LEED BD+C: Hospitality.** Buildings dedicated to hotels, motels, inns, or other businesses within the service industry that provide transitional or short-term lodging with or without food.
- **LEED BD+C: Healthcare.** Hospitals that operate twenty-four hours a day, seven days a week and provide inpatient medical treatment, including acute and long-term care.
- **LEED BD+C: Homes and Multifamily Low rise.** Single-family homes and multi-family residential buildings of one to three stories. Projects three to five stories may choose the Homes rating system that corresponds to the ENERGY STAR program in which they are participating.
- **LEED BD+C: Multifamily Midrise.** Multifamily residential buildings of four or more occupied stories above grade. The building must have 50% or more residential space. These buildings may also use LEED BD+C: New Construction.

LEED for Interior Design and Construction (ID+C). Interior spaces that are a complete interior fit-out. In addition, at least 60% of the project's gross floor area must be complete by the time of certification.

- **LEED ID+C: Commercial Interiors.** Interior spaces dedicated to functions other than retail or hospitality.
- **LEED ID+C: Retail.** Interior spaces used to conduct the retail sale of consumer product goods. Includes both direct customer service areas (showroom) and preparation or storage areas that support customer service.
- **LEED ID+C: Hospitality.** Interior spaces dedicated to hotels, motels, inns, or other businesses within the service industry that provide transitional or short-term lodging with or without food.

LEED for Building Operations and Maintenance (O+M). Buildings that are fully operational and occupied for at least one year. The project may be undergoing improvement work or little to no construction. Must include the entire building's gross floor area in the project.

- **LEED O+M: Existing Buildings.** Existing buildings that do not primarily serve K–12 educational, retail, data centers, warehouses and distribution centers, or hospitality uses.
- **LEED O+M: Retail.** Existing buildings used to conduct the retail sale of consumer product goods. Includes both direct customer service areas (showroom) and preparation or storage areas that support customer service.
- **LEED O+M: Schools.** Existing buildings made up of core and ancillary learning spaces on K–12 school grounds. May also be used for higher education and non-academic buildings on school campuses.
- **LEED O+M: Hospitality.** Existing buildings dedicated to hotels, motels, inns, or other businesses within the service industry that provide transitional or short-term lodging with or without food.
- **LEED O+M: Data Centers.** Existing buildings specifically designed and equipped to meet the needs of high density computing equipment such as server racks, used for data storage and processing. LEED O+M: Data Centers only addresses whole building data centers.
- **LEED O+M: Warehouses & Distribution Centers.** Existing buildings used to store goods, manufactured products, merchandise, raw materials, or personal belongings (such as self-storage).
- **LEED O+M: Multifamily.** Existing multifamily projects with 20 or more units, either a single building with 20 units or multiple buildings within the same complex (e.g., 20 townhomes).

…LEED Rating System Descriptions

LEED for Neighborhood Development (ND). New land development projects or redevelopment projects containing residential uses, nonresidential uses, or a mix. Projects may be at any stage of the development process, from conceptual planning through construction. It is recommended that at least 50% of total building floor area be new construction or major renovation. Buildings within the project and features in the public realm are evaluated.

- **LEED ND: Plan.** Projects in conceptual planning or master planning phases, or under construction.
- **LEED ND: Built Project.** Completed development projects.

Man-Hour Conversions

Minutes		Hours
1	=	0.017
2	=	0.033
3	=	0.050
4	=	0.067
5	=	0.083
6	=	0.100
7	=	0.117
8	=	0.133
9	=	0.150
10	=	0.167
11	=	0.183
12	=	0.200
13	=	0.217
14	=	0.233
15	=	0.250

Minutes		Hours
16	=	0.267
17	=	0.283
18	=	0.300
19	=	0.317
20	=	0.333
21	=	0.350
22	=	0.367
23	=	0.383
24	=	0.400
25	=	0.417
26	=	0.433
27	=	0.450
28	=	0.457
29	=	0.483
30	=	0.500

Minutes		Hours
31	=	0.517
32	=	0.533
33	=	0.550
34	=	0.567
35	=	0.583
36	=	0.600
37	=	0.617
38	=	0.633
39	=	0.650
40	=	0.667
41	=	0.683
42	=	0.700
43	=	0.717
44	=	0.733
45	=	0.750

Minutes		Hours
46	=	0.767
47	=	0.783
48	=	0.800
49	=	0.817
50	=	0.833
51	=	0.850
52	=	0.867
53	=	0.883
54	=	0.900
55	=	0.917
56	=	0.933
57	=	0.950
58	=	0.967
59	=	0.983
60	=	1.000

T568A and T568B Wiring Color Codes (ANSI/TIA-568)

T568A		T568B	
Pin #	**Wire Color**	**Pin #**	**Wire Color**
1	White/Green	1	White/Orange
2	Green	2	Orange
3	White/Orange	3	White/Green
4	Blue	4	Blue
5	White/Blue	5	White/Blue
6	Orange	6	Green
7	White/Brown	7	White/Brown
8	Brown	8	Brown

Locking Wiring Devices

2-Pole, 3-Wire

Wiring Diagram	NEMA ANSI	Receptacle Configuration	Rating
	ML2 C73.44		15 A 125 V
	L5-15 C73.42		15 A 125 V
	L5-20 C73.72		20 A 125 V
	L6-15 C73.74		15 A 250 V
	L6-20 C73.75		20 A 250 V
	L6-30 C73.76		30 A 250 V
	L7-15 C73.43		15 A 277 V
	L7-20 C73.77		20 A 277 V
	L8-20 C73.79		20 A 480 V
	L9-20 C73.81		20 A 600 V

3-Pole, 4-Wire

Wiring Diagram	NEMA ANSI	Receptacle Configuration	Rating
	L14-20 C73.83		20 A 125/250 V
	L14-30 C73.84		30 A 125/250 V
	L15-20 C73.85		20 A 3ϕ 250 V
	L15-30 C73.86		30 A 3ϕ 250 V
	L16-20 C73.87		20 A 3ϕ 480 V
	L16-30 C73.88		30 A 3ϕ 480 V
	L17-30 C73.89		30 A 3ϕ 600 V

3-Pole, 3-Wire

Wiring Diagram	NEMA ANSI	Receptacle Configuration	Rating
	ML3 C73.30		15 A 125/250 V
	L10-20 C73.96		20 A 125/250 V
	L10-30 C73.97		30 A 125/250 V
	L11-15 C73.98		15 A 3ϕ 250 V
	L11-20 C73.99		20 A 3ϕ 250 V
	L12-20 C73.101		20 A 3ϕ 480 V
	L12-30 C73.102		30 A 3ϕ 480 V
	L13-30 C73.103		30 A 3ϕ 600 V

4-Pole, 4-Wire

Wiring Diagram	NEMA ANSI	Receptacle Configuration	Rating
	L18-20 C73.104		20 A 3ϕY 120/208 V
	L18-30 C73.105		30 A 3ϕY 120/208 V
	L19-20 C73.106		20 A 3ϕY 277/480 V
	L20-20 C73.108		20 A 3ϕY 347/600 V

4-Pole, 5-Wire

Wiring Diagram	NEMA ANSI	Receptacle Configuration	Rating
	L21-20 C73.90		20 A 3ϕY 120/208 V
	L22-20 C73.92		20 A 3ϕY 277/480 V
	L23-20 C73.94		20 A 3ϕY 347/600 V

Miscellaneous VDV Symbols					
Device	**Abbrev.**	**Symbol**	**Device**	**Abbrev.**	**Symbol**
Cable Antenna Outlet (CATV) 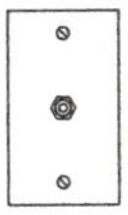	CAO		Plywood Backboard 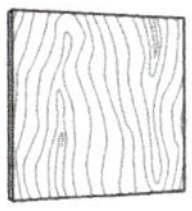		
Master Antenna Outlet (MATV) 	MAO		Telephone Undercarpet Flat Conductor 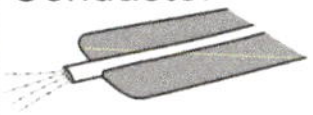	UCT	---UCT---
Equipment Cabinet 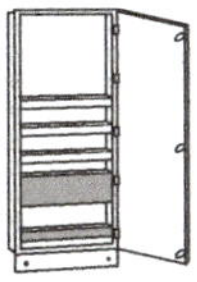	EQC		Data Undercarpet Flat Conductor 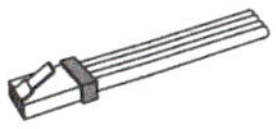	UCD	---UCD---
Equipment Rack 	EQR	 Wall Mounted	Power Pole 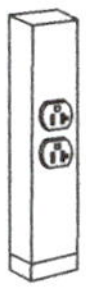	PP	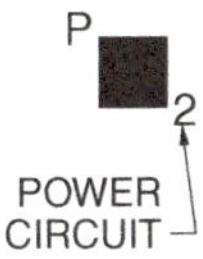
Equipment Rack 	EQR	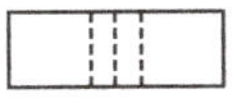 Freestanding	Telecom Pole 	TP	
Terminal Cabinet 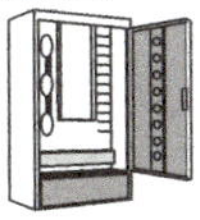	TC		Telecom Power Pole 	TPP	

Maximum Backbone Distance			
Cable Type	**Main to Horizontal Cross-Connect***	**Main to Intermediate Cross-Connect***	**Intermediate to Horizontal Cross-Connect***
Copper (Voice)	2624	1640	984
Multimode fiber	6560	5575	984
Single-mode fiber	9840	8855	984

* in ft

Maximum Work Area Cable Length		
Length of Horizontal Cable*	**Maximum Length of Work Area Cable***	**Maximum Combined Length of all Patch and Equipment Cords***
295	16	33
279	30	46
262	44	59
246	57	72
230	72	89

* in ft

Network Device Symbols

Network devices are represented in network diagrams by symbols that indicate their respective functions. The telecommunications industry has not standardized these symbols, but many are used so commonly that they have become de facto standards. Variations are commonly used to denote devices with special features or a combination of network functions.

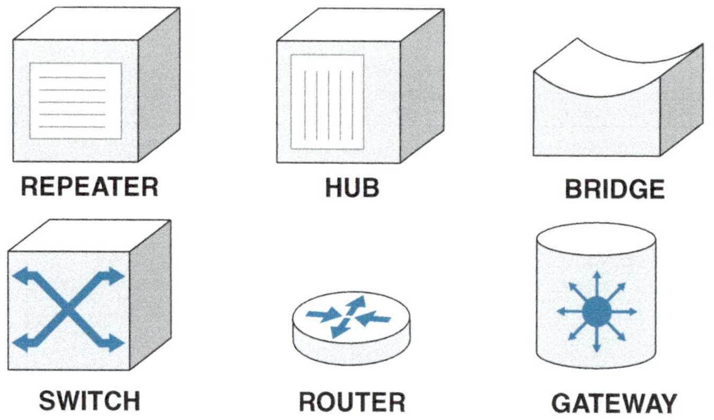

Fiber-Optic Minimum Bend Radius		
	No Load Condition	**Maximum Load**
Intrabuilding 2 or 4 Fiber	25 mm	50 mm
Intrabuilding Riser	10 × OD	15 × OD
Interbuilding Riser	10 × OD	20 × OD

Nonlocking Wiring Devices

2-Pole, 3-Wire

Wiring Diagram	NEMA ANSI	Receptacle Configuration	Rating
	5-15 C73.11		15 A 125 V
	5-20 C73.12		20 A 125 V
	5-30 C73.45		30 A 125 V
	5-50 C73.46		50 A 125 V
	6-15 C73.20		15 A 250 V
	6-20 C73.51		20 A 250 V
	6-30 C73.52		30 A 250 V
	6-50 C73.53		50 A 250 V
	7-15 C73.28		15 A 277 V
	7-20 C73.63		20 A 277 V
	7-30 C73.64		30 A 277 V
	7-50 C73.65		50 A 277 V

4-Pole, 4-Wire

Wiring Diagram	NEMA ANSI	Receptacle Configuration	Rating
	18-15 C73.15		15 A 3ϕY 120/208 V
	18-20 C73.26		20 A 3ϕY 120/208 V
	18-30 C73.47		30 A 3ϕY 120/208 V
	18-50 C73.48		50 A 3ϕY 120/208 V
	18-60 C73.27		60 A 3ϕY 120/208 V

3-Pole, 3-Wire

Wiring Diagram	NEMA ANSI	Receptacle Configuration	Rating
	10-20 C73.23		20 A 125/250 V
	10-30 C73.24		30 A 125/250 V
	10-50 C73.25		50 A 125/250 V
	11-15 C73.54		15 A 3ϕ 250 V
	11-20 C73.55		20 A 3ϕ 250 V
	11-30 C73.56		30 A 3ϕ 250 V
	11-50 C73.57		50 A 3ϕ 250 V

3-Pole, 4-Wire

Wiring Diagram	NEMA ANSI	Receptacle Configuration	Rating
	14-15 C73.49		15 A 125/250 V
	14-20 C73.50		20 A 125/250 V
	14-30 C73.16		30 A 125/250 V
	14-50 C73.17		50 A 125/250 V
	14-60 C73.18		60 A 125/250 V
	15-15 C73.58		15 A 3ϕ 250 V
	15-20 C73.59		20 A 3ϕ 250 V
	15-30 C73.60		30 A 3ϕ 250 V
	15-50 C73.61		50 A 3ϕ 250 V
	15-60 C73.62		60 A 3ϕ 250 V

Fiber-Optic Cable Transmission Performance			
Cable Type*	**Cable Type Wavelength†**	**Maximum Attenuation‡**	**Minimum Transmission Capacity§**
50/125 Multimode	850	3.5	500
	1300	1.5	500
62.5/125 Multimode	850	3.5	160
	1300	1.5	500
Single-mode inside	1310	1.0	N/A
Plant cable	1550	1.0	N/A
Single-mode outside plant cable	1310	0.5	N/A
	1550	0.5	N/A

Note: The manufacturer's documentation on the fiber's performance can be used to demonstrate compliance with the above performance requirements.
* in mm
† in nm
‡ in db/km
§ in MHz × km

Fiber-Optic Test Equipment Summary					
Type of Test	**Equipment**	**Investment**	**Accuracy**	**Customer Requirements**	**Properties Tested**
Visual	Visual fault locator	\$300–\$400	Good	Never	Continuity, polarity
Attenuation	Optical test set	\$4K–\$8K	Better	Always	Actual link loss, continuity, polarity
Faults	OTDR	\$8K–\$17K	Best	On occasion	Signature traces, length measurements events analysis (splice and connector) troubleshooting

NEC® Table 770.179 Cable Markings	
Cable Marking	**Type**
OFNP	Nonconductive optical fiber plenum cable
OFCP	Conductive optical fiber plenum cable
OFNR	Nonconductive optical fiber riser cable
OFCR	Conductive optical fiber riser cable
OFNG	Nonconductive optical fiber general-purpose cable
OFCG	Conductive optical fiber general-purpose cable
OFN	Nonconductive optical fiber general-purpose cable
OFC	Conductive optical fiber general-purpose cable

NFPA 70 NEC®

VDV System Abbreviations

Equipment

110	Twisted pair termination block
ADF	Area distribution facility
BDF	Building distribution frame or facility
BEF	Building entrance frame or facility
CAB	Telecom cabinet or enclosure
CONN	Connector
CSC	Copper splice closure
CVE	Controlled environmental vault
ER	Equipment room
FDF	Fiber distribution facility
FF	Front facing
FS	Fiber shelf/fiber termination panel
FSC	Fiber-optic splice enclosure
HH	Handhole
IDC	Internet data center
IDF	Intermediate distribution frame or facility
MDF	Main distribution frame
MH	Manhole; maintenance hole
MPOE	Minimum point of entry; main point of entry
OCEF	Optical cable entrance facility
OTDR	Optical time domain reflectometer
PAV	Pavement
PBB	Primary bonding busbar
PC	Plastic conduit
PG	Pair group
POP	Point of presence
PR	Pair
PVC	Polyvinyl chloride
R/W	Right-of-way
SBB	Secondary bonding busbar
SC	Splice closure
SCS	Structured cabling system
SER	Serial
SMR	Surface-mounted raceway
SS	Fiber splice shelf
TEBC	Telecom equipment bounding busbar
TC	Telecom closet
TCC	Telecom conduit
TCH	Telecom conduit sleeve, horizontal
TCR	Telecom horizontal and vertical riser conduit
TCT	Telecom cable tray
TEC	Telecom entrance conduit
TEL	Telephone
TELECOM	Telecommunications
TERM	Terminal
TP	Twisted pair
TPB	Telecom pull box
TR	Telecom room
TSL	Telecom wall or floor slot
TSV	Telecom conduit sleeve, vertical

Wire and Cable

AFMW	24 AWG bonded fill flooded twisted cable
ARMM	24 AWG riser armored bonded multipair cable
AWG	American wire gauge
CAT3	Category 3 twisted pair copper cable
CAT6	Category 6 twisted pair copper cable
CM	Communications cable
CMP	Communications plenum cable
CMR	Communications riser cable
COAX	Coaxial cable
FO	Fiber optic
HDPE	High-density polyethylene
LTFF	Loose tube filled and flooded
MDPE	Medium-density polyethylene
MM	Loose tube filled and flooded
MPP	Multipurpose plenum cable
OFC	Optical fiber conductive cable
OFCP	Optical fiber conductive plenum cable
OFCR	Optical fiber conductive riser cable
OFNR	Optical fiber nonconductive riser cable
OFN	Optical fiber nonconductive cable
OFNP	Optical fiber nonconductive plenum cable
SM	Single-mode fiber-optic cable
STP	Shielded twisted pair
TB	Tight buffered
TBC	Telecom bounding cable
UTP	Unshielded twisted pair

Glossary

A

abbreviation: A letter or series of letters or words denoting a complete word.

access control system: A security system that controls door locks.

acronym: A word formed from the first letter of each of the successive parts or words of a compound term.

addendum: The addition of omitted or updated information to a set of construction drawings, technical specifications, or a contract.

air-core cable: A cable in which the conductors are surrounded by an air space within the jacket.

alien crosstalk: The crosstalk noise experienced by two or more adjacent twisted-pair cables.

alternate product: A product acceptable to the engineer or specifier for which the contractor submits alternate pricing.

aluminum-polyethylene (ALPETH) cable: A cable covered with separate aluminum and polyethylene jackets.

aluminum-polyvinyl-chloride (ALVYN) sheath: An aluminum-polyvinylchloride (PVC) cable sheath used for riser cable where a flame-retardant sheath is required.

American National Standards Institute (ANSI): A private nonprofit organization that oversees the development of voluntary consensus standards for products, services, processes, systems, and personnel in the United States.

Americans with Disabilities Act (ADA) of 1990: An antidiscrimination law that prohibits employers from denying employment to applicants based on their physical disabilities.

application assurance warranty: An inclusion to a standard product warranty that is designed to provide the end user with additional assurances of performance.

applications assurance warranty: A manufacturer warranty that ensures that standards-based technology designed to be used with a specific system is supported by the manufacturer.

apprentice: An individual training to learn a trade through a combination of on-the-job training and related classroom instruction in an apprenticeship program.

apprenticeship program: A training program that provides practical hands-on training accompanied by theoretical classroom instruction.

architect: A person licensed and qualified to perform architectural services such as analyzing a construction project, creating the project design and specifications, and managing the project.

architect's scale: A triangular scale, 1′ in length, with six graduated surfaces.

architect's supplemental instruction (ASI): A document that depicts and clarifies minor changes to a print or other project document.

ARMM cable: A color-coded riser cable that consists of bare copper insulated with foam polyethylene (PE) and a polyvinyl-chloride (PVC) jacket.

ASIS International (ASIS): A professional organization for security professionals that issues certifications, standards, and guidelines for the security profession.

audiovisual over internet protocol (AV/IP): The transmission of audiovisual data over a network, such as a LAN, a wide area network (WAN), or the internet.

audiovisual (A/V) system: A digital voice-data-video system that is a combination of sound and video elements.

authority having jurisdiction (AHJ): The organization, office, or individual responsible for approving the equipment and materials used for electrical wiring installation.

auxiliary disconnect outlet (ADO): A device located within the tenant or living unit used to terminate the backbone cable.

B

barrier: A physical impediment that is intended to block passage.

Bell cable: A VDV cable specified for outdoor use by private companies and Regional Bell Operating Companies (RBOCs).

bid: An offer from a contractor, subcontractor, or supplier to perform construction work, provide a construction service, or provide materials or supplies at a stated price.

bid form: A document that contractors are required to use when bidding a project. It includes space to enter project identification, prices, quantities, substitutions, and bidder contact information.

blown fiber: A type of fiber-optic installation method in which fiber-optic cables are blown through preinstalled conduit or tubes using air pressure to float the cable through the conduit or tube.

buffer: A fiber-optic cable layer that provides protection for optical fiber within a cable.

building diagram: A simple view of the services and media types installed from one point to another point, from one building to another building.

building entrance terminal: A cable entrance point where an input cable between buildings is terminated.

Building Industry Consulting Service International (BICSI): A professional telecommunications standards and certification organization.

building information modeling (BIM): The process of digitally representing the physical and functional properties of a facility.

C

cabinet: An enclosure used for terminating telecommunications cables, wiring, and connection devices that has a hinged cover and is usually installed flush-mounted to a wall.

cable manager: A device used to secure, organize, and manage cables in a data center or equipment room.

cementitious material: A material that can harden like cement.

Certified Protection Professional (CPP) certification: A professional certification for individuals with a minimum of seven years of security experience in the physical security field, three of which must be in responsible charge of a security function, who have successfully passed a board certification licensing exam.

change order: A written form that details any variation from the original plans and specifications of a project as directed by the project owner or designer.

cladding: A layer of glass fused to a glass fiber-optic core to aid light transmission and to act as protection for the core.

code: A regulation and requirement devised to maintain uniformity of work and proper standards of procedure.

commissioning: The process of planning, documenting, scheduling, testing, adjusting, verifying, and training in order to operate a facility as a fully functional system per the owner's project requirements.

compartmentation: The deliberate division of a building into sections or units.

computer-aided design (CAD): The design and drafting of a component, member, or structure using a computer.

computer room air conditioning (CRAC) unit: An HVAC device that monitors and maintains the temperature, air distribution, and humidity in a network room, space, or data center.

construction change directive (CCD): A mechanism for directing a contractor to perform additional work outside of the original contract documents and when the time and cost of the work is not agreed upon between the owner and general contractor.

construction documents: A set of instructions created by a designer and used by a contractor to estimate project costs and build a project.

Construction Specifications Institute (CSI): An organization that develops standardized construction specifications.

contract document: A set of instructions created by a designer and used by a contractor to estimate project costs and build a project.

contractor guarantee: A written assurance issued by a general contractor that furnished materials and equipment are of acceptable quality and workmanship as well as free from defects.

core: The smallest section of a glass fiber strand and is used to carry light waves.

cyber security: The convergence of the technologies, processes, and practices designed to protect networks, devices, programs, and data from attack, damage, or unauthorized access. Also known as information technology security.

D

data center: A centralized location where computing resources critical to an organization are maintained in a highly controlled physical environment.

descriptive specification: A type of performance specification that provides detailed requirements for material properties and workmanship, fabrication, and installation.

design: The process of taking an abstract idea and developing it into a working solution.

design-assist (DA) procurement method: A construction project procurement method in which one or more subcontractors are retained, prior to the completion of design, to assist the architect, engineer, or construction manager in the development of the design and construction documents.

design development (DD): The design phase that further develops and refines a schematic design.

design narrative: A brief description of a project and its coordination process.

detail drawing: A drawing that shows a small part of a plan, elevation, or section view at an enlarged scale.

detailed takeoff estimate: An estimate in which each individual construction component is accounted for, or taken off.

digital electricity: A method of transferring high amounts of power over nonpower cabling.

digital estimating method: An estimating method in which quantity takeoff and pricing calculation is accomplished using electronic spreadsheets and integrated database systems.

distributed antenna system (DAS): A network of low- and high-powered antennas distributed vertically and horizontally throughout a building.

distribution device: A hardware device used for the distribution of VDV systems throughout a dwelling or commercial structure.

E

elevation: An orthographic view of a vertical surface without allowance for perspective.

endothermic fireproofing material: A fireproofing material that releases water vapor when exposed to high temperatures.

engineer's scale: A measuring and scaling tool that is graduated in the decimal system.

enterprise data center: A private facility operated for the sole use of supporting a single organization.

entrance facility (EF): The point where outside public and private network service provider cables can enter a building and be consolidated.

equipment room (ER): A centralized room or space for telecommunications equipment that serves the occupants of a building or multiple buildings within a campus environment.

estimating: The process of determining labor and material costs required to complete a project.

estimator: An individual who performs quantity takeoff and pricing to determine the cost of a construction project.

F

field superintendent: The general contractor's site representative who is responsible for the day-to-day operations and overall supervision of the construction job site.

fire alarm system: A building system used to notify building occupants of a possible fire upon the detection of smoke, heat, or both.

fireproofing: The process of applying protective material to a structural or finish member to increase the fire resistance of the member and protect it from failure in the event of excessive heat or fire.

fire rating: The measure of resistance of a material or component to failure when exposed to fire and is expressed as the number of hours a material or component will retain its integrity.

firestop: Any material, device, or assembly of parts installed after a penetration of a fire-rated wall, ceiling, or floor area to prevent passage of flame, smoke, or gases through the rated barrier.

firestopping: The process of applying and installing a material or member that seals open construction to inhibit the spread of fire, smoke, and fumes in a structure.

foreman: A field supervisor who is responsible for directing each labor crew and activity on a project but is not considered company management.

fusion splice: A permanent splice accomplished by the application of heat with a temperature high enough to melt the ends of two sections of glass fiber.

fusion splicer: A device that uses a high-voltage electrical arc between two electrodes to heat and melt the ends of glass fibers in order to join them and form a continuous strand.

G

Gantt chart: A type of bar chart in which project tasks are represented by a bar or line, with one end representing the start date and the other end representing the completion date.

gateway: A device that connects two or more dissimilar networks.

gel-filled cable: A cable that has a gel-type substance inside its jacket that encases its conductors.

general contractor (GC): An individual or commercial firm that is hired to provide goods and services for the undertaking of construction activities.

H

handhole: A buried nonmetallic box, in which the lid is flush with the surface of the ground, that provides access, housing, and protection for cable splices, branches, and termination points.

HDBaseT: A firmware-based consumer electronics connectivity technology that enables a single standard CAT 5e or CAT 6 LAN cable with a standard RJ-45 connector to simultaneously carry uncompressed 10.2 Gbps high-definition video and audio, 100 Mbps of data, and 100 W of power over a distance of 100 m (328′).

HDBaseT Alliance: A standards organization that promotes, advances, and standardizes HDBaseT technology in a variety of industries by forming an alliance with manufacturers and other organizations.

I

index-matching material: A material in liquid, paste, film, or gel form with a refractive index nearly equal to the core index of a fiber that is used to reduce reflections from a fiber end face.

information technology (IT) equipment: The hardware, software, and networks that are used by an organization to input, store, process, output, and disperse data and information.

information technology security: *See* cyber security.

Institute Electrical and Electronics Engineers (IEEE): A technical professional society that produces standards for telecommunications, local area networks, and computing.

insulation displacement connector (IDC): A type of telecommunications connector in which the wire is punched down into a metal holder that strips the insulation on the wire and allows electrical contact to be made.

interbuilding riser system: A system of telecommunications cables that connect one building to another.

intercom: An internal communication system that allows direct intercommunication with another connected device within a building, building complex, campus, factory, or residential dwelling.

intermediate distribution frame (IDF): A metal rack designed to connect cables and is located in an equipment room or telecommunications room.

International Electrotechnical Commission (IEC): An organization that develops international safety standards for electrical equipment.

International Organization for Standardization (ISO): A developer of voluntary international standards and facilitates world trade by providing common standards between nations.

intrabuilding riser system: Telecommunications cables that are part of the building subsystem that connect one telecommunications room (TR) to another within the same building.

intrusion detection and notification system: A security system that monitors, detects, and sends notifications in the event of a breach or another security event in commercial, institutional, and industrial facilities in order to protect life and property.

intumescent fireproofing material: A fireproofing material that expands rapidly when exposed to high temperatures, closing voids left by burning or melting construction materials.

invitation for bid (IFB): A request for contractors, subcontractors, and suppliers to submit pricing for services and/or materials for a project.

L

laser-optimized (LO) multimode fiber: A high-capacity multimode fiber designed for short-wave 10 Gb optical transmission.

LC connector: A small form factor connector that is half the size of a standard connector.

Leadership in Energy and Environmental Design™ (LEED®): An international green building certification system developed by the USGBC.

ledger sheet: A grid on a sheet of paper consisting of rows and columns into which item descriptions, locations, code numbers, quantities, and costs are entered.

limited-power (LP) cable: A cable that can carry both power and data in a single cable to supply power to equipment such as IP video surveillance cameras, intelligent LED lighting systems, and devices that use PoE or digital electricity.

loose-tube buffer: A fiber-optic cable construction in which the glass is carried loosely in a buffer that has a diameter greater than the diameter of the fiber.

M

main distribution frame (MDF): An apparatus that connects the telephone lines coming into the building from outside to the internal lines connected to the equipment inside the building.

MasterFormat: A uniform system of numbers and titles for organizing information concerning construction requirements, products, and activities into a standard sequence.

mechanical splice: A fiber-optic splice in which the ends of two different fibers are mechanically connected.

milestone: A recognizable event that represents a meaningful completion or identifiable point in a project.

model building code: A complete set of requirements developed for the design and construction industries to provide a minimum standard for the health, safety, and welfare of building occupants.

multifiber push-on (MPO) connector: A fiber-optic connector for ribbon cables with a capacity of 8 to 72 fibers that is designed to provide multifiber connectivity in one connector to support high bandwidth and high-density cabling system applications.

multitenant data center: A data center operated by a third party for the purpose of serving multiple enterprise tenants.

N

National Electrical Code® (NEC®): A standard on practices for the design and installation of electrical VDV systems.

National Electrical Manufacturers Association (NEMA): A national organization that provides standardization for electrical equipment.

National Fire Protection Association® (NFPA®): A national organization that provides guidance in assessing the hazards of the products of combustion.

network video recorder (NVR): A video recorder that uses an IP camera and server to view, record, and play back an IP video stream over a network.

NFPA 101®, Life Safety Code®: A standard used for strategies to protect building occupants based on building construction, protection, and occupancy features that minimize the effects of fire and related hazards.

NFPA 70E®, Standard for Electrical Safety in the Workplace®: A standard that covers the electrical safety requirements of workers performing electrical work.

O

Occupational Safety and Health Administration (OSHA): A federal regulatory agency that develops and publishes safety and occupational health regulations and has legal authority to inspect workplaces, enforce regulations, and propose fines and penalties for violations.

optical time domain reflectometer (OTDR): A test instrument used to measure fiber-optic cable attenuation (signal loss).

one-cord test method: A fiber-optic cable test method that allows for testing of a fiber-optic link from end-to-end and includes losses from all connectors.

outside plant (OSP) distribution system: A telecommunications infrastructure designed for installation outside any structure.

owner: An individual, a group, or another entity that enters into a contract with designers, contractors, and other professionals to obtain goods and services provided by the construction industry.

owner's representative: An individual who represents the business interests of the owner, especially if the owner is a group of individuals or a separate corporation.

P

pathway: A set of devices or spaces used to route telecommunications cables from a telecommunications room or equipment room to a work area.

percentage estimate: A type of general-knowledge estimate formulated by using data from similar projects with known completion costs and calculating the completion cost against the entire cost of the finished building.

performance-based specification: A document that outlines the intended results of a project design.

Physical Security Professional (PSP) certification: A professional certification for individuals with a minimum of four years of experience in the physical security field who have successfully passed a board certification licensing exam.

pixel (picture element): The smallest unit area of a video screen that can be turned on or off or varied in intensity.

plenum: A building space that facilitates air circulation for HVAC systems by providing pathways for airflows typically greater than atmospheric pressure.

point-to-point cabling system: A data center cabling system composed of jumper fiber cables that are used to connect one switch, server, or storage unit directly to another switch, server, or storage unit.

polyethylene (PE): A thermoplastic material with electrical insulating properties.

power-limited fire protection (PLFA) cable: A nonplenum-rated cable used for fire alarm installations. PLFA cables are not suitable for installation in risers, ducts, plenums, or other environmental air spaces.

Power over Ethernet (PoE): A networking feature that allows network cables to carry electrical power over an existing data connection with a single CAT 5e or CAT 6 cable.

premises: A space occupied by an authorized user in a building on continuous or contiguous property that is not separated by public road or highway.

primary bonding busbar (PBB): A busbar that serves as the central connection point between the grounding electrode system for a building electrical service and the grounding and bonding system for telecommunications.

product option: One of several listed products in the specification.

project manager (PM): The contractor's representative who oversees the overall planning and execution of a construction project.

product substitution: A request to substitute one product for another product described in a specification.

project specification: A written supplement to construction prints that provides additional construction information.

public safety network: A network used by emergency services to prevent or respond to incidents that can harm or endanger persons or property.

punch list: A list compiled by the owner and/or lead designer during the walk-through of a completed project that indicates items that must be repaired or completed before project finalization.

purchase order (PO): A document with the details of a purchase that is often coded with a specific number.

R

registered communications distribution designer (RCDD): An individual with education, training, and expertise in the design, implementation, and integration of telecommunications/data transport systems and infrastructure.

request for information (RFI): A document used for clarifying scope and the coordination of project elements that arise during the construction process.

resolution: The raw number of pixels contained within an image.

ribbon fiber: A type of fiber-optic cable in which multiple fibers are clad in a single, flat, ribbon-like cable.

riser cable: A telecommunications cable and associated hardware that runs from a telecommunications closet, or intermediate distribution frame (IDF), to a main distribution frame (MDF) or to another telecommunications room (TR).

room-scheduling system: A digital calendar scheduling system used in commercial environments to allocate space and resources for upcoming events and meetings.

S

safety inspection: The process of critically examining and evaluating the safety and occupational health hazards of a job site.

schedule: A table or detailed list on a print that provides information about specific building components and equipment such as doors, windows, and mechanical equipment.

schematic design (SD): The initial design phase that defines the general scope and conceptual design of a project.

secondary bonding busbar (SBB): A busbar located in the telecommunications room (TR) or equipment room (ER) of a building that connects to the primary bonding busbar (PBB).

service provider data center: An off-site data center that is leased by a service provider to an organization.

shop drawing: A detailed drawing or set of drawings that shows how building elements are to be fabricated or installed and contains other pertinent information such as construction materials, finishes, and dimensions.

smoke barrier: A continuous membrane, either vertical or horizontal, such as a wall, floor, or ceiling assembly that is designed and constructed to restrict the movement of smoke.

specification: Written information included with a set of prints.

spreadsheet: A computer program that uses cells organized into rows and columns to perform various calculations when formulas and numbers are entered.

standard: An accepted reference or practice.

standard connector (SC): A fiber-optic cable connector that uses a push-pull latching mechanism. Female MPO connectors have holes in the ferrule to accept the alignment pins from the male connector.

storage area network (SAN): A dedicated network that provides access to servers, local area networks, and shared storage devices.

student response system (SRS): A digital educational platform that allows instructors to electronically pose questions and gather student responses during classroom (in-house or virtual) instruction.

subcontractor: An individual or commercial firm that specializes in a particular trade or field of construction.

sustainable building: The practice of constructing a building using processes that are environmentally responsible and resource-efficient throughout the life cycle of the building.

T

takeoff: The practice of reviewing contract documents, including the prints and specifications, to determine the quantities of material to be included in a bid.

telecommunications bonding backbone (TBB): A cable that is connected to a primary bonding busbar in the telecommunications entrance facility, to an earth ground in the electrical entrance facility, and to structural steel on each floor.

telecommunications bonding conductor (TBC): A conductor that bonds a primary bonding busbar to the grounding electrode system for a building electrical service.

Telecommunications Industry Association (TIA): An international trade group that represents several hundred telecommunications companies.

telecommunications riser: The infrastructure that connects the tenant spaces of a building from the local exchange carrier presence in the entrance facility to each floor.

telecommunications room (TR): An enclosed space that houses telecommunications equipment, cable terminations, and cross-connect cabling used to service work areas on the same floor of a building.

through-penetration firestop system: A firestop system that protects penetrations that pass completely through a barrier such as a wall or a floor.

tight buffer: A fiber-optic cable construction where each glass fiber is buffered tightly by a protective thermoplastic coating to a diameter of 900 μm.

title block: The area of a print that contains information about the contents of the print.

traditional estimating method: An estimating method in which material quantities are calculated and entered onto ledger sheets for pricing.

transceiver: An electronic device that can send and receive signals on the Ethernet media system.

two-cord test method: A fiber-optic cable test method used when the connector on the power meter is different from the connector on the fiber link to be tested.

U

Underwriters Laboratories (UL): An independent organization that tests equipment and products to see if they conform to national codes and standards.

unit-price agreement: A construction contract that is based on the cost of work or per item of material.

V

video surveillance system: A system that controls security cameras.

voice-data-video (VDV) cabling system: A system that is organized into a hierarchy of wiring termination and interconnection structures.

W

wayfinding sign: An interactive touch-screen display.

wideband multimode fiber (WBMMF): A fiber-optic cable that can transmit up to four times the amount of data that could be transmitted on a single multimode fiber-optic cable.

wireless access point (WAP): A device connected to a wired network that connects wireless communication devices to the same network.

Index

Page numbers in italic refer to figures.

M

N

O